P9-BTY-226

Molecular Biology Intelligence Unit

DNA Conformation and Transcription

Takashi Ohyama, Ph.D.
Department of Biology
Faculty of Science and Engineering
Konan University
Kobe, Japan

Landes Bioscience / Eurekah.com
Georgetown, Texas
U.S.A.

Springer Science+Business Media
New York, New York
U.S.A.

DNA Conformation and Transcription

Molecular Biology Intelligence Unit

Landes Bioscience / Eurekah.com
Springer Science+Business Media, Inc.

ISBN: 0-387-25579-6 Printed on acid-free paper.

Springer Science+Business Media, Inc., 233 Spring Street, New York, New York 10013, U.S.A.
http://www.springeronline.com

Please address all inquiries to the Publishers:
Landes Bioscience / Eurekah.com, 810 South Church Street, Georgetown, Texas 78626, U.S.A.
Phone: 512/ 863 7762; FAX: 512/ 863 0081
http://www.eurekah.com
http://www.landesbioscience.com

Printed in the United States of America.

9 8 7 6 5 4 3 2 1

Library of Congress Cataloging-in-Publication Data

DNA conformation and transcription / [edited by] Takashi Ohyama.
p. ; cm. -- (Molecular biology intelligence unit)
Includes bibliographical references and index.
ISBN 0-387-25579-6
1. DNA--Conformation. 2. Genetic transcription. I. Ohyama, Takashi. II. Series: Molecular biology intelligence unit (Unnumbered)
[DNLM: 1. DNA--chemistry. 2. DNA--metabolism. 3. Nucleic Acid Conformation. 4. Transcription, Genetic. QU 58.5 D62907 2005]
QP624.5.S78D58 2005
611'.018166--dc22

2005012636

CONTENTS

Part V
Chromatin Infrastructure in Transcription: Roles of DNA Conformation and Properties

EDITOR

Takashi Ohyama
Department of Biology
Faculty of Science and Engineering
Konan University
Kobe, Japan
Email: ohyama@konan-u.ac.jp
Chapters 3, 5, 13

CONTRIBUTORS

Jaime F. Angulo
Laboratoire de Génétique
de la Radiosensibilité
Département de Radiobiologie
et de Radiopathologie
Direction des Sciences du Vivant
Commissariat à l'Energie Atomique
Centre d'Etudes de Fontenay-Aux-Roses
Fontenay-Aux-Roses, France
Email: jaime.angulo-mora@cea.fr
Chapter 6

Munehiko Asayama
Laboratory of Molecular Genetics
School of Agriculture
Ibaraki University
Ibaraki, Japan
Email: asam@mx.ibaraki.ac.jp
Chapter 3

Denis S.F. Biard
Laboratoire de Génétique
de la Radiosensibilité
Département de Radiobiologie
et de Radiopathologie
Direction des Sciences du Vivant
Commissariat à l'Energie Atomique
Centre d'Etudes de Fontenay-Aux-Roses
Fontenay-Aux-Roses, France
Email: denis.biard@cea.fr
Chapter 6

Søren Brunak
Center for Biological Sequence Analysis
Biocentrum-DTU
The Technical University of Denmark
Lyngby, Denmark
Email: brunak@cbs.dtu.dk
Chapter 14

Andrew V. Colasanti
Wright-Rieman Laboratories
Rutgers University
Piscataway, New Jersey, U.S.A.
Email: andrewc@rutchem.rutgers.edu
Chapter 2

Mensur Dlakić
Department oicf Microbiology
Montana State University
Bozeman, Montana, U.S.A.
Email: mdlakic@montana.edu
Chapter 14

Claudio O. Gualerzi
Department of Biology MCA
University of Camerino
Camerino (MC), Italy
Email: claudio.gualerzi@unicam.it
Chapter 4

Alan Herbert
Framingham Heart Study Genetics
Laboratory
Department of Neurology
Boston University School of Medicine
Boston, Massachusetts, U.S.A.
Email: aherbert@bu.edu
Chapter 7

Susumu Hirose
Department of Developmental Genetics
National Institute of Genetics
Mishima, Japan
Email: shirose@lab.nig.ac.jp
Chapter 10

Kuniharu Matsumoto
Department of Developmental Genetics
National Institute of Genetics
Mishima, Japan
Email: kunmatsu@lab.nig.ac.jp
Chapter 10

Philippe Mauffrey
Laboratoire de Génétique
de la Radiosensibilité
Département de Radiobiologie
et de Radiopathologie
Direction des Sciences du Vivant
Commissariat à l'Energie Atomique
Centre d'Etudes de Fontenay-Aux-Roses
Fontenay-Aux-Roses, France
Email: philippe.mauffrey@cea.fr
Chapter 6

Laurent Miccoli
Laboratoire de Génétique
de la Radiosensibilité
Département de Radiobiologie
et de Radiopathologie
Direction des Sciences du Vivant
Commissariat à l'Energie Atomique
Centre d'Etudes de Fontenay-Aux-Roses
Fontenay-Aux-Roses, France
Email: laurent.miccoli@cea.fr
Chapter 6

Wilma K. Olson
Wright-Rieman Laboratories
Rutgers University
Piscataway, New Jersey, U.S.A.
Email: olson@rutchem.rutgers.edu
Chapter 2

Ghislaine Pinon-Lataillade
Laboratoire de Génétique
de la Radiosensibilité
Département de Radiobiologie
et de Radiopathologie
Direction des Sciences du Vivant
Commissariat à l'Energie Atomique
Centre d'Etudes de Fontenay-Aux-Roses
Fontenay-Aux-Roses, France
Email: ghislaine.lataillade@cea.fr
Chapter 6

Cynthia L. Pon
Department of Biology MCA
University of Camerino
Camerino (MC), Italy
Email: cynthia.pon@unicam.it
Chapter 4

Vladimir N. Potaman
Center for Genome Research
Institute of Biosciences and Technology
Texas A&M University System Health
Sciences Center
Houston, Texas, U.S.A.
Email: VPotaman@ibt.tamhsc.edu
Chapter 1

Richard R. Sinden
Laboratory of DNA Structure
and Mutagenesis
Center for Genome Research
Institute of Biosciences and Technology
Texas A&M University System Health
Sciences Center
Houston, Texas, U.S.A.
Email: RSinden@ibt.tamhsc.edu
Chapter 1

Stefano Stella
Department of Biology MCA
University of Camerino
Camerino (MC), Italy
Email: stefano.stella@unicam.it
Chapter 4

Michael Y. Tolstorukov
Laboratory of Experimental
and Computational Biology
National Cancer Institute, NIH
Bethesda, Maryland, U.S.A.
Email: tolstorm@mail.nih.gov
Chapter 2

Andrew A. Travers
MRC Laboratory of Molecular Biology
Cambridge, U.K.
Email: aat@mrc-lmb.cam.ac.uk
Chapter 11

Karen Usdin
Genomic Structure and Function Section
Laboratory of Molecular
and Cellular Biology
National Institute of Diabetes
Digestive and Kidney Diseases
National Institutes of Health
Bethesda, Maryland, U.S.A.
Email: ku@helix.nih.gov
Chapter 9

David W. Ussery
Center for Biological Sequence Analysis
Biocentrum-DTU
The Technical University of Denmark
Lyngby, Denmark
Email: Dave@CBS.dtu.DK
Chapter 14

Michael W. Van Dyke
Department of Molecular
and Cellular Oncology
University of Texas M.D. Anderson
Cancer Center
Houston, Texas, U.S.A.
Email: mvandyke@mdanderson.org
Chapter 8

Michael A. Weiss
Department of Biochemistry
Case Western Reserve University
Cleveland, Ohio, U.S.A.
Email: michael.weiss@case.edu
Chapter 12

Fei Xu
Wright-Rieman Laboratories
Rutgers University
Piscataway, New Jersey, U.S.A.
Email: feix@rutchem.rutgers.edu
Chapter 2

Victor B. Zhurkin
Laboratory of Experimental
and Computational Biology
National Cancer Institute, NIH
Bethesda, Maryland, U.S.A.
Email: zhurkin@nih.gov
Chapter 2

PREFACE

The human genome project was virtually completed in 2003, the same year in which the 50th anniversary of the discovery of the DNA double helix came around. This project has opened the door to the genomic era. Nowadays, we have easy access to genome sequence databases for more than 100 organisms. However, despite such remarkable progress in genome science, we are still far from a clear understanding of how genomic DNA is packaged without entanglement into a nucleus, how genes are wrapped up in chromatin, how chromatin structure is faithfully inherited from mother to daughter cells, and how the differential expression of genes is enabled in a given cell type. Exploring and answering these questions constitutes one of the next frontiers in the 21st century.

Multifarious DNA structures are found in the genome, i.e., curved DNA structures found in regions of periodically occurring A-tracts, triplex structures composed of homopurine/homopyrimidine regions, quadruplex structures made up of guanine-rich sequences, left-handed DNA helix (Z-DNA) formed by alternating purine-pyrimidine sequences, cruciform structures formed by inverted repeats, and so forth. The implication of these structures for DNA packaging and gene expression has long been argued. In the meantime, much circumstantial evidence and several lines of direct evidence have been presented, and we are beginning to appreciate how these structures provide additional structural and functional dimensions to chromatin organization and gene expression. However, to the best of my knowledge, there is no book in which the fruits of the studies performed to date have been compiled, in order to shed light on the roles of DNA conformation in transcription.

The challenge of this book is to collect these results, and it is intended to serve as a source of information. The book is not only aimed at specialists, but also at students and non-specialists who have no prior knowledge of this field. Contributors from the respective fields describe, in 14 chapters, the history, up-to-date topics, what has been clarified, and what is still to be discovered. These chapters are grouped into five closely related parts. Part I presents fundamental knowledge on DNA structure, which will simply guide the reader into the 'DNA world'. Readers will see that there are many DNA structures in genomes besides the DNA described by James D. Watson and Francis H. C. Crick, which is now called B-form DNA, and will further understand how these structures are constructed and what possible biological functions they may have. In Part II, the role of curved DNA is discussed. Among the multifarious DNA structures, curved DNA has been most widely and intensively studied in relation to transcriptional regulation. The roles of this structure in both prokaryotic and eukaryotic transcription are discussed, and the proteins that bind curved DNA and their functional roles are described. Part III focuses on the role of Z-DNA, triplex, quadruplex and supercoiled DNA. In addition, the relationship between DNA conformation, transcriptional defects and human diseases is discussed, with Fragile X mental retardation syndrome and Friedreich's ataxia used as examples. Highlighted in Part IV are the roles of architectural transcription factors and protein-induced DNA bends in gene regulation. Furthermore, the mechanism of mammalian male sex determination is discussed as a relevant biological phenomenon. Finally, Part V focuses on how the structural and mechanical properties of DNA influence nucleosome positioning and chromatin organization, and discusses how these effects influence gene expression.

"This structure has novel features which are of considerable biological interest." This is the second (and the last) sentence in the first paragraph of Watson and Crick's historical paper, "Molecular structure of nucleic acids—a structure for deoxyribose nucleic acid" (Nature 171, 737-738, 1953). Now, the time has come to solve the riddle that is 'written' in DNA conformation, as another point of considerable biological interest regarding the DNA molecule.

I acknowledge the farsightedness of Ron Landes at Landes Bioscience who commissioned me to edit this book. I am particularly thankful to the authors for their scholarly contributions, and for their patience during the editing and the production of the book. I am deeply indebted to Junko Ohyama, my wife, for helping me in every aspect of this endeavor. I am also thankful to Toru Higashinakagawa at Waseda University for his advice, and to Cynthia Conomos and Sara Lord at Landes Bioscience. Members of my laboratory have tolerated my preoccupation with this book project. I am also very grateful to them.

Takashi Ohyama
Kobe, Japan
September 2004

Part I
Multifarious DNA Structures Found in Genomes

Chapter 1

DNA:
Alternative Conformations and Biology

Vladimir N. Potaman and Richard R. Sinden

Abstract

Local structural transitions from the common B-DNA conformation into other DNA forms can be functionally important. This chapter describes the structures of DNA forms called alternative DNA conformations that are different from the canonical B-DNA helix. Also discussed are the requirements for the formation of alternative DNA structures, as well as their possible biological roles. The formation of non-B-DNA within certain sequence elements of DNA can be induced by changes in environmental conditions, protein binding and superhelical tension. Several lines of evidence indicate that alternative DNA structures exist in prokaryotic and eukaryotic cells. The data on their involvement in replication, gene expression, recombination and mutagenesis continues to accumulate.

Introduction

Genetic information is generally stored in long double-stranded DNA molecules. Hydrogen bonding between nucleobases keeps the complementary DNA strands organized into a right-handed helical structure called B-DNA. Structural transitions into other DNA forms can occur within certain sequence elements of DNA and these can be functionally important. Several non-B-DNA structures (oftentimes called unusual or alternative DNA structures) can be important for interactions with proteins involved in replication, gene expression and recombination. They may also play different roles in the formation of nucleosomes and other supramolecular structures involving DNA. DNA sequences characterized as "random" or "mixed sequence" typically only form A-DNA or B-DNA. Special sequence characteristics or defined symmetry elements are required to form alternative structures such as left-handed Z-DNA, cruciforms, intramolecular triplexes, quadruplex DNA, slipped-strand DNA, parallel-stranded DNA, and unpaired DNA structures.[1] Together with variations in DNA supercoiling, local alternative structures provide enormous potential for autoregulation of DNA functions. This chapter will briefly review major alternative DNA structures and their potential involvement in biology.

B-DNA and A-DNA

Structure

Table 1 lists structural parameters for three structural families of DNA helices. B-DNA is the term given for the canonical right-handed DNA helix that is the most common form of DNA. Canonical B-DNA is a double helix made of two antiparallel strands that are held together via hydrogen bonding in the A•T and G•C base pairs (Fig. 1). One helical turn of B-DNA contains about 10.5 base pairs that are buried inside the helix and are almost

DNA Conformation and Transcription, edited by Takashi Ohyama. ©2005 Eurekah.com and Springer Science+Business Media.

Table 1. Structural parameters of DNA helices

Structural Parameter		A-DNA	B-DNA	Z-DNA
Direction of helix rotation		Right handed	Right handed	Left handed
Residue per helical turn		11	10.5	12
Axial rise per residue		2.55 Å	3.4 Å	3.7 Å
Pitch (length) of the helix		28.2 Å	34 Å	44.4 Å
Base pair tilt		20°	-6°	7°
Rotation per residue		32.7°	34.3°	-30°
Diameter of helix		23 Å	20 Å	18 Å
Configuration	dA, dT, dC	anti	anti	anti
of glycosidic bond	dG	anti	anti	syn
Sugar Pucker	dA, dT, dC	C3' endo	C2' endo	C2' endo
	dG	C3' endo	C2' endo	C3' endo

perpendicular to the helical axis. DNA exists as a cylinder of 20 Å in diameter with two grooves, a major and a minor groove, spiraling around the cylinder. In B-DNA the distance between the bases (rise) is 3.4 Å. Studies of oligonucleotide duplexes in crystals showed significant sequence-dependent variability of the structural parameters listed in Table 1 that define the structure of the B-DNA helix. In bent DNA, for example, certain B-DNA parameters add up over a length of several base pairs to produce a permanently curved DNA helix. A- and Z-DNA are also double-helical but the spatial arrangement of base pairs differs significantly from that for B-DNA. Other DNA structures may have regions of unpaired strands or be composed of three and even four strands.

A-DNA has 11 base pairs per helical turn, base pairs are tilted to about 20°, with respect to the helical axis, the grooves are not as deep as those in B-DNA, the sugar pucker is C3' endo compared to C2' endo for B-DNA, and the base pairs are shifted to the helix periphery which

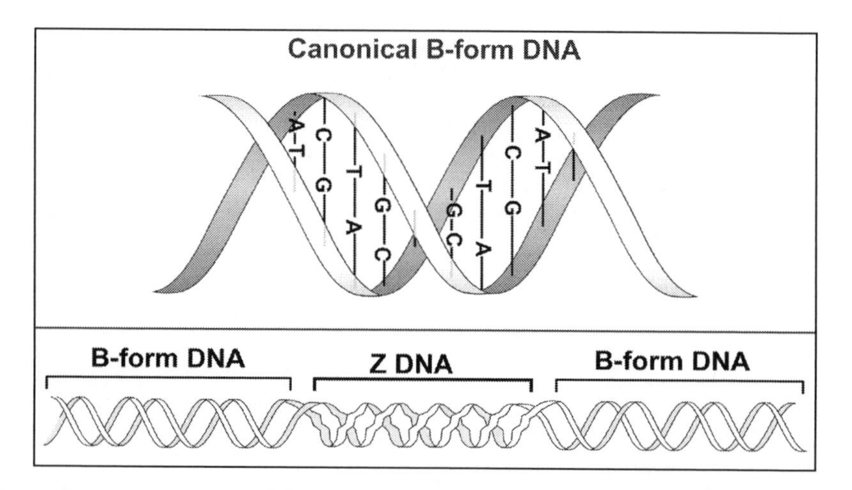

Figure 1. B-DNA and Z-DNA. Top) B-DNA exists as a right-handed helix with about 10.5 bp per helical turn. The bases (denoted by the letters) are organized in the central region of the helix with the phosphate backbone (denoted by the ribbon) distributed on the outside of the helix. B-DNA has two grooves: a major, the wider, groove and a minor, the narrower, groove. Bottom) Left-handed Z-DNA can exist within regions of a B-DNA helix. A Z-DNA helical region is shown flanked by B-DNA regions.

creates a 9 Å hole in the helix center. In A-DNA the average rise is 2.55 Å. A-DNA and B-DNA have different patterns of bound cations and water molecules[2-4] that result in different stability conditions for these structures. B-DNA is stable under a broad variety of conditions, whereas A-DNA has been observed under conditions of reduced water content, such as in DNA fibers at 75% relative humidity or in solutions containing organic solvents or high salt concentrations.[5-7] The ease of conversion from B-DNA into the A-form is somewhat sequence-specific,[3,8] being more difficult in sequences containing 5'-AA-3' steps, and easiest in sequences containing 5'-CC-3' and 5'-ACT-3' steps.[8] Because of the variability in structural parameters for different di- and trinucleotides within the helix, B- and A-DNA actually consist of families of conformations. Experiments with oligonucleotide duplexes in crystals have shown that B- and A-DNA do not represent deep local energetic minima and that a number of intermediate structures may form upon a mild change in conditions.[9-11]

Biological Relevance of A-DNA

Biochemical, crystallographic and computer simulation analyses of the A-DNA structure and protein-DNA complexes indicate that an A-like DNA conformation may either form upon binding of certain proteins to DNA, or be an important intermediate step in forming the strongly distorted DNA conformation observed within at least some complexes with proteins.[12-14] Several examples below illustrate these structural roles of A-DNA in biological processes.

TBP-Binding

Nanosecond scale molecular dynamics simulations in water using two different starting structures show that the DNA oligomer, GCGTATATAAAACGC, which contains a target site for the TATA-box binding protein (TBP), adopts an A-like conformation in the region of the TATA-box and undergoes bending related to that seen within the complex with the TBP.[15] This is consistent with A-DNA being an important intermediate step in forming a strongly distorted DNA structure observed within its complex with TBP in crystals.[9]

CAP Binding

The *Escherichia coli* cyclic AMP receptor protein (CAP) has two symmetrically related inverted recognition elements separated by a spacer whose length may be either 6 or 8 bp (e.g., TGTGAxxxxxxTCACA).[16] CAP binding induces DNA bending with DNA remaining in the B-form when the spacer is 6 bp. For the 8 bp spacer, an additional transition into the A-form is necessary to shorten the distance between TGTGA sites for CAP binding.[17]

Complexes with Polymerases

The B-to-A transition may occur in DNA complexes with enzymes that cut or seal at the (O3'-P) phosphodiester linkage. The transition is necessary to expose atoms of the sugar-phosphate backbone, such as the 3'-oxygen ordinarily buried within the chain backbone, for enzymatic attack.[14] A polymerase-induced A-DNA conformation has been identified in crystallographic studies of HIV reverse transcriptase bound to DNA.[18] The function of a conformational switch from the B-form to an underwound A-form DNA at the polymerase active site may provide discrimination between correct and incorrect base pairing[19] because of a lower sequence-dependent structural variability in A-DNA compared with B-DNA. A-DNA in the vicinity of the DNA polymerase active site may improve the base pair fit in the nascent template-primer duplex and increase a reliability of proofreading thereby contributing to the fidelity of synthesis.[13]

Protection from DNA Damage

A-DNA stabilization by a group of proteins from sporulating bacteria *Bacillus subtilis* has been described.[20] Nucleobases in A-DNA are an order of magnitude less susceptible to UV damage compared with B-DNA.[21] Therefore, the conformational change on protein binding in the spores may be responsible for the well-known resistance of DNA in spores to UV damage.[20]

DNA Supercoiling

In most biological systems DNA is normally negatively supercoiled. Supercoiling is a property of topologically closed DNA molecules (those in which the free rotation of the DNA ends is restrained).[1,22] Through changes in twisting and writhing, supercoiling makes the molecular shape and helix structure of DNA remarkably dynamic. The most important topological property of supercoiled DNA is its linking number, Lk, which is an integer number of times one strand crosses the other in a planar projection. Due to the continuity of DNA strands, the linking number can only change when at least one strand is cut by chemicals, ionizing radiation, or enzymes and then sealed. DNA topology is described by the equation

$$Lk = Tw + Wr$$

where Tw is the number of twists or double helical turns, and Wr is the number of supercoils or writhes. For a covalently closed molecule, Lk must remain constant but Tw and Wr can change simultaneously. For relaxed DNA,

$$Lk = Lk_o = N/10.5$$

where N is the number of base pairs in DNA, and 10.5 is the average number of base pairs per helical turn. Usually DNA isolated from cells is negatively supercoiled, such that $(Lk\text{-}Lk_o) < 0$. DNA with $Lk < Lk_o$ is said to be underwound in terms of the number of helical turns. Such a state of DNA underwinding results in a torsional tension in the DNA double helix. The deficit of helical twists is compensated for by DNA supertwisting into the right-handed supercoils. The lack of one helical turn results in one supercoil. The level of supercoiling is characterized by the term "superhelical density" or σ, where

$$\sigma = 10.5\ \tau/N$$

where τ is the number of titratable (measurable) supercoils (and 10.5 and N are as defined above). Besides existing as interwound supertwists, negative supercoils can exist as left-handed toroidal coils that can be represented, for example, by DNA wrapping around a protein. Negatively supercoiled DNA contains free energy since the underwinding creates a high energy state. The free energy of supercoiling is given by the following relationship

$$\Delta G = (1100\ RT/N)(Lk\text{-}Lk_o)^2$$

where R is the gas constant, T is the temperature in degrees Kelvin, and N is as defined above. The free energy of supercoiling can be used to locally unpair the DNA helix and drive the formation of alternative DNA structures (see below) or unwind DNA for interaction with transcription or replication proteins.

In vivo most DNA is negatively supercoiled. This is easily understood for circular molecules such as plasmids and bacterial chromosomes in which the free rotation of DNA strands is restrained. Circular bacterial chromosomes are long enough to be additionally subdivided into smaller topological domains. In fact, the 2.9 Mb *E. coli* chromosome is organized into about 45 independent domains in vivo.[23] For linear DNA to exist in a supercoiled state, it must be organized into one or more topological domains. Eukaryotic chromosomes may form independent loops stabilized by the interaction of specific DNA regions with proteins attached to the nuclear matrix. In addition, RNA polymerase can define topological domains in eukaryotic cells.[24]

Linking number in vivo is regulated by enzymes called topoisomerases that transiently break and reseal the DNA double helix. Type I topoisomerases break only one strand of the DNA, allowing one strand to rotate around the other. Type II topoisomerases break and reseal both DNA strands. Correspondingly, the linking number changes in increments of 1 and 2 for type I and type II topoisomerases, respectively. In bacterial cells the level of supercoiling is carefully maintained by topoisomerase I, that relaxes supercoils, and topoisomerase II (gyrase), that introduces negative supercoils. In bacterial cells about half of the free energy from DNA supercoiling (called unrestrained supercoiling) is available for biological reactions, while the other half is presumably restrained by virtue of stable left-handed toroidal coiling around proteins. On average in bulk eukaryotic DNA, supercoils are restrained by the organization into nucleosomes. However, DNA in individual genes can contain unrestrained negative supercoiling.[24-27] Transient changes in the level of supercoiling can be caused by proteins tracking through the DNA. In particular, the movement of an RNA polymerase during transcription generates waves of negative supercoiling behind and positive supercoiling in front of the enzyme.[28]

The state of DNA supercoiling may be important for the regulation of cell functions in a number of ways. (i) The energy from DNA supercoiling can be used to facilitate the opening of the promoter or origin of replication regions by RNA polymerase or replication proteins.[1] (ii) DNA supercoiling may facilitate functional enhancer-promoter communication over a large distance, probably by bringing the enhancer and promoter in the plectonemically wound DNA into close proximity.[29] (iii) The supercoil-induced formation of alternative structures in the regulatory regions may also influence protein binding. One particular example, albeit an artificially created system, is a down-regulation of transcription from an inverted repeat-containing promoter where the cruciform formation possibly prevents an assembly of transcription machinery.[30] An example of transcriptional up-regulation is a likely supercoil-driven Z-DNA formation in the Rous sarcoma virus promoter that prevents nucleosome formation and facilitates access of transcription proteins to the gene regulatory regions.[31]

Supercoil-Induced DNA Structures and Their Biological Roles

While DNA mostly has a seemingly random distribution of nucleobases in the sequence, defined order sequences may rather frequently occur. These include inverted repeats that can form cruciforms, mirror repeats that may adopt intramolecular triplex DNA conformations, and direct repeats, that can form slipped mispaired structures, and $(GC)_n$ and $(GT)_n$ tracts that can form Z-DNA.

Cruciform Structure

An inverted repeat or a palindrome is a DNA sequence that reads the same from the 5' to 3' in either strand. For example, many type II restriction enzyme sites are palindromic. To form a cruciform the *inter*strand hydrogen bonds in the inverted repeat must be broken and *intra*strand hydrogen bonds then established between complementary bases in each single strand, thus forming two hairpin-like arms with 3-4 unpaired bases at their tips (Fig. 2A). As a whole, the cruciform consists of two rather long duplex DNA arms, and two comparatively short hairpin arms which form a four-way junction. The structure of the four-way junction is such that the nucleobases in and around the junction are fully involved in base pairing.[32] Cruciforms can form in topologically closed molecules where they use energy from DNA supercoiling to melt the center of the inverted repeat, allowing the intrastrand hairpin nucleation.[1,32] The thermodynamic stability of the cruciform comes from relaxation of one negative supercoil per 10.5 bp of DNA sequence that converts into the cruciform. The propensity for cruciform formation increases in longer inverted repeats that relax more supercoils than shorter ones. It also depends on temperature and the base composition of the inverted repeat, most importantly, in its center, in accordance with a requirement of partial DNA melting before the hairpin base pairing. Although schematically the cruciform is usually shown as having a cross shape as in the schematic representation in Figure 2, such an extended structure is favored only under the low-salt conditions, where electrostatic repulsion between phosphates pushes all four cruciform arms apart (Fig. 2B). Under physiologically relevant salt conditions, where the phosphates are partially shielded and repulsion is reduced, the cruciform adopts an X-type structure with unequal inter-arm angles as seen in the AFM image in Figure 2C.[33] The extended cruciform is rather stiff, as judged from little fluctuation of the inter-arm angles, whereas the X-type cruciform has a pronounced mobility of the hairpin arms observed by atomic force microscopy in liquid.[33] The distribution of inverted repeats in eukaryotic DNA is nonrandom and they are clustered at or near genetic regulatory regions, which suggests that they are important biologically.[34-36]

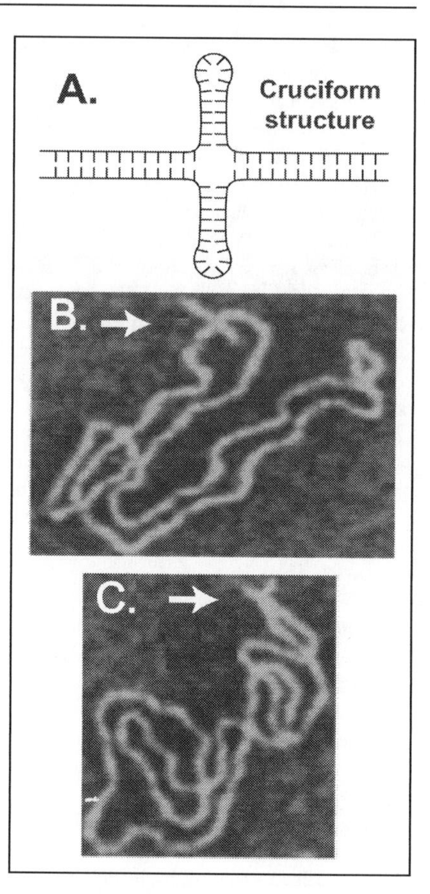

Figure 2. Cruciform structures. A) A schematic representation of a cruciform is shown. An unpaired loop of 3-4 bp typically exists at the tip of the cruciform arms. B) Atomic force microscopy (AFM) image of a 106 bp inverted repeat extruded into a cruciform with 53 bp arms in the *extended* conformation in supercoiled plasmid DNA. C) AFM image of a 106 bp inverted repeat extruded into a cruciform with 53 bp arms in the *X-type* conformation.

Z-DNA Structure

Left-handed Z-DNA has been mostly found in alternating purine-pyrimidine sequences $(CG)_n$ and $(TG)_n$.[37] Z-DNA is thinner (18 Å) than B-DNA (20 Å), the bases are shifted to the periphery of the helix, and there is only one deep, narrow groove equivalent to the minor groove in B-DNA. In contrast to B-DNA where a repeating unit is 1 base pair, in Z-DNA the repeating unit is 2 bp. For Z-DNA in $(CG)_n$ sequences the twist angle for a CpG step is 9°, whereas it is 51° for the GpC step, totaling 60° in the 2 bp repeating unit. The helix repeat in Z-DNA is 12 bp/turn and an average rise is 3.7 Å/bp, compared with 10.5 bp/turn and 3.4 Å/bp in B-DNA. The backbone follows a zigzag path as opposed to a smooth path in B-DNA. The sugar and glycosidic bond conformations alternate: C2' endo in *anti* dC or dT and C3' endo in *syn* dG or dA. Electrostatic interactions play a crucial role in Z-DNA formation. Because of the zigzag backbone path, some phosphate groups are closer and electrostatic repulsion between them is greater than in B-DNA. Therefore, Z-DNA is stabilized by high salt concentrations or polyvalent cations that shield interphosphate repulsion better than monovalent cations. Other factors also contribute to Z-DNA stability. If an alternating purine-pyrimidine sequence occurs in a circular DNA molecule, DNA supercoiling is a major driving force for Z-DNA formation. Z-DNA formation unwinds DNA about two supercoils per 12 bp of DNA. The junctions between the B- and Z-DNA in supercoiled DNA span several base pairs in which nucleobases behave as if they were unpaired. In particular, they are partially reactive to single-strand specific chemicals. A computer analysis of over one million base pairs of human DNA, containing 137 complete genes identified 329 potential Z-DNA-forming sequences.[38]

Like inverted repeats, potential Z-DNA-forming sequences have a distinctly nonrandom distribution with a strong bias toward locations near the site of transcription initiation.

Triplex DNA Structure

When the hydrogen bonds in the A•T and G•C base pairs in canonical B-form DNA are formed, several hydrogen bond donor and acceptor groups in nucleobases remain unused. Each purine base has two such groups on the edges that are exposed in the major groove. These groups can be used to form base triads that are unit blocks of triple-stranded (triplex) DNA that consists of the B-form double helix and the third strand bound in the major groove.[39-42] The third strand bases form the so-called Hoogsteen-type hydrogen bonds with purines in the B-form duplex. Energetically favorable triplexes have duplex pyrimidines (Py) and purines (Pu) segregated in complementary strands (Py•Pu duplex). For a snug fit in the duplex major groove, the third strands are made of either only pyrimidines (Py•Pu•Py triplex), or mostly purines with a fraction of pyrimidines (Py•Pu•Pu triplex). In the Py•Pu•Py triplex, the usual base triads are T•A•T and C•G•C$^+$ (cytosine is protonated and this requires pH < 5). In the Py•Pu•Pu triplex the usual triads are T•A•A and C•G•G, and less frequently T•A•T. Triplex DNA may form intermolecularly, between a duplex target and a third oligonucleotide strand. It may also form intramolecularly in supercoiled DNA within a Py•Pu sequence of mirror repeat symmetry. For this, half of the mirror repeat Py•Pu sequence unpairs and one of the unpaired strands folds back and binds as a third strand to purines in the repeat's double-stranded half. The resulting local structure contains three notable features: a triple-stranded region; a fourth, unpaired strand; and a short (3-4 nt) stretch of unpaired bases in the fold-back strand (Fig. 3A). The Py•Pu•Py triplex/single strand combination is termed H-DNA to reflect the necessity of cytosine protonation in the C•G•C$^+$ triads.[43] By analogy, the Py•Pu•Pu triplex/single strand combination is termed H'-DNA.[44] Similar to the cruciform, H (H')-DNA may only form under torsional stress in a topologically closed DNA (Fig. 3B).[41,42] Among other factors that promote H (H')-DNA are longer lengths of Py•Pu mirror repeats and the presence of multivalent cations.[42,45] The presence of single-stranded regions provides the DNA molecule with local increased flexibility akin to a hinge, which is incidentally another reason for calling the structure H-DNA. However, the angle between the outgoing duplex arms in the H-DNA structure fluctuates over a smaller range than in the X-type cruciform.[46] Analysis of the genomic databases showed that in eukaryotes mirror repeated sequences occur more frequently than statistically expected.[34,36] In the human genome, H-DNA-forming sequences may occur as frequently as 1 in 50,000 bp, whereas in the *E. coli* genome they are not abundant.[34]

Search for Unusual DNA Structures in Vivo

Numerous attempts have been undertaken to show the formation of supercoil-induced alternative DNA structures in living cells. The differential chemical susceptibility of double and single-stranded DNA regions in cruciforms and H-DNA as well as of structural junctions in Z-DNA has been exploited to probe for the formation of alternative structures in vivo. Using OsO_4 reactivity with unpaired thymines, the cruciforms,[47] Z-DNA[48] and H-DNA[49] were detected in supercoiled plasmids propagated in *E. coli* cells. The differences in photochemical reactivity of TA dinucleotides with psoralen (reactive in double-stranded DNA but not in single-stranded DNA or in the junctions between the B- and Z-DNA regions) were used to show the formation of cruciforms, Z-DNA, and H-DNA in *E. coli*.[50,51] H'-DNA formation in *E. coli* was also detected by chloroacetaldehyde reactivity with unpaired adenines and cytosines.[52] Elevated plasmid supercoiling in *E. coli* was interpreted as a combination of (i) supercoil relaxation by the formation of Z-DNA or cruciforms and (ii) a compensatory supercoiling increase by DNA gyrase.[53,54] The analysis of sites differentially susceptible to DNA methylase in B- and Z-DNA showed that $(GC)_n$ sequences in plasmids or integrated in the *E. coli* chromosome form Z-DNA in vivo.[55,56] Monoclonal antibodies were raised that recognize structural features of either cruciforms, Z-DNA, or triple-stranded DNA. These were then used to

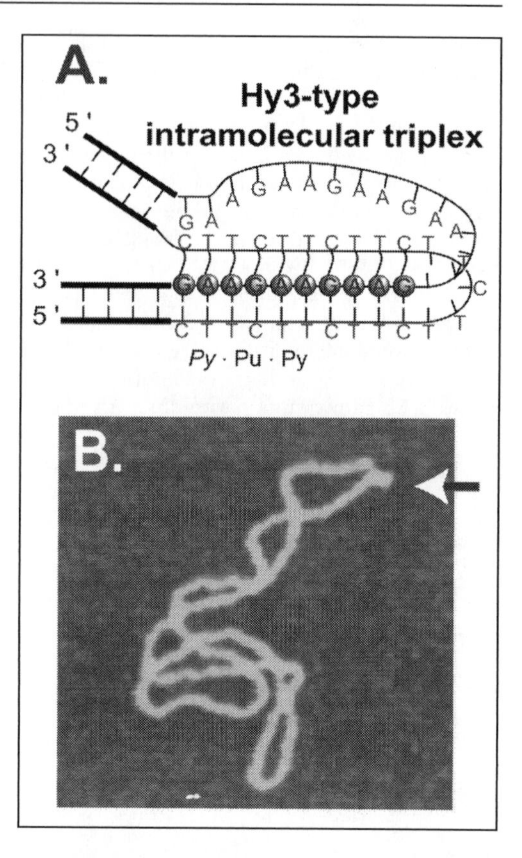

Figure 3. Intramolecular triplex DNA. A) A schematic representation of an Hy-3 type intramolecular triplex is shown, formed within a mirror repeat sequence of $(GAA)_n$•$(TTC)_n$. B) AFM image of a 46 Pu•Py tract from a region of the human *PKD1* gene in a triplex conformation within a plasmid DNA.

probe eukaryotic chromosomal DNA for the structures in question. Local structural transitions into cruciforms,[57] Z-DNA[58] and triplex DNA[59] were detected by immunofluorescence. Thus, several lines of evidence indicate the presence of alternative DNA structures in prokaryotic and eukaryotic cells. The existence of proteins that specifically bind to alternative DNA structures also supports the notion of H-, Z- and cruciform DNA formation in vivo.

Cruciform, Z-DNA and H-DNA-Binding Proteins

An integral part of our understanding of the biological roles of alternative DNA structures comes from the identification of proteins that specifically interact with these structures. A number of proteins bind to different structural elements in cruciforms.[35] They include HMG proteins, a replication initiation protein RepC, the cruciform binding protein CBP,[35] and four-way junction resolvases.[60,61] Among the Z-DNA-binding proteins are the highly specific binders, such as Zα domain-contatining proteins ADAR1 and ESL3,[37,62] and relatively low specific proteins, such as HMG proteins, zeta crystalline and type III intermediate filament proteins.[63] Proteins that bind to triple-stranded DNA have been identified in the HeLa cell extracts and keratinocyte cDNA expression library.[64-66] In addition, several proteins that bind single-stranded Py or Pu sequences have been partially characterized.[42,67-69]

Possible Biological Roles of Supercoil-Driven Alternative Structures

Many similar biological roles for alternative DNA structures including cruciforms, Z-DNA, and H-DNA have been proposed.[1] This is perhaps not surprising because these sequences often occur in the regulatory regions of genes that may use different structures for the same purposes. The dependence of all three structures on DNA supercoiling as well as the preference of structures to form in certain locations of topological domains also add to the apparent similarity in their functions.

Modulation of Supercoiling

The extent of supercoiling is known to affect transcription, recombination, and replication such that an optimum DNA topology may be required for these processes.[1,70] The formation of cruciforms, Z-DNA and H-DNA may cause partial relaxation of excessive superhelicity in a topological domain. Specific cases of DNA replication and gene expression have been described that depend on superhelicity changes induced by the formation of cruciforms, Z-DNA and H-DNA.[30,35,71]

Nucleosome Exclusion

DNA wrapping around histones in nucleosomes interferes with the protein binding to promoters and origins of replication.[72] Nucleosome formation, on the one hand, and the formation of cruciforms, Z-DNA and triplex DNA, on the other hand, are mutually exclusive.[31,73-75] Thus, the alternative structure-forming DNA sequences may expose nucleosome-free DNA, making them accessible to transcription, replication and recombination proteins.

Positioning of Sequence Elements, Molecular Switch

Supercoiled DNA at physiological ionic strength forms a plectonemic superhelix in which distant parts of the double helix are intertwined. The slithering motion of one duplex region on the other results in a wide distribution of distances between any two pre-selected remote sites. Similar to strongly bent DNA,[76] the X-type cruciforms and H-DNA tend to occupy the apical positions in plectonemic DNA structures[33,46] and therefore, may specifically position distant DNA sites. This was first realized for H-DNA whose fold-back structure seemed suitable for bringing remote sequence elements into close proximity. In agreement with this idea, increased recombination rates were observed when homologous sequences were separated by H-DNA-forming elements.[77,78] It is likely that the X-type cruciforms may also position DNA elements for recombination or for promoter-enhancer interactions. Moreover, cruciform transitions between the X-type and extended conformations may serve to switch between the favorable and unfavorable arrangements of interacting DNA sites.[79]

Roles in Transcription

Analyses of genomic databases show that sequences capable of forming cruciforms, Z- and H-DNA are frequently found around transcription initiation sites.[34,36,38] The formation of alternative DNA structures in these sequences may influence transcription by changing the supercoiling levels within a domain thereby changing the energy cost for protein-DNA binding. The formation of an alternative structure may also alter interactions between transcription factors bound to different sites due to a change in their spatial positioning. At least two of the structures, cruciforms and H-DNA, may spatially organize DNA around their formation sites so that certain DNA segments are brought into close proximity.[33,46] Gene expression may also depend on protein binding to unusual DNA structures. For example, poly(ADP-ribose) polymerase (PARP) may bind to the junction-containing DNA structures such as cruciforms.[80] Repressive PARP binding to potential cruciforms in a promoter of its own gene and dissociation upon DNA strand break-induced autoribosylation are parts of the mechanism of autoregulation of PARP expression.[80,81] In another example, in the human proenkephalin gene switching of a region of DNA between the linear and cruciform form provides a mechanism of gene regulation.[82] More correlations of transcrption with the formation of non-B-DNA structures are discussed in detail in other chapters of this book.

Roles in Replication

One of the well-studied effects of alternative structures on replication is a block to polymerases due to template folding, which was shown for cruciforms/hairpins[83,84] and H-DNA.[85-87] Unless unwound by the replication accessory proteins, including helicases,[88] polymerization blocks may result in genetic mutations that lead to the development of human diseases, such as polycystic kidney disease and Friedrich ataxia. Single-stranded parts of the cruciform and H-DNA may serve as recognition elements for the replication initiation proteins.[35,89] Protein binding may also be directed to the four-way junction of the cruciform to initiate replication as shown for CBP in HeLa cells.[35]

Roles in Recombination

There are several relationships between the formation of alternative structures and DNA recombination. Consistent with an idea of sequence positioning by a fold-back structure of H-DNA, facilitated recombination was observed between distant elements separated by the Py•Pu tract.[77,78] Several models of Z-DNA assisted recombination have been proposed.[90] DNA strand exchange during recombination requires initial duplex-duplex interaction. For this, exposed N7 and C8 of guanosines in one Z-DNA duplex are available for interaction with another Z-DNA duplex so as to initiate recombination. During the synapsis step in homologous recombination a paranemic joint, a nascent heteroduplex where strands from different DNA molecules base pair without breaking them, can be formed from the alternating left-handed and right-handed turns.

Slipped-Strand DNA

If a region of DNA contains a block of several nucleotides that repeats many times, there are multiple opportunities for the formation of base pairs in an out-of-register or "slipped" fashion. A slipped-strand DNA (S-DNA) structure forms when a section of the repeating duplex unwinds so that one region of the direct repeat forms the Watson-Crick base pairs with another region of the repetitive sequence forming two loop-out regions in opposite strands (Fig. 4A).[1,91] The likelihood of DNA slippage increases with increasing length of the repeats and increasing potential for partial base pairing in the looped-out single strand. The out-of-register base pairing is more probable in the GC-rich repeats because they have a better propensity for nucleation of the double-stranded structure than the average 50% GC flanking sequences if the DNA strands are temporarily separated and then allowed to re-form the duplex. Interruptions in the direct repeat tracts significantly reduce the number of possible out-of-register configurations and, therefore, the probability of S-DNA formation.

Biological Significance

S-DNA has been of considerable interest in the last decade. Fourteen genetic neurodegenerative diseases and three fragile sites have been associated with the expansion of $(CTG)_n$•$(CAG)_n$, $(CGG)_n$•$(CCG)_n$, or $(GAA)_n$•$(TTC)_n$ repeat tracts. Different models have been proposed for the expansion of triplet repeats, most of which presume the formation of alternative DNA structures in repeat tracts. One of the most likely structures, S-DNA, can stably and reproducibly form within the GC-rich triplet repeat sequences, $(CTG)_n$•$(CAG)_n$, $(CGG)_n$•$(CCG)_n$. In fact, given that the loops of the slipped out arms are complementary, good evidence exists that there is a further conformational transition to a folded slipped strand structure (Fig. 4B), as formed by the model slipped strand structure shown in Figure 4C. S-DNA may be involved in triplet repeat mutagenesis in several ways, such as a simple primer/template misalignment or reiterative synthesis, involving repetitive slippage events. More details on the S-DNA structure and its role in the triplet repeat mutagenesis may be found in recent reviews.[91,92]

Slipped misalignment during DNA replication is very important in spontaneous frameshift mutagenesis. In 1966, Streisinger et al proposed a model that explained frameshift mutations within runs of a single base by a slippage of the nascent DNA strand on the replication template strand.[93] Since the genetic code is read as triplets, adding or deleting a single base shifts the reading frame of all bases downstream of the mutation. As a result, part of the mRNA encodes amino acids that are different from those in the wild-type protein. Further work has shown that direct repeats and more complex DNA repeats often contribute to frameshift

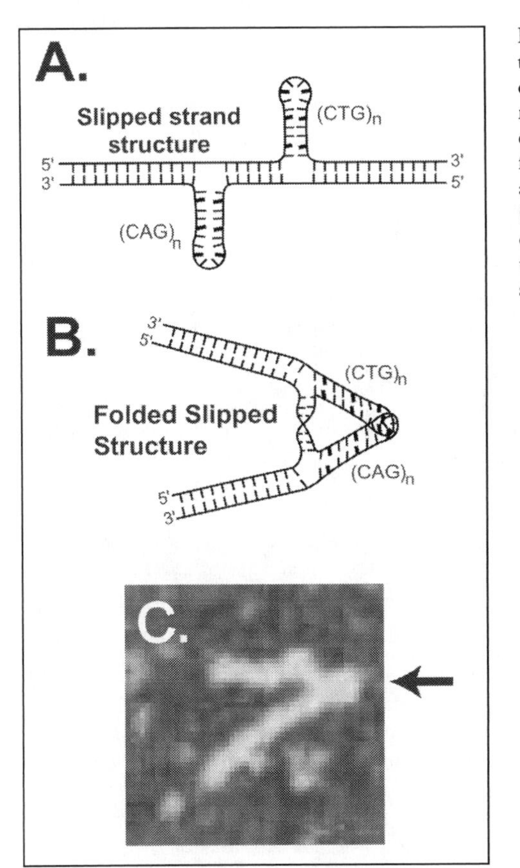

Figure 4. Slipped-strand DNA. A) A representation of slipped-strand DNA, formed within a tract of $(CTG)_n{\bullet}(CAG)_n$ is shown. B) The complementary loops of the two slipped-strand structure can form interstrand hydrogen bonds, forming a folded slipped strand structure. C) An AFM image of a DNA molecule containing $(CTG)_{23}$ and $(CAG)_{23}$ hairpin arms, separated by 30 bp, in opposite strands. The molecule is folded with a thick DNA region that likely represents a folded slipped-strand structure.[92]

mutagenesis.[94-96] The hairpin-forming sequences intervening the repeats have been shown to stabilize S-DNA and promote mutations.

DNA Unwinding Elements

DNA unwinding elements (DUE) have been identified in both prokaryotic and eukaryotic DNA sequences (see ref. 1, for review). DUEs are AT-rich sequences about 30-100 bp long. They have little sequence similarity except for being AT-rich. Under torsional stress, unwinding of the double helix occurs first in AT-rich sequences, therefore, DUEs can be maintained as unpaired DNA regions in the presence of negative supercoiling (Fig. 5A). In the presence of Mg^{2+}, DUEs tend to remain double-stranded and other regions (such as inverted repeats) unwind to partially relieve superhelical tension. Thus, the ability of DUEs to form denaturation bubbles may be dependent on the level of unrestrained supercoiling and the local ionic environment in cells.

Biological Significance

DUEs are commonly associated with replication origins and chromosomal matrix attachment regions. DUEs are a common feature of DNA replication origins in *E. coli* and yeast.[97] Replication from the yeast origins shows a correlation between the extent of DNA unwinding and the proficiency of the DUEs as replication origins. Similarly, DNA unwinding is also required at the *E. coli* origin of replication. The AT-rich DUEs are also found in at least some mammalian origins.[98] Thus, DUEs seem to fulfill a primary requirement for the initiation of DNA replication in all systems, which is the formation of an unpaired region in DNA where

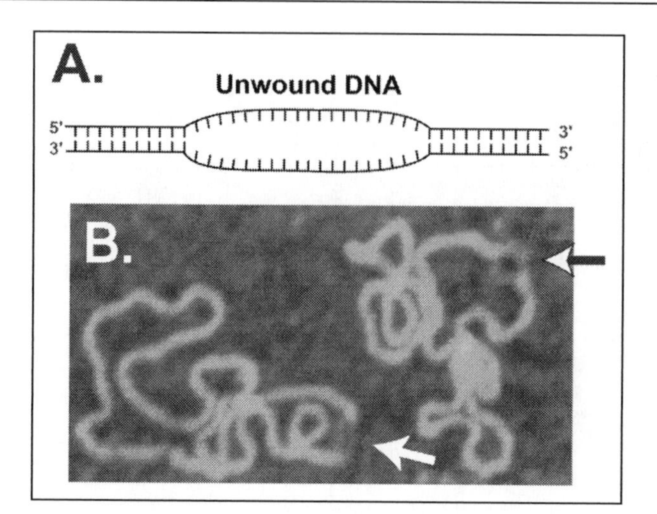

Figure 5. Unwound (unpaired) DNA. A) A representation of an unpaired region within a B-DNA helical region is shown. This structure, typically called unwound DNA, is formed under superhelical tension within A+T rich DNA regions. B) Unwound regions in supercoiled plasmids, formed within $(ATTCT)_{23}$•$(AGAAT)_{23}$ tracts from the human *SCA10* gene.

the replication complex assembles. Recent studies showed that an AT-rich repetitive sequence $(ATTCT)_n$•$(AGAAT)_n$, whose spontaneous length expansion has been associated with the development of the disease, spinocerebellar ataxia type 10, has the properties of DUE (Fig. 5B).[99] Under superhelical stress, the repeating sequence preferentially unpairs and may potentially bind the proteins of the replication complex. Unscheduled initiation of replication from the false origin in combination with a possible primer/template slippage in the repeated sequence may produce longer than expected products of DNA replication. These may be incorporated into the repeat tract leading to the expansion of repeat length and eventually to the development of the disease.[99]

Conclusion

Local structural transitions from the common B-DNA conformation into other DNA forms can be functionally important. Such transitions within certain sequence elements of DNA can be induced by changes in environmental conditions, protein binding and superhelical tension. Several lines of evidence indicate that alternative DNA structures may exist in prokaryotic and eukaryotic cells. The data on their involvement in replication, gene expression, recombination and mutagenesis continues to accumulate.

References

1. Sinden RR. DNA Structure and Function. San Diego: Academic Press, 1994.
2. Cheatham TE III, Kollman PA. Insight into the stabilization of A-DNA by specific ion association: spontaneous B-DNA to A-DNA transitions observed in molecular dynamics simulations of $d[ACCCGCGGGT]_2$ in the presence of hexaamminecobalt(III). Structure 1997; 5:1297-1311.
3. Feig M, Pettitt BM. A molecular simulation picture of DNA hydration around A- and B-DNA. Biopolymers 1998; 48:199-209.
4. Egli M. DNA-cation interactions: quo vadis? Chem Biol 2002; 9:277-286.
5. Fuller W, Wilkins MHF, Wilson HR et al. The molecular configuration of deoxyribonucleic acid. IV. X-ray diffraction of the A form. J Mol Biol 1965; 12:60-80.
6. Ivanov VI, Minchenkova LE, Minyat EE et al. The B to A transition of DNA in solution. J Mol Biol 1974; 87:817-833.
7. Nishimura Y, Torigoe C, Tsuboi M. Salt induced B-A transition of poly(dG)•poly(dC) and the stabilization of A form by its methylation. Nucleic Acids Res 1986; 14:2737-2748.

8. Ivanov VI, Minchenkova LE. The A-form of DNA: in search of the biological role. Mol Biol (Mosk) 1994; 28:1258-1271; Engl Transl 1995; 28:780-788.
9. Guzikevich-Guerstein G, Shakked Z. A novel form of the DNA double helix imposed on the TATA-box by the TATA-binding protein. Nat Struct Biol 1996; 3:32-37.
10. Vargason JM, Henderson K, Ho PS. A crystallographic map of the transition from B-DNA to A-DNA. Proc Natl Acad Sci USA 2001; 98:7265-7270.
11. Dickerson RE, Ng HL. DNA structure from A to B. Proc Natl Acad Sci USA 2001; 98:6986-6988.
12. Nekludova L, Pabo CO. Distinctive DNA conformation with enlarged major groove is found in Zn-finger-DNA and other protein-DNA complexes. Proc Natl Acad Sci USA 1994; 91:6948-6952.
13. Timsit Y. DNA structure and polymerase fidelity. J Mol Biol 1999; 293:835-853.
14. Lu XJ, Shakked Z, Olson WK. A-form conformational motifs in ligand-bound DNA structures. J Mol Biol 2000; 300:819-840.
15. Flatters D, Young M, Beveridge DL et al. Conformational properties of the TATA-box binding sequence of DNA. J Biomol Struct Dyn 1997; 14:757-765.
16. Barber AM, Zhurkin VB, Adhya S. CRP-binding sites: evidence for two structural classes with 6-bp and 8-bp spacers. Gene 1993; 130:1-8.
17. Ivanov VI, Minchenkova LE, Chernov BK et al. CRP-DNA complexes: inducing the A-like form in the binding sites with an extended central spacer. J Mol Biol 1995; 245:228-240.
18. Jacobo-Molina A, Ding J, Nanni RG et al. Crystal structure of human immunodeficiency virus type 1 reverse transcriptase complexed with double-stranded DNA at 3.0 Å resolution shows bent DNA. Proc Natl Acad Sci USA 1993; 90:6320-6324.
19. Kiefer JR, Mao C, Braman JC et al. Visualizing DNA replication in a catalytically active *Bacillus* DNA polymerase crystal. Nature 1998; 391:304-307.
20. Mohr SC, Sokolov NV, He CM et al. Binding of small acid-soluble spore proteins from *Bacillus subtilis* changes the conformation of DNA from B to A. Proc Natl Acad Sci USA 1991; 88:77-81.
21. Becker MM, Wang Z. B-A transitions within a 5 S ribosomal RNA gene are highly sequence-specific. J Biol Chem 1989; 264:4163-4167.
22. Vologodskii AV. Topology and Physics of Circular DNA. Boca Raton: CRC Press, 1992.
23. Sinden RR, Pettijohn DE. Chromosomes in living *Escherichia coli* cells are segregated into domains of supercoiling. Proc Natl Acad Sci USA 1981; 78:224-228.
24. Kramer PR, Fragoso G, Pennie W et al. Transcriptional state of the mouse mammary tumor virus promoter can affect topological domain size in vivo. J Biol Chem 1999; 274:28590-28597.
25. Ljungman M, Hanawalt PC. Presence of negative torsional tension in the promoter region of the transcriptionally poised dihydrofolate reductase gene in vivo. Nucleic Acids Res 1995; 23: 1782-1789.
26. Jupe ER, Sinden RR, Cartwright IL. Specialized chromatin structure domain boundary elements flanking a *Drosophila* heat shock gene locus are under torsional strain in vivo. Biochemistry 1995; 34:2628-2633.
27. Kramer PR, Sinden RR. Measurement of unrestrained negative supercoiling and topological domain size in living human cells. Biochemistry 1997; 36:3151-3158.
28. Liu LF, Wang JC. Supercoiling of the DNA template during transcription. Proc Natl Acad Sci USA 1987; 84:7024-7027.
29. Liu Y, Bondarenko V, Ninfa A et al. DNA supercoiling allows enhancer action over a large distance. Proc Natl Acad Sci USA 2001; 98:14883-14888.
30. Horwitz MS, Loeb LA. An *E. coli* promoter that regulates transcription by DNA superhelix-induced cruciform extrusion. Science 1988; 241:703-705.
31. Krajewski WA. Enhancement of transcription by short alternating C•G tracts incorporated within a Rous sarcoma virus-based chimeric promoter: in vivo studies. Mol Gen Genet 1996; 252: 249-254.
32. Murchie AI, Lilley DM. Supercoiled DNA and cruciform structures. Methods Enzymol 1992; 211:158-180.
33. Shlyakhtenko LS, Potaman VN, Sinden RR et al. Structure and dynamics of supercoil-stabilized DNA cruciforms. J Mol Biol 1998; 280:61-72.
34. Schroth GP, Ho PS. Occurrence of potential cruciform and H-DNA forming sequences in genomic DNA. Nucleic Acids Res 1995; 23:1977-1983.
35. Pearson CE, Zorbas H, Price GB et al. Inverted repeats, stem-loops, and cruciforms: significance for initiation of DNA replication. J Cell Biochem 1996; 63:1-22.
36. Cox R, Mirkin SM. Characteristic enrichment of DNA repeats in different genomes. Proc Natl Acad Sci USA 1997; 94:5237-5242.
37. Herbert A, Rich A. Left-handed Z-DNA: structure and function. Genetica 1999; 106:37-47.
38. Schroth GP, Chou PJ, Ho PS. Mapping Z-DNA in the human genome. Computer-aided mapping reveals a nonrandom distribution of potential Z-DNA-forming sequences in human genes. J Biol Chem 1992; 267:11846-11855.

39. Wells RD, Collier DA, Hanvey JC et al. The chemistry and biology of unusual DNA structures adopted by oligopurine•oligopyrimidine sequences. FASEB J 1988; 2:2939-2949.
40. Htun H, Dahlberg JE. Topology and formation of triple-stranded H-DNA. Science 1989; 243:1571-1576.
41. Frank-Kamenetskii MD, Mirkin SM. Triplex DNA structures. Annu Rev Biochem 1995; 64:65-95.
42. Soyfer VN, Potaman VN. Triple-Helical Nucleic Acids. New York: Springer, 1996.
43. Lyamichev VI, Mirkin SM, Frank-Kamenetskii MD. Structures of homopurine-homopyrimidine tract in superhelical DNA. J Biomol Struct Dyn 1986; 3:667-669.
44. Kohwi Y, Kohwi-Shigematsu T. Magnesium ion-dependent triple-helix structure formed by homopurine-homopyrimidine sequences in supercoiled plasmid DNA. Proc Natl Acad Sci USA 1988; 85:3781-3785.
45. Potaman VN, Sinden RR. Stabilization of intramolecular triple/single-strand structure by cationic peptides. Biochemistry 1998; 37:12952-12961.
46. Tiner WJ Sr, Potaman VN, Sinden RR et al. The structure of intramolecular triplex DNA: atomic force microscopy study. J Mol Biol 2001; 314:353-357.
47. McClellan JA, Boublikova P, Palecek E et al. Superhelical torsion in cellular DNA responds directly to environmental and genetic factors. Proc Natl Acad Sci USA 1990; 87:8373-8377.
48. Rahmouni AR, Wells RD. Stabilization of Z DNA in vivo by localized supercoiling. Science 1989; 246:358-363.
49. Karlovsky P, Pecinka P, Vojtiskova M et al. Protonated triplex DNA in *E. coli* cells as detected by chemical probing. FEBS Lett 1990; 274:39-42.
50. Zheng GX, Kochel T, Hoepfner RW et al. Torsionally tuned cruciform and Z-DNA probes for measuring unrestrained supercoiling at specific sites in DNA of living cells. J Mol Biol 1991; 221:107-122.
51. Ussery DW, Sinden RR. Environmental influences on the in vivo level of intramolecular triplex DNA in *Escherichia coli*. Biochemistry 1993; 32:6206-6213.
52. Kohwi Y, Malkhosyan SR, Kohwi-Shigematsu T. Intramolecular dG•dG•dC triplex detected in *Escherichia coli* cells. J Mol Biol 1992; 223:817-822.
53. Haniford DB, Pulleyblank DE. The in vivo occurrence of Z DNA. J Biomol Struct Dyn 1983; 1:593-609.
54. Haniford DB, Pulleyblank DE. Transition of a cloned $d(AT)_n$-$d(AT)_n$ tract to a cruciform in vivo. Nucleic Acids Res 1985; 13:4343-4363.
55. Jaworski A, Hsieh WT, Blaho JA et al. Left-handed DNA in vivo. Science 1987; 238:773-777.
56. Lukomski S, Wells RD. Left-handed Z-DNA and in vivo supercoil density in the *Escherichia coli* chromosome. Proc Natl Acad Sci USA 1994; 91:9980-9984.
57. Ward GK, Shihab-el-Deen A, Zannis-Hadjopoulos M et al. DNA cruciforms and the nuclear supporting structure. Exp Cell Res 1991; 195:92-98.
58. Nordheim A, Lafer EM, Peck LJ et al. Negatively supercoiled plasmids contain left-handed Z-DNA segments as detected by specific antibody binding. Cell 1982; 31:309-318.
59. Agazie YM, Burkholder GD, Lee JS. Triplex DNA in the nucleus: direct binding of triplex-specific antibodies and their effect on transcription, replication and cell growth. Biochem J 1996; 316(Pt 2):461-466.
60. Sharples GJ. The X philes: structure-specific endonucleases that resolve Holliday junctions. Mol Microbiol 2001; 39:823-834.
61. Constantinou A, Chen XB, McGowan CH et al. Holliday junction resolution in human cells: two junction endonucleases with distinct substrate specificities. EMBO J 2002; 21:5577-5585.
62. Kim YG, Muralinath M, Brandt T et al. A role for Z-DNA binding in vaccinia virus pathogenesis. Proc Natl Acad Sci USA 2003; 100: 6974-6979.
63. Li G, Tolstonog GV, Traub P. Interaction in vitro of type III intermediate filament proteins with Z-DNA and B-Z-DNA junctions. DNA Cell Biol 2003; 22:141-169.
64. Kiyama R, Camerini-Otero RD. A triplex DNA-binding protein from human cells: purification and characterization. Proc Natl Acad Sci USA 1991; 88:10450-10454.
65. Guieysse AL, Praseuth D, Helene C. Identification of a triplex DNA-binding protein from human cells. J Mol Biol 1997; 267:289-298.
66. Ciotti P, Van Dyke MW, Bianchi-Scarra G et al. Characterization of a triplex DNA-binding protein encoded by an alternative reading frame of loricrin. Eur J Biochem 2001; 268:225-234, and references therein.
67. Hildebrandt M, Lacombe ML, Mesnildrey S et al. A human NDP-kinase B specifically binds single-stranded poly-pyrimidine sequences. Nucleic Acids Res 1995; 23:3858-3864.
68. Brunel F, Zakin MM, Buc H et al. The polypyrimidine tract binding (PTB) protein interacts with single-stranded DNA in a sequence-specific manner. Nucleic Acids Res 1996; 24:1608-1615.

69. Farokhzad OC, Teodoridis JM, Park H et al. CD43 gene expression is mediated by a nuclear factor which binds pyrimidine-rich single-stranded DNA. Nucleic Acids Res 2000; 28:2256-2267.
70. Wang JC, Lynch AS. Transcription and DNA supercoiling. Curr Opin Genet Dev 1993; 3:764-768.
71. Kato M, Shimizu N. Effect of the potential triplex DNA region on the in vitro expression of bacterial β-lactamase gene in superhelical recombinant plasmids. J Biochem 1992; 112:492-494.
72. Simpson RT. Nucleosome positioning: occurrence, mechanisms, and functional consequences. Prog Nucleic Acid Res Mol Biol 1991; 40:143-184.
73. Nickol J, Martin RG. DNA stem-loop structures bind poorly to histone octamer cores. Proc Natl Acad Sci USA 1983; 80:4669-4673.
74. Casasnovas JM, Azorin F. Supercoiled induced transition to the Z-DNA conformation affects the ability of a $d(CG/GC)_{12}$ sequence to be organized into nucleosome-cores. Nucleic Acids Res 1987; 15:8899-8918.
75. Westin L, Blomquist P, Milligan JF et al. Triple helix DNA alters nucleosomal histone-DNA interactions and acts as a nucleosome barrier. Nucleic Acids Res 1995; 23:2184-2191.
76. Laundon CH, Griffith JD. Curved helix segments can uniquely orient the topology of supertwisted DNA. Cell 1988; 52:545-549.
77. Kohwi Y, Panchenko Y. Transcription-dependent recombination induced by triple-helix formation. Genes Dev 1993; 7:1766-1778.
78. Rooney SM, Moore PD. Antiparallel, intramolecular triplex DNA stimulates homologous recombination in human cells. Proc Natl Acad Sci USA 1995; 92:2141-2144.
79. Shlyakhtenko LS, Hsieh P, Grigoriev M et al. A cruciform structural transition provides a molecular switch for chromosome structure and dynamics. J Mol Biol 2000; 296:1169-1173.
80. Soldatenkov VA, Chasovskikh S, Potaman VN et al. Transcriptional repression by binding of poly(ADP-ribose) polymerase to promoter sequences. J Biol Chem 2002; 277:665-670.
81. Oei SL, Herzog H, Hirsch-Kauffmann M et al. Transcriptional regulation and autoregulation of the human gene for ADP-ribosyltransferase. Mol Cell Biochem 1994; 138:99-104.
82. Spiro C, McMurray CT. Switching of DNA secondary structure in proenkephalin transcriptional regulation. J Biol Chem 1997; 272:33145-33152.
83. Bedinger P, Munn M, Alberts BM. Sequence-specific pausing during in vitro DNA replication on double-stranded DNA templates. J Biol Chem 1989; 264:16880-16886.
84. Hacker JK, Alberts BM. The rapid dissociation of the T4 DNA polymerase holoenzyme when stopped by a DNA hairpin helix. A model for polymerase release following the termination of each Okazaki fragment. J Biol Chem 1994; 269:24221-24228.
85. Baran N, Lapidot A, Manor H. Formation of DNA triplexes accounts for arrests of DNA synthesis at $d(TC)_n$ and $d(GA)_n$ tracts. Proc Natl Acad Sci USA 1991; 88:507-511.
86. Dayn A, Samadashwily GM, Mirkin SM. Intramolecular DNA triplexes: unusual sequence requirements and influence on DNA polymerization. Proc Natl Acad Sci USA 1992; 89:11406-11410.
87. Potaman VN, Bissler JJ. Overcoming a barrier for DNA polymerization in triplex-forming sequences. Nucleic Acids Res 1999; 27:e5.
88. Kopel V, Pozner A, Baran N et al. Unwinding of the third strand of a DNA triple helix, a novel activity of the SV40 large T-antigen helicase. Nucleic Acids Res 1996; 24:330-335.
89. Liu G, Malott M, Leffak M. Multiple functional elements comprise a mammalian chromosomal replicator. Mol Cell Biol 2003; 23:1832-1842.
90. Blaho JA, Wells RD. Left-handed Z-DNA and genetic recombination. Prog Nucleic Acid Res Mol Biol 1989; 37:107-126.
91. Pearson CE, Sinden RR. Trinucleotide repeat DNA structures: dynamic mutations from dynamic DNA. Curr Opin Struct Biol 1998; 8:321-330.
92. Sinden RR, Potaman VN, Oussatcheva EA et al. Triplet repeat DNA structures and human genetic disease: dynamic mutations from dynamic DNA. J Biosci 2002; 27(Suppl 1):53-65.
93. Streisinger G, Okada Y, Emrich J et al. Frameshift mutations and the genetic code. Cold Spring Harb Symp Quant Biol 1966; 31:77-84.
94. Ripley LS. Frameshift mutation: determinants of specificity. Annu Rev Genet 1990; 24:189-213.
95. Sinden RR, Wells RD. DNA structure, mutations, and human genetic disease. Curr Opin Biotechnol 1992; 3:612-622.
96. Strauss BS. Frameshift mutation, microsatellites and mismatch repair. Mutat Res 1999; 437: 195-203.
97. Miller CA, Umek RM, Kowalski D. The inefficient replication origin from yeast ribosomal DNA is naturally impaired in the ARS consensus sequence and in DNA unwinding. Nucleic Acids Res 1999; 27:3921-3930, and references therein.
98. Berberich S, Trivedi A, Daniel DC et al. In vitro replication of plasmids containing human *c-myc* DNA. J Mol Biol 1995; 245:92-109.
99. Potaman VN, Bissler JJ, Hashem VI et al. Unpaired structures in SCA10 $(ATTCT)_n \bullet (AGAAT)_n$ repeats. J Mol Biol 2003; 326:1095-1111.

CHAPTER 2

Sequence-Dependent Variability of B-DNA: An Update on Bending and Curvature

Victor B. Zhurkin, Michael Y. Tolstorukov, Fei Xu, Andrew V. Colasanti and Wilma K. Olson

Abstract

DNA bending is universal in biology—both the storage and the retrieval of information encoded in the base-pair sequence require significant deformations, particularly bending, of the double helix. The A-tract curvature, which modulates these processes, has thus been a subject of long-standing interest. Here we describe the ongoing evolution of models developed to account for the sequence-dependent bending and curvature of DNA, namely the AA-wedge, junction, and flexible anisotropic dimer models. We further show that recent high-resolution NMR structures of DNA A-tracts are consistent with crystallographically observed structures, and that the combined data provide a realistic basis for describing the behavior of curved DNA in solution.

Introduction

The phenomenon of DNA 'curvature' (or intrinsic bending) in solution was first observed in the kinetoplast DNA (k-DNA) of the trypanosome *Leischmania tarentolae*.[1,2] Since these reports, more than 2,000 original papers and over 100 reviews have been published on the subject of 'DNA Bending and Curvature'. Nevertheless, until very recently, the structural mechanisms which underlie the A-tract-induced curvature of DNA have been a topic of heated debate.[3-6] For example, detailed models based on known DNA crystal structures failed to account for gel retardation and cyclization rates. Conversely, simplistic models offered to rationalize the solution data contradicted the X-ray structures. No short overview can profess to be complete in such circumstances.

Here we aim at laying the basic stereochemical groundwork for the next chapters. We pay only limited attention to the physico-chemical origins of DNA curvature, such as the comparative roles of the spine of hydration, bound cations, and water activity in stabilizing the A-tract conformation—for a review see refs. 6-8. Instead, we summarize the evolution of structural concepts over the past 20-30 years, highlighting the erratic history of the field and recent key findings which appear to settle earlier differences of opinion.

We emphasize the truly remarkable progress to date in understanding the sequence-dependent behavior of DNA (at least, in vitro). In this historical context, we see, in fact, that current uncertainties regarding DNA conformation in solution are small. In terms of the bending angles at the base-pair level, a very high precision of 1-2° has been achieved. We conclude with a brief mention of open problems in the interpretation of experimental data, and possible ways to resolve them (e.g., accounting for DNA flexibility in the analysis of polyacrylamide gel electrophoresis (PAGE) data).

DNA Conformation and Transcription, edited by Takashi Ohyama.

B-DNA Family of Forms

We deal exclusively with B-DNA, defined phenomenologically as a family of structural forms, which are:

i. stabilized in aqueous solution of relatively low ionic strength;
ii. linked with one another by means of non-cooperative transitions;
iii. separated from the A and Z forms by cooperative transitions.[9]

According to both X-ray and NMR measurements, the B-DNA forms are right-handed duplexes, with deoxyribose sugars puckered predominantly in a C2′-endo conformation (see below). The B form is universal in the sense that it can accommodate any DNA sequence. On the other hand, certain sequences, such as poly(dA•dT), have been observed only in a B-like conformation, and not in the A and Z forms described in Chapter 1.

'Typical' B-DNA variability is presented in Figure 1. These examples illustrate the known variation of B-DNA helical structure as a function of sequence[10] and environment.[9,11,12] The changes in base-pair inclination shown here correspond approximately to thermal fluctuations of the duplex under 'standard' conditions (see below). Any realistic explanation of the DNA curvature phenomenon must take this variability into account.

Sequence-Dependent Anisotropic Bending of DNA

Initially, the DNA duplex was described as an ideal isotropic rod, with elastic properties independent of sequence. The measured persistence length of ~500 Å[13] corresponds to thermal fluctuations of the bending angle of ~5°.[14] Later, once it became clear how tightly DNA is packed in chromatin, the initial isotropic representation of the double helix was rightfully questioned,[15] and the concept of DNA anisotropy was introduced.[14,16] The latter idea implies that bending across the grooves (in the Tilt direction) is much less pronounced than bending into the grooves (in the Roll direction), see Figure 2A. The variation of Tilt is hindered by stacking interactions between the bases and stereochemical constraints of the backbone, both of which resist the stretching of one strand and the compression of the other. This feature, first predicted by energy calculations,[16,17] was later confirmed once a number of protein-DNA complexes had been crystallized.[18,19] Arguably, DNA anisotropy reveals itself most strikingly in the recently solved SWI/SNF-independent (Sin) mutant nucleosomes,[20] where sharp bends with Roll of magnitude 20-25° ('mini-kinks') are alternately directed into the minor and major grooves at regular 5-6 bp increments along the DNA (Fig. 2B).

The concept of anisotropy is complemented by the notion of sequence-dependent variability of DNA, and bending in particular (Fig. 2A).[21-23] Overall, the DNA bending preferences follow a simple rule: the purine-pyrimidine (RY) and AA•TT dimers bend predominantly into the minor groove, whereas the pyrimidine-purine (YR) and GG•CC dimers bend more frequently toward the major groove. Historically, this rule originates from Calladine's[24] steric clash model, which was used to rationalize the alternation of positive and negative Roll at sequential base-pair steps in the Dickerson-Drew dodecamer structure.[25] Later, a somewhat modified version of this idea was corroborated by energy calculations[26] and used to account for DNA bending.[27,28] Despite its simplicity, the 'YR/RY rule' still holds for numerous B-DNA[10] and protein-DNA[29] crystal structures.

The pyrimidine-purine (YR) dimers deserve special attention. In addition to being the most anisotropic dimers (their equilibrium Roll angle has the largest absolute value), they are also the most flexible among all dimeric steps.[26,29-31] The bending and 'bendability' of YR steps are particularly notable in protein-DNA complexes, where these dimers are often severely kinked and serve as targets for intercalation.[32] The YR distortions are believed to be operative in indirect recognition and also appear to be important for the wrapping of DNA in large nucleoprotein complexes.[18,19]

The Sin mutant nucleosomes[20] mentioned above, provide a remarkable example of the DNA structural variability in nucleo-protein complexes (Fig. 2C). As is clear from the color-coded mosaic of Roll angles in these structures, the tight association of DNA with the core of histone proteins requires a regularly alternating pattern of (red $B \rightarrow B^A$ and blue $B \rightarrow B^C$) DNA deformation. These nucleosomes have a common DNA template, containing six A-tracts highlighted by blue lines (Fig. 2C). Three of the A-tracts bend exactly into the minor groove, one exactly into the major groove, and two show mixed (but minor-groove dominant) bending. Thus, we see that the nucleosomal structure takes advantage of the slight minor-groove bending preferences of the A-tracts, but the imperfect spacing of A-tracts in the crystallized sequences precludes a uniform conformational response.

The notion of anisotropic, sequence-dependent bending of DNA presented above provides a useful perspective for comprehending the controversies over A-tract bending and a framework for interpretation of recent high-resolution structures.

A-Tract Curvature: AA-Wedge, Purine-Clash, and Junction Models

The A-tract story began with the seminal 1980 work of Trifonov and Sussman[33] who discovered the periodicity of AA•TT dimers in genomic DNA and tied this observation to the DNA packaging in chromatin. Specifically, they argued that the AA•TT dimer had an intrinsic 'wedge-like' shape, which when repeated in phase with the helical periodicity of the duplex would introduce systematic intrinsic bending in DNA. (The Roll and Tilt components of the AA•TT wedge, however, were not initially specified.) Therefore, when DNA curvature was subsequently found in studies of k-DNA containing periodically repeated A_5- and A_6-tracts, the observation was interpreted by Marini et al[1,2] as a manifestation of the AA-wedges in solution.

On the other hand, the crystal structure of the Dickerson-Drew dodecamer, $d(CGCGAATTCGCG)_2$, which had been resolved by that time,[25] was in apparent contradiction with the idea of AA wedges. Successive A•T base pairs were nearly coplanar in the central AATT tetramer of the dodecamer, and the only distortions of any significance were found in the CG and GC dimeric steps at the ends of the molecule. The interpretation, by Calladine, of the structural irregularities in the dodecamer in terms of 'purine-clashes',[24] in combination with the results of energy calculations,[26] led to an alternative explanation of k-DNA curvature.[27,28] According to this scheme, the YR and RY dimers are responsible for intrinsic DNA bending, and the A-tracts remain essentially 'straight' (i.e., the A•T base pairs are perpendicular to the duplex axis). Indeed, the k-DNA sequence is organized in such a way that the A_5 and A_6 runs are interrupted by pyrimidine-rich segments of 4-5 bp, e.g., CCC-A_5-TGTC-A_6-TAGGC-A_6-TGCC-A_5. The local bends in CA dimers (which are directed into the major groove) are separated by 9-11 bp, as are the bends at AT and AC steps (which are directed toward the minor groove). Thus, it was imagined that these local bends (YR and RY) would accumulate and produce a significant global bend. The 'purine-clash' models were refuted, however, when it was shown[34-37] that they don't distinguish between 'strongly' curved and 'slightly' distorted DNA sequences (for comparison of the early models see the reviews of Tan and Harvey[38] and Sundaralingam and Sekharudu[39]).

Crucial gel electrophoresis experiments carried out by Hagerman,[34,35] Diekmann,[36] and Koo, Wu, and Crothers[37] established three important features of curved DNA:

i. Properly phased A-tracts are indispensable for 'strong' DNA curvature (e.g., substitution of AAGAA for A_5 diminishes the effect drastically).
ii. A-tract orientation is important (A_4T_4-induced bending differs from that of T_4A_4).
iii. Flanking sequences have a limited influence on the magnitude of DNA curvature (the gel retardation associated with the GA_5G sequence is 10-15% less than that for CA_5C).

To account for these results, Ulanovsky and Trifonov [40] refined the AA-wedge model, specifying, for the first time, values of the Roll and Tilt angles at AA•TT dimeric steps, and Crothers and coworkers introduced their 'junction' model.[37,41] Although Wu and Crothers[42] noted that they could not exclude a wedge-like 'smooth' bending mechanism, they expressed a preference for junction bending "because it leads to more interesting hypotheses as a focus for further work." Introduction of yet another model of A-tract curvature, however, led to a certain

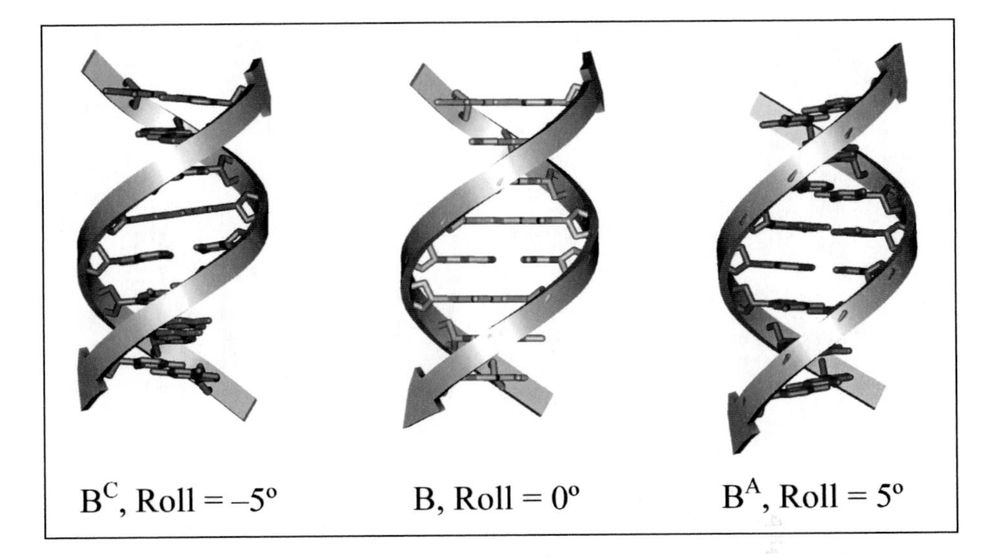

Figure 1. Schematic representation of DNA structural variability within the B family of forms. Three regular conformations are shown. B^C (blue): Twist = 38°, Roll = -5° (intermediate substate between the B and C forms); B (gray): Twist = 36°, Roll = 0° (canonical B-DNA); B^A (red): Twist = 34°, Roll = 5° (intermediate substate between the canonical A and B forms). Note that DNA twisting is strongly correlated with the inclination of base pairs since duplex unwinding is accompanied by an increase in the Roll angle.[10] For definition of the base-pair step parameters (Twist, Tilt, and Roll) see reference 80. To calculate these parameters from the published coordinates, and to generate atomic coordinates for the given Twist, Tilt, and Roll values, the CompDNA/3DNA software was used.[10,81] For clarity, the propeller twist and buckle angles are set to be zero (here and in Figs. 3 and 5).

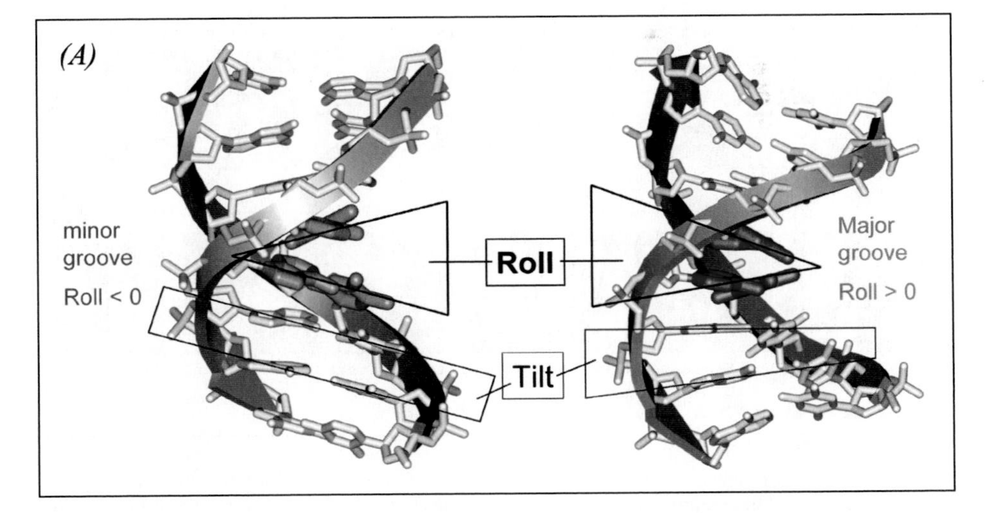

Figure 2A. Anisotropic sequence-dependent bending of DNA. The images are based on the crystallographic structure of the nucleosome.[79] Although the DNA distortions represent a larger variation of structural parameters than in free B-DNA, the duplex remains within the B-family of forms. Here and below, the color coding is the same as that used in Figure 1: red for Roll > 0 (bending into the major groove); blue for Roll < 0 (bending into the minor groove).

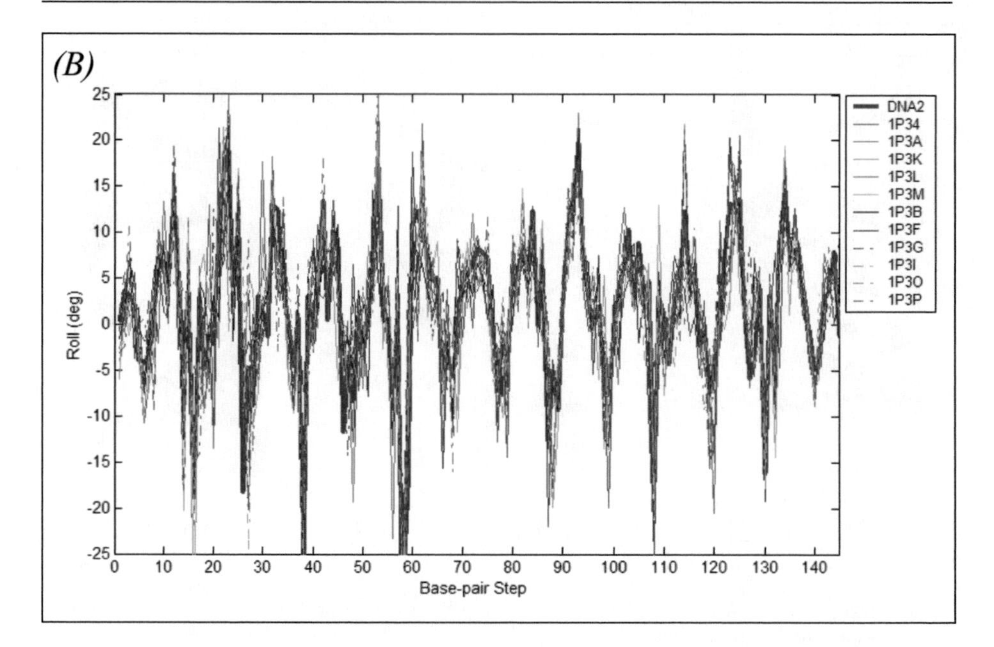

Figure 2B. Sequential variation of Roll along the DNA in 12 high-resolution nucleosome structures with a common DNA template.[20] Except for DNA2, which denotes the unmodified reference structure, the labels refer to the Protein Data Bank (PDB) entries of core particle structures with point mutations of histone H3 or H4. The composite plots highlight the 'mini-kinks' (i.e., large positive or negative Roll every 5-6 bp) which bring about the tight folding of DNA.

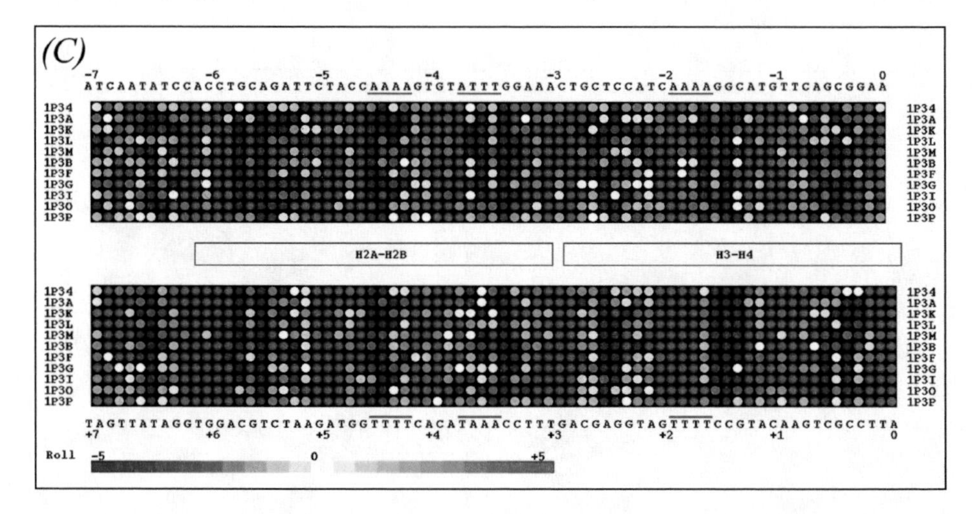

Figure 2C. The color-coded mosaic reveals the periodic interconversion of local conformation between states of negative (blue) and positive (red) Roll, i.e., B^C and B^A forms, in the 11 Sin mutant nucleosomes described in B; the PDB entries are given on the left and right sides. The 146 bp sequence, written along the upper and lower edges of the image, is divided into two fragments to emphasize the imperfect symmetry of the structures. Base-pair positions are expressed in terms of the number of helical turns away from the dyad (located at 0). A-tracts are highlighted by blue lines. Histone proteins in contact with DNA are noted in boxes at the observed sites of interaction.

'dualism' of interpretation. Both models were equally successful in accounting for PAGE data (but both were equally unacceptable from the crystallographic point of view—see below).

The principal difference between the wedge and junction models is the conjecture made on the 'nature' of interactions stabilizing the A-tract geometry. The AA-wedge model is based on the 'first approximation' that the average conformation of any dimeric step (e.g., AC•GT or AA•TT) is independent of its neighbors. In particular, the AA•TT dimer is believed to have the same distorted conformation in the context of both CAAC•GTTG and AAAA•TTTT. (As shown below, this hypothesis is not confirmed by X-ray and NMR structures.) By contrast, the 'junction' model is based on the assumption that an A-tract (made up of four or more consecutive adenines in the same strand) is stabilized in a 'specific' conformation which is somewhat different from the canonical B form. The latter idea builds upon the concept of 'junction bending' originated by Selsing et al[43] in their construction of a stereochemically optimal B/A junction. In other words, the AA-wedge model is a nearest-neighbor dimeric model, while the junction model postulates cooperative interactions along the DNA chain, which make A-tracts different from other sequences.

This difference between the two models leads to differences in the description of DNA deformation. The wedge model considers the dimeric step as the elementary structural unit of a duplex, and the 'wedge angles' accordingly describe transitions from the i-th to the $(i+1)$-st base pair (coordinate frames are assigned to each base pair). By contrast, the 'junction' model ignores possible irregularities within the A-tracts and non-A-tracts, and only considers the 'effective' deformations at the 5′- and 3′-ends of the A-tracts (see Fig. 3).

Subsequent modifications of the wedge model, in which all 16 dimers are considered[44] do not change the basic tenets, (i) that deviation from base-pair co-planarity occurs predominantly in the AA•TT steps and (ii) that the A-tract occurs naturally in a conformation similar to the B^C form in Figure 1. The wedge model also incorporates sequence-dependent values of Twist, which are based on known solution properties of DNA.[45]

The 'overall' DNA bend of an A-tract is directed approximately into the minor groove in the center of the run of A's, the bending vector being shifted somewhat toward the 3′-end of the fragment.[37,41] Among various A_n-tracts, the bend angle is probably the largest for $n = 6$, in as much as in this case the gel retardation is the strongest.[41] The bend angle for the A_6-tract has been estimated to be 17-21° from cyclization experiments.[46] In fact, the experimental bend angle estimates differ by roughly two-fold, ranging from 13.5° (based on the analysis of 2D scanning force microscopy images[47]) to 28° (based on early PAGE-circularization data[48]). Recent topological measurements of supercoiled DNA by Lutter and coworkers find the A-tract bend angle to be 22° at room temperature.[49] Thus, we consider a value $20 \pm 2°$ to be the best current estimate of the DNA bending angle per A_6-tract (under 'standard' conditions). As mentioned above, both the AA-wedge model and the junction model ascribe this intrinsic bending to a specific conformation of the A-tract, with the AA-dimers rolled into the minor groove and the base pairs inclined with respect to the local DNA axis (Fig. 3A,B).

It should be noted that the introduction of a 20° bend per A-tract requires only relatively small distortions in local structure. The Roll angles in the A-tract need not differ any more than 5-6° from those of 'random', mixed-sequence DNA. That is, the expected difference between the two structures does not exceed their thermal fluctuations under 'standard' conditions. (These fluctuations correspond to the difference between B^C and B, or between B and B^A forms in Fig. 1). This is yet another reason why the stereochemical mechanisms of A-tract curvature have proved to be evasive for so long.

Non-A-Tract Model

The crystallographic B-DNA structures provide a completely different perspective on the problem. The base pairs in A-tracts of known structures are essentially coplanar, with Roll(AA) ≈ 0° (B form in Fig. 1; see Fig. 4A and Table 1). By contrast, the base pairs are inclined in the non-A-tracts, with Roll positive (B^A form in Fig. 1). Following this 'non-A-tract'

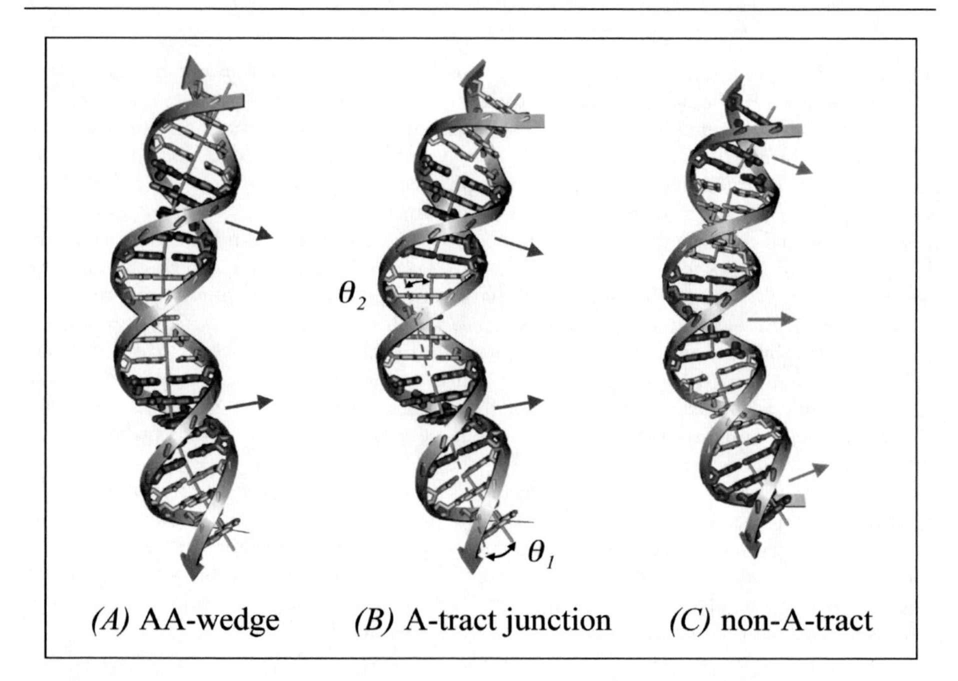

Figure 3. Three alternative models for the curved $C_5A_5C_5A_5C_5$ fragment: A) AA-wedge model; B) junction model; C) non-A-tract model. Blue arrows: rolling of AA-dimers into the minor groove (A, B); red arrows: rolling of non-AA-dimers into the major groove (C). A) The A-tract generated from non-coplanar dimer steps is shown in blue (Roll(AA) = -6.5°, Tilt(AA) = 3.2°). Dimeric step parameters are taken from Bolshoy et al;[44] the bend angle per A-tract is 28° according to this model. B) A-tracts (in blue) have a regular B^C conformation (Fig. 1) and non-A-tracts (in gray) a canonical B conformation. The helical axis of each fragment is shown. The angles between these axes are the 'junctions angles'. (Note that adjacent base pairs at the junctions are slightly unstacked to increase the visual effect.) The θ_1 and θ_2 angles in the junction model do not necessarily imply that there are sharp distortions at the ends of the A-tracts.[37] Rather, these parameters are 'virtual' angles introduced to simplify description of the DNA trajectory. Instead of using the 10 Roll and 10 Tilt angles of the 'wedge' model for each helical turn of DNA, only two 'effective' angles are used, θ_1 and θ_2. (For the Roll and Tilt components of θ_i see Koo et al.[46]) C) A-tracts (in gray) have a B conformation and non-A-tracts (in red) a B^A conformation (Fig. 1). This scheme corresponds to the non-A-tract model suggested by Olson, Calladine, Dickerson, and colleagues.[3,50,51]

model suggested by Olson,[50] Calladine,[51] Dickerson,[3] and colleagues, curved DNA resembles the image shown in Figure 3C. Importantly, the direction of DNA bending and its magnitude are the same as that predicted by the A-tract junction model (Fig. 3B). Thus, based on PAGE or cyclization data, it is impossible to distinguish between the two models. Only high-resolution techniques like NMR can help to solve the conundrum.

High-Resolution NMR Structures of A-Tract DNA

New techniques taking advantage of the partial orientation of macromolecules in a liquid crystalline medium, and subsequent measurement of the residual dipole couplings make possible an unprecedentedly high accuracy in the determination of DNA conformation in solution.[52,53] Using this approach, Bax and coworkers[53,54] found that the A•T base pairs in the central AATT tetramer of the Dickerson-Drew dodecamer are practically coplanar: Roll(AA) ≈ Tilt(AA) ≈ -1°. Moreover, the central part of the new NMR structure differs no more from the 11 solved X-ray structures of the same sequence than the crystallographic models differ from one another (Fig. 4). The NMR and X-ray conformations of the dodecamer,

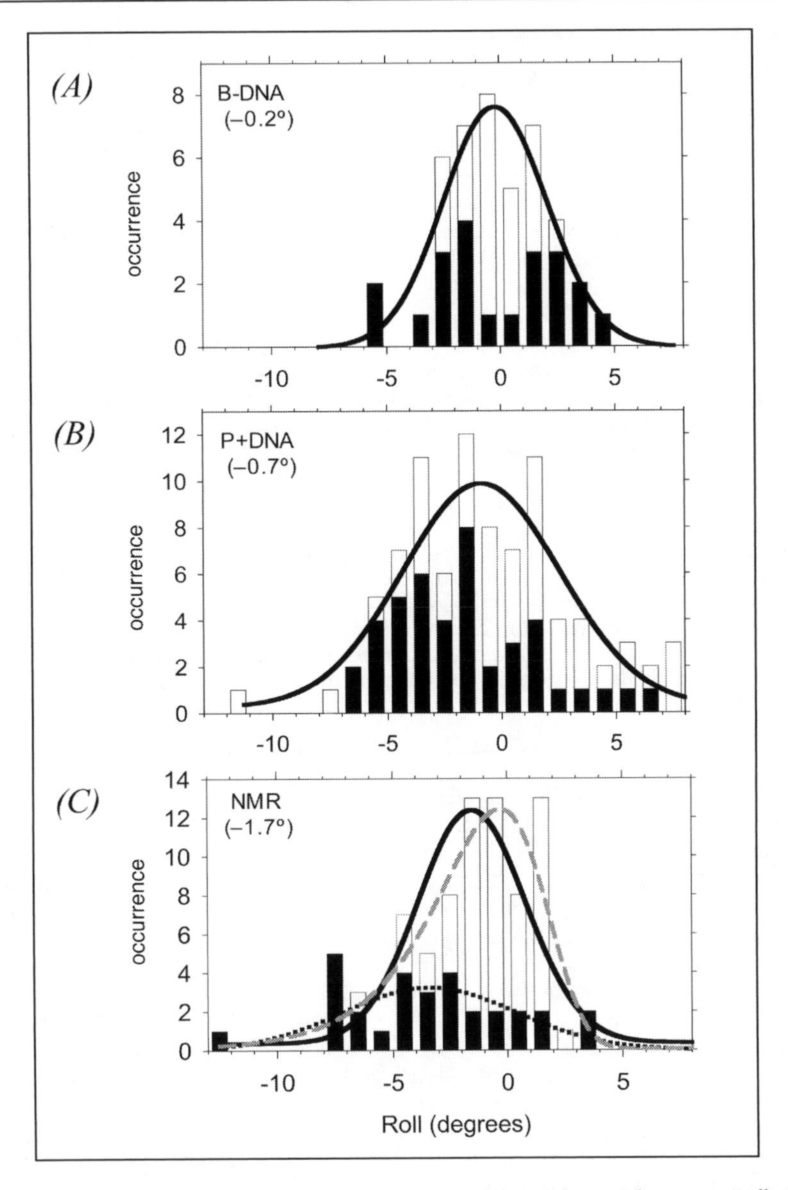

Figure 4. Distribution of Roll angles of A-tracts in solution and the solid state. The average Roll values are given in parentheses. A) Roll values in B-DNA X-ray structures. Solid bars: A_n-tracts ($n = 4$ to 6) in structures with resolution 2.6 Å and better. Open bars: AA•TT dimers in eleven Dickerson-Drew dodecamers containing a central AATT motif. Curves in A and B: Gaussian approximation of the cumulative A_n and AATT distributions. B) Roll values in protein-DNA crystal complexes (with resolution 2.6 Å and better). Solid bars: A_n-tracts ($n = 4$ to 6); open bars: the AA•TT dimers in A_nT_m-tracts ($n+m \geq 4$, $n < 4$, $m < 4$), such as AATT and AAAT. DNA fragments distorted by intercalation (e.g., in complexes with TBP and IHF) are omitted from consideration. C) Roll(AA) values in the NMR-resolved oligomers, $GGCA_4CGG$[56] (solid bars) and $GGCA_6CGG$[55] (open bars). For each sequence, the 10 best structures were taken (PDB entries 1NEV and 1FZX, respectively). Curves: Gaussian approximations of the distribution of Roll in the A_4 structure (dotted line) and in both NMR structures (solid line); broken line: 'skewed' non-Gaussian approximation of the distribution in the two NMR structures.

Table 1. The AA·TT dimeric step parameters in X-ray and NMR structures

	N	Twist (°)	Tilt (°)	Roll (°)
B-DNA, A_n and AATT	43	35.8 (2.7)	-0.1 (2.7)	-0.2 (2.2)
B-DNA, A_2 and A_3	14	33.3 (5.1)	-2.8 (1.5)	3.2 (5.4)
P+DNA, A_n and A_nT_m	89	35.8 (3.1)	-0.8 (2.8)	-0.7 (3.7)
P+DNA, A_n-tracts	43	36.1 (3.3)	-0.2 (2.2)	-1.7 (3.2)
NMR, A_n-tracts	80	37.2 (2.1)	-0.7 (2.8)	-1.7 (2.9)
NMR, A_2 and A_3	30	35.6 (1.1)	-2.2 (1.6)	3.0 (2.5)

B-DNA and P+DNA: crystallographic structures of B-DNA and protein-DNA complexes, with resolution 2.6 Å and better. NMR: high resolution structures from the Lu[55] and Crothers[56] groups; the ten lowest energy structures are taken from each of the PDB entries: 1FZX and 1NEV (A_n-tracts) and 1G14 (A_2 and A_3). For each dataset, the average and root-mean-square deviations (in parentheses) are given; *N* is the number of dimeric steps in a dataset. A_n-tracts are consecutive runs of adenines in one strand, $n = 4$ to 6. Accordingly, A_nT_m are runs of *n* adenines followed by *m* thymines, $n+m = 4$ to 6. A_2 and A_3 are 'isolated' adenine dimers or trimers, which don't belong to the A_n- or A_nT_m-tracts described above. For example, the tetramer TTAA contains two dimers AA·TT, whereas AATT contains no 'isolated' AA·TT dimers (because AATT belongs to the A_nT_m set). Note that in the two 'A_2' and 'A_3' sets (B-DNA and NMR) the average Roll is positive (~3°), while in the other sets Roll is negative. The used Nucleic Acid Database (NDB) entries are described below. B-DNA, A_n-tracts, 21 steps: bdj081, bdl006, bdl047, bdl015 (-1 and -2). B-DNA, AATT, 22 steps: eleven structures of the Dickerson-Drew dodecamer. B-DNA, A_2 and A_3: bd0033, bd0034, bd0051, bdj019, bdj031, bdj055, bdl059. P+DNA, A_n-tracts: pd0045, pd0050, pd0125, pd0187, pd0189, pd0314, pdr015, pdr056, pdt033, pdt038, pdt040.

however, differ at the terminal CGCG tetramers, but this is a natural consequence of the packing interactions in the crystal phase.

Two other oligomers recently resolved by NMR contain central A_4- and A_6-tracts, which are flanked by GC-rich termini.[55,56] (For brevity, the oligomer $GGCA_4CGG$[55] is denoted the A_4-decamer, and $GGCA_6CGG$[56] the A_6-dodecamer.) These sequences were selected by analogy with naturally curved k-DNA[1,2] and synthetic DNA fragments which produce the strongest retardation in PAGE experiments.[41,46] The solution structures of the two oligomers are close to the crystallographic data (Fig. 4 and Table 1). In terms of the average Roll angle, the A-tracts in solution differ from those found in B-DNA crystals by 1.5°, and from protein-bound A-tracts by a mere 1.0° (Table 1). Note that in this selection, both A_n-tracts and A_nT_m-tracts are included to make the crystallographic datasets more representative. If, however, the protein-bound A_n-tracts are considered separately from the A_nT_m-tracts, then the average Roll(A_n) in the X-ray complexes is equal to -1.7°, that is, a value identical to the mean value found in solution (Table 1). For B-DNA crystal structures, A_n- and A_nT_m-tracts are indistinguishable in terms of the average Roll value.

In short, 'pure' A_n-tracts (without A_nT_m) in solution are characterized by the same inclination of base pairs as those in protein-DNA co-crystals (on average), but differ somewhat from the A-tracts in 'free' B-DNA crystals (of course, this conclusion is based on limited statistics). A similar tendency was observed earlier, when we compared DNA twisting under the same conditions. In that case, the structure of DNA in solution[45] was also closer to that in protein-DNA co-crystals[29] than to the structure of 'pure' crystallographic B-DNA.[10] Apparently, the protein environment in co-crystals is closer to 'standard' solvent conditions than the environment in B-DNA crystals. (If the crystallographically observed A-tract conformation is, indeed, affected by the presence of the dehydrating agent MPD (2-methyl-2,4-pentanediol)[57,58] then, as we see here, the effect does not exceed 1.5° in Roll.)

Comparison of the NMR-resolved A-tract structures with existing models shows that the A-tract in the A_4-decamer[56] is quite irregular (the Roll(AA) angles differ by 6°). That is, the duplex is somewhat distorted at each dimeric step, rather than in one or two selected locations. Indeed, the Crothers group describes the overall duplex structure as a 'delocalized' bend[56] rather than a junction bend of the type anticipated in their earlier work.

The other important result following from the comparison of X-ray and NMR data is clear evidence in favor of cooperative interactions stabilizing a 'specific' A-tract conformation. Both in B-DNA crystals and in solution, isolated adenine dimers have a positive Roll of ~3° (Table 1), which makes them closer to 'random' sequence DNA than to an A-tract. This contradicts one of the major AA-wedge model postulates—namely, that the AA-dimer and the A-tract have identical (or very similar) conformations. The other postulate—that the A_n•T_n sequences are unique—is remarkably confirmed.[33,59] Ulanovsky and Trifonov[40] correctly predicted the qualitative effect—the preference of the A-tracts (compared to random-sequence DNA) for negative Roll—before there were any 'solid' data, but the absolute magnitude of the predicted AA-wedge, ~9°, is far too large (Table 1). (Current best estimates of the mean AA wedge are actually closer to the coplanar base-pair geometry predicted in the non-A-tract models cited above.) Note also that the slight negative Tilt(AA) observed by X-ray and NMR (Table 1) indicates that the thymines are opened and the adenines closed[16]—a trend opposite to the wedge model.[40,44] The 'cooperative interactions', which stabilize a 'specific' A-tract conformation different from random-sequence B-DNA (see above), are likely to be solvent-mediated interactions in the grooves (e.g., the minor groove hydration spine associated with crystallized A-tracts and the major groove cation binding found in the structures of GC-rich sequences). We do not discuss these physico-chemical effects here and refer the reader to published reviews.[6-8]

Further examination of the NMR-resolved structures shows that the A_6-dodecamer[55] (average Roll(AA) = -0.8°) is closer to the crystallographic B-DNA A-tracts than is the A_4-decamer[54] (Roll(AA) = -3.3°). In Figure 4C, the 'black' bars corresponding to the A_4-decamer are more widely distributed than the 'white' ones associated with A_6 (so far, it is unclear whether this scattering reflects real DNA flexibility, or a lack of NMR-restraints in the computer simulations used to deduce the structure). Overall, the two structures produce a lopsided distribution of Roll(AA) angles 'skewed' toward negative values (broken line curve in Fig. 4C).

If supported by future solution structures, this 'skewness' would corroborate our idea[6,60] of the asymmetric bending of AA-sequences (preferentially into the minor groove). This kind of asymmetry is important, because it increases the 'effective' bend angle measured in cyclization[60] and loop formation[61,62] experiments, where high-energy DNA distortions are operative in achieving a 'closed' DNA configuration.

Modeling of A-Tract Curvature Based on NMR and X-Ray Data

Below, we offer a simple qualitative model of DNA bending based on current estimates of the sequence-dependent Roll angles in solution. This 'model' is not meant to explain all available data—rather, it is given to illustrate how by using NMR and X-ray data, one can realistically account for the magnitude and directionality of the A-tract bending observed in PAGE and cyclization experiments. As noted above, the Roll angles in the A-tract need not differ more than 5-6° from those of 'random', mixed-sequence DNA.

Our model is based on three important features of the two A-tract NMR structures[55,56] discussed above:

i. On average, Roll(AA) is less than Roll(non-AA).
ii. There is a significant positive Roll(CA) at the 5′-end of the A-tract (a finding consistent with both X-ray observations[19] and molecular simulations, i.e., Monte Carlo[30,31] and molecular dynamics[63,64] calculations).
iii. The 3′-end of the A-tract is characterized by a negative Tilt(AC) ≈ -5° (which brings the adenine and cytosine close to one another).

Accordingly, we adopt the following Roll values: Roll(AA) = -2° (comparable to -1.7° in Fig. 4C); Roll(CA) = 8° (the same as in the A_6-dodecamer[55]); Roll(MN) = 3° for all other dimers (in agreement with the average Roll = 2.7° found in protein-DNA cocrystals[29] and the mean Roll = 4.6° reported in earlier NMR-resolved DNA structures[65]). The Tilt angle is taken to be zero for all dimers (see below), and the Twist angles are taken from Kabsch et al.[45] The latter values are not critical for short DNA fragments of 30-40 bp, but do matter in the description of the overall shape of 100-200 bp DNA fragments,[29] an issue beyond the scope of the present article.

The bend angle per A_6-tract associated with the model is 19° (Fig. 5A), a value consistent with the 20 ± 2° experimental interval.[46,49] Furthermore, the overall DNA bending vector is directed approximately along the dyad axis of the fourth A•T pair in the A-tract (highlighted in green in Fig. 5A), a result also in agreement with the experimental data mentioned above.[37,41] The predicted overall bend angle is not very sensitive to the value of Roll(CA): reduction of this Roll by 5° (from 8° to 3°) diminishes the global bend angle by only 2° (from 19° to 17°). The large positive Roll(CA), denoted C-A1 in Figure 5A, is critical to the directionality of the DNA bend, i.e., the shift in directionality toward the 3′-end of the A-tract. The importance of the CA step in the bending of A-tracts has long been appreciated, both in experimental[66,67] and computational[30,63] studies, but its key role in bending directionality has not been previously pointed out.

The Tilt at the 3′-end of the A-tract is more critical to the magnitude of bending (A6-C in Fig. 5A). For example, if Tilt(AC) = -5° as in the NMR-resolved A_4- and A_6-structures,[55,56] then the overall bend angle increases up to 24°. That is, the Tilt at the 3′-end of the A-tract is roughly additive with the sum of the Roll(AA) angles. We do not include this Tilt(AC) in the present model, however, because it is likely to be related to the effect of the cytosine sugar switch, and as such it has to be treated by a model which explicitly incorporates DNA flexibility (see below).

Unresolved Problems: Sugar Switching and Overall DNA Flexibility

Below, we emphasize several as yet unanswered questions related to the mechanisms of DNA bending and curvature.

Sugar Switching and DNA Bending

NMR studies of DNA oligonucleotide duplexes show that 90-95% of the deoxyribose sugars attached to purines (A and G) remain in the B-like C2′-endo conformation, or S domain, in solution.[68] This conformational propensity is less clear-cut in the case of pyrimidines (especially cytosine): in some examples, up to 35-40% of the sugar rings are switched to the C3′-endo conformation (N domain).[54,69] The conformational preference of the cytosine sugar for the C3′-endo form is also predicted in energy minimization[70] and ab initio quantum mechanical studies[71] and is seen in the distribution of conformational states in high-resolution crystal structures.[72] It should be noted, however, that detection of the sequence-dependent N/S equilibrium in solution is an extremely sophisticated and time-consuming procedure, so that even in the recent well resolved NMR structures of DNA A-tracts[55,56] the question of deoxyribose interconversion has not yet been addressed. For example, in the A_6-dodecamer[55] the sugar rings of the terminal cytosines apparently adopt an unfavorable O1′-endo conformation. Such states are suggestive of fast, undetectable N↔S interconversions in cytosine, with the N-population high enough to shift the average sugar pucker toward the O1′-endo form. The sugar switching is evidently a complicated context-depending phenomenon, requiring further investigation.

Using high frequency antiphase NMR spectroscopy, in the Sarma group[69] it was demonstrated, in the context of CAAAC, that the 3′-terminal cytosine is especially prone to sugar switching. Such deformation may have important consequences in terms of DNA bending. The inter-phosphate distance along the DNA chain is known to be shorter in the A form (with sugars in the C3′-endo conformation) than in the B form (with C2′-endo puckered rings).

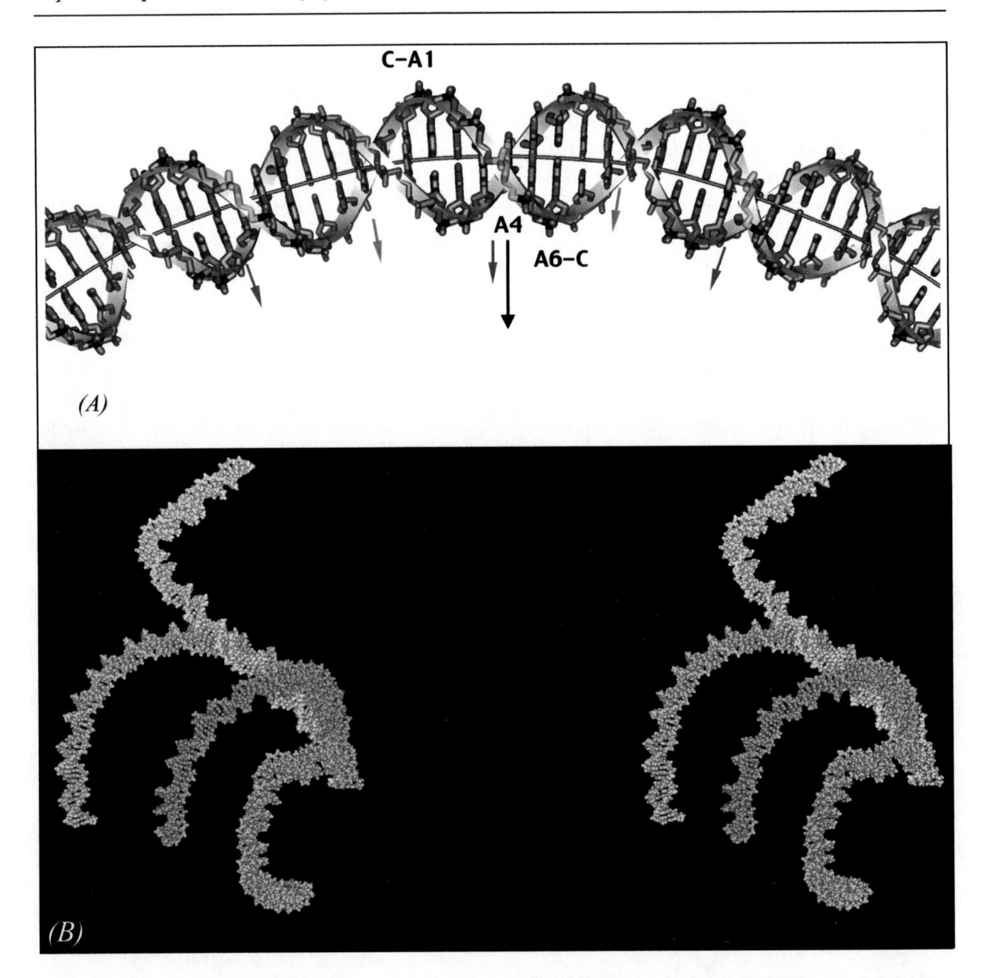

Figure 5. Curved $(A_6C_5A_6C_4)_n$ fragment constructed on the basis of current X-ray and NMR data. A) Static 'equilibrium' conformation is bent by 19° per helical turn, as a result of the negative Rolls of the AA-dimers (blue arrows directed into the minor groove) and the positive Rolls of the non-AA-dimers (red arrows directed into the major groove). The overall DNA bending vector (black arrow) is directed approximately along the dyad axis of the A•T pair marked A4 and shown in green. The AA•TT dimers are shown in blue, and the GG•CC dimers in red. Specific Roll and Tilt values are given in the text, and the procedure to calculate the 'overall' DNA bend angle is described in the literature.[30] The dimeric step C-A1 (at the 5′-end of the A-tract) is characterized by a positive Roll,[30,55,56,64] and the step A6-C (at the 3′-end of the A-tract) by a negative Tilt.[55,56,73] B) Effect of thermal fluctuations on DNA curvature: a 'bouquet' of representative instantaneous chain configurations is shown in stereo. The 100 bp fragments, superimposed at their right termini, were generated by Monte Carlo simulations.[60] Fluctuations in Roll, Tilt, and Twist correspond to a persistence length of 500 Å.[13] The 'equilibrium' conformation (in blue and yellow) is the same as in A.

Therefore, it is natural to expect that S→N sugar switching would introduce local compression in the sugar-phosphate backbone, and that this compression would, in turn, induce DNA bending toward the perturbed sugar. Indeed, this is exactly what has been predicted in energy calculations of the CAAAC fragment mentioned above.[73] Specifically, an isolated sugar switch was found to introduce a negative Tilt (-2° to -7°) in the same direction as that recently observed in the A_4- and A_6-oligomers.[55,56] Such a Tilt corresponds to bending the DNA axis 'down' at the A6-C position in Figure 5A—that is, to an increase in the overall A-tract bending as described above. Interestingly, a similar S→N sugar switch has also been observed at the 3′-end of an A-tract in the context of CA_5C.[73]

The cytosine S and N conformers are in dynamic equilibrium in solution; therefore, the aforementioned Tilt angle at the 3′-end of the A-tract should be considered within the framework of a dynamic model of DNA bending. Such models are only beginning to emerge. Hopefully, in the near future several more A-tract containing duplexes will be studied as exhaustively as the Dickerson-Drew dodecamer has been studied in the Bax group.[54] This would provide the information necessary for developing a dynamic model of DNA bending, which includes, among other features, sugar ring interconversion.

Global DNA Flexibility

There is no good reason why the gel electrophoresis properties of 150-200 bp DNA fragments have been interpreted for so long in terms of 'static' models rather than the deformable DNA segments commonly used to account for the ring closure rates of chain molecules of the same length. As noted above, DNA curvature is brought about by local effects of the same magnitude, or even less than, the thermal fluctuations of the base-pair step parameters (e.g., static bends vs. angular deformations of ~5° in the case of Roll). Thus, from the point of view of statistical mechanics, a 'static' model of DNA curvature is absurd.

To evaluate the effect of fluctuations on DNA curvature, one can use either all-atom chain simulations (Monte Carlo[30,31] and molecular dynamics[63,64]) or a simplified Twist-Roll-Tilt model[60] corresponding to the observed bending and torsional rigidities of DNA. The latter 'flexible anisotropic dimer' model[60] was used to illustrate the dramatic effect of room-temperature fluctuations on the shape of naturally curved DNA (Fig. 5B). Clearly, many of the states of a curved DNA chain are extended by such fluctuations, that is, 'real curved' DNA is longer than the idealized chain produced by a static model. Conversely, 'moderately' curved DNA fragments in reality are more compact than their 'static' model representation.[6,60] Therefore, current quantitative estimates of the sequence-dependent bending angles of DNA, based on a static interpretation of PAGE data, may be biased.

Although critical evaluation of the static model has been presented in numerous reviews,[6,31,64,74] this oversimplified representation of the DNA chain still dominates the literature. Only within the last year have Mohanty and colleagues begun to develop a theoretical framework for the interpretation of PAGE data in terms of the sequence-dependent conformation and deformability of DNA.[75,76]

DNA Curvature and Gene Regulation

Realistic modeling of the looping and other perturbations of DNA critical for its biological functioning requires knowledge (as accurate as possible) of the numerical values of the parameters which govern the sequence-dependent 'mechanical' properties of the double helix, such as the equilibrium Twist, Roll, and Tilt angles, the torsional and bending stiffness, etc. The recent progress in determination of reliable three-dimensional DNA structures in solution and the growing number of DNA structural examples in the solid phase bode well for improvements in current knowledge-based force fields, i.e., the mechanical parameters used in modeling long chains.[29] As new data accumulate it will become possible to include, in addition to the mean values and variance of base-pair step parameters, the sequence context of the dimeric building blocks, the dependence of the base-pair steps on backbone structure, and the role of local environmental variables on conformational state. Such information will make it possible to model large protein-DNA assemblages where the duplex is severely distorted.

Genome-Wide A-Tract Distribution in Bacteria

In *E. coli* and other bacteria, there is an increased number of A-tracts (compared to random DNA sequences with the same base composition). These A-tracts are organized in ~100 bp clusters, in which the A-tracts are phased with a period of ~11 bp.[77] Frequently, two or three phased A-tracts are located inside a cluster, and as such they produce a net bend of 40-60°.

Although a 60° deflection is a modest bend, preexisting curvature of this magnitude would reduce the energy of elastic deformation required for DNA loop formation (180° bending) by more than half. The 3 kcal/mole decrease in free energy of the curved segment would increase the probability of looping by ~100 fold. (Note that this estimate is valid only for linear, spatially unconstrained DNA molecules; Tolstorukov and Zhurkin, unpublished data.)

Importantly, these A-tract clusters are observed throughout bacterial genomes, both in coding[77] and regulatory sequences.[78] In addition to reducing the energy required for DNA compaction, the A-tract clusters may serve as binding sites for the bacterial nucleoid-associated proteins that have propensities for curved DNA. Thus, A-tract clusters may be operative in the compaction of bacterial genomes, providing intrinsic DNA curvature and increasing the stability of DNA complexes with architectural proteins.

Conclusion

We are now in a 'transition' period in which X-ray and NMR-based models of DNA curvature are beginning to evolve. The original concepts, such as the preference for A-tracts to bend into the minor groove (compared to mixed-sequence DNA), hold. The new 'knowledge-based' models operate with real base-pair steps, as opposed to the hypothetical components of early wedge and junction models, and incorporate the deformability encoded in known structures.

Acknowledgements

We are grateful to Ad Bax for valuable discussions, and to David Beveridge, Tali Haran, and Karolin Luger for sharing unpublished data. Support of this work through USPHS grant GM20861 and the Program in Mathematics and Molecular Biology based at Florida State University (predoctoral fellowship support to AVC) is also gratefully acknowledged. Space limitations forced us to refer frequently to reviews rather than to primary references.

References

1. Marini JC, Levene SD, Crothers DM et al. Bent helical structure in kinetoplast DNA. Proc Natl Acad Sci USA 1982; 79:7664-7668. Correction ibid 1983; 80:7678.
2. Marini JC, Levene SD, Crothers DM et al. A bent helix in kinetoplast DNA. Cold Spring Harb Symp Quant Biol 1983; 47:279-283.
3. Goodsell DS, Kaczor-Grzeskowiak M, Dickerson RE. The crystal structure of C-C-A-T-T-A-A-T-G-G. Implications for bending of B-DNA at T-A steps. J Mol Biol 1994; 239:79-96.
4. Dickerson RE, Goodsell DS, Kopka ML. MPD and DNA bending in crystals and in solution. J Mol Biol 1996; 256:108-125.
5. Haran TE, Kahn JD, Crothers DM. Sequence elements responsible for DNA curvature. J Mol Biol 1994; 244:135-143.
6. Olson WK, Zhurkin VB. Twenty years of DNA bending. In: Sarma RH, Sarma MH, eds. Biological Structure and Dynamics. Vol. 2. Schenectady: Adenine Press, 1996:341-370.
7. Crothers DM, Shakked Z. DNA bending by adenine-thymine tracts. In: Neidle S, ed. Oxford Handbook of Nucleic Acid Structures. London: Oxford University Press, 1999:455-470.
8. Hud NV, Plavec J. A unified model for the origin of DNA sequence-directed curvature. Biopolymers 2003; 69:144-158.
9. Ivanov VI, Minchenkova LE, Minyat EE et al. Cooperative transitions in DNA with no separation of strands. Cold Spring Harb Symp Quant Biol 1983; 47:243-250.
10. Gorin AA, Zhurkin VB, Olson WK. B-DNA twisting correlates with base pair morphology. J Mol Biol 1995; 247:34-48.
11. Ivanov VI, Minchenkova LE, Schyolkina AK et al. Different conformations of double-stranded nucleic acid in solution as revealed by circular dichroism. Biopolymers 1973; 12:89-110.
12. Ivanov VI, Lysov YP, Malenkov GG et al. Conformational possibilities of double-helical DNA. Studia Biophys 1976; 55: 5-13.
13. Hagerman PJ. Flexibility of DNA. Annu Rev Biophys Biophys Chem 1988; 17:265-286.
14. Schellman JA. Flexibility of DNA. Biopolymers 1974; 13:217-226.
15. Crick FH, Klug A. Kinky helix. Nature 1975; 255:530-533.
16. Zhurkin VB, Lysov YP, Ivanov VI. Anisotropic flexibility of DNA and the nucleosomal structure. Nucleic Acids Res 1979; 6:1081-1096.

17. Ulyanov NB, Zhurkin VB. Flexibility of complementary dinucleotide phosphates. A Monte Carlo study. Mol Biol (Engl transl) 1982; 16:857-867.
18. Suzuki M, Yagi N. Stereochemical basis of DNA bending by transcription factors. Nucleic Acids Res 1995; 23:2083-2091.
19. Dickerson RE. DNA bending: the prevalence of kinkiness and the virtues of normality. Nucleic Acids Res 1998; 26:1906-1926.
20. Muthurajan UM, Bao Y, Forsberg LJ et al. Crystal structures of histone Sin mutant nucleosomes reveal altered protein-DNA interactions. EMBO J 2004; 23:260-271.
21. Klug A, Jack A, Viswamitra MA et al. A hypothesis on a specific sequence-dependent conformation of DNA and its relation to the binding of the *lac*-repressor protein. J Mol Biol 1979; 131:669-680.
22. Dickerson RE. Base sequence and helix structure variation in B- and A-DNA. J Mol Biol 1983; 166:419-441.
23. Drew HR, Travers AA. DNA bending and its relation to nucleosome positioning. J Mol Biol 1985; 186:773-790.
24. Calladine C.R. Mechanics of sequence-dependent stacking of bases in B-DNA. J Mol Biol 1982; 161:343-352.
25. Dickerson RE, Drew HR. Structure of a B-DNA dodecamer. Influence of base sequence on helix structure. J Mol Biol 1981; 149:761-786.
26. Ulyanov NB, Zhurkin VB. Sequence-dependent anisotropic flexibility of B-DNA. A conformational study. J Biomol Struct Dyn 1984; 2:361-385.
27. Hagerman PJ. Evidence for the existence of stable curvature of DNA in solution. Proc Natl Acad Sci USA 1984; 81:4632-4636.
28. Zhurkin VB. Sequence-dependent bending of DNA and phasing of nucleosomes. J Biomol Struct Dyn 1985; 2:785-804.
29. Olson WK, Gorin AA, Lu X-J et al. DNA sequence-dependent deformability deduced from protein-DNA crystal complexes. Proc Natl Acad Sci USA 1998; 95:11163-11168.
30. Zhurkin VB, Ulyanov NB, Gorin AA et al. Static and statistical bending of DNA evaluated by Monte Carlo simulations. Proc Natl Acad Sci USA 1991; 88:7046-7050.
31. Olson WK, Zhurkin VB. Modeling DNA deformations. Curr Opin Struct Biol 2000; 10:286-297.
32. Werner MH, Gronenborn AM, Clore GM. Intercalation, DNA kinking, and the control of transcription. Science 1996; 271:778-784.
33. Trifonov EN, Sussman JL. The pitch of chromatin DNA is reflected in its nucleotide sequence. Proc Natl Acad Sci USA 1980; 77:3816-3820.
34. Hagerman PJ. Sequence dependence of the curvature of DNA: a test of the phasing hypothesis. Biochemistry 1985; 24:7033-7037.
35. Hagerman PJ. Sequence-directed curvature of DNA. Nature 1986; 321:449-450.
36. Diekmann S. Sequence specificity of curved DNA. FEBS Lett 1986; 195:53-56.
37. Koo H-S, Wu H-M, Crothers DM. DNA bending at adenine•thymine tracts. Nature 1986; 320:501-506.
38. Tan RK, Harvey SC. A comparison of six DNA bending models. J Biomol Struct Dyn 1987; 5:497-512.
39. Sundaralingam M, Sekharudu YC. Sequence directed DNA bending and curvature. An overview. In: Olson WK, Sarma MH, Sarma RH, Sundaralingam M, eds. Structure and Expression. Vol. 3. DNA Bending and Curvature. Schenectady: Adenine Press, 1988:9-23.
40. Ulanovsky LE, Trifonov EN. Estimation of wedge components in curved DNA. Nature 1987; 326:720-722.
41. Koo H-S, Crothers DM. Calibration of DNA curvature and a unified description of sequence-directed bending. Proc Natl Acad Sci USA 1988; 85:1763-1767.
42. Wu H-M, Crothers DM. The locus of sequence-directed and protein-induced DNA bending. Nature 1984; 308:509-513.
43 Selsing E, Wells RD, Alden CJ et al. Bent DNA: visualization of a base-paired and stacked A-B conformational junction. J Biol Chem 1979; 254:5417-5422.
44. Bolshoy A, McNamara P, Harrington RE et al. Curved DNA without A-A: experimental estimation of all 16 DNA wedge angles. Proc Natl Acad Sci USA 1991; 88:2312-2316.
45. Kabsch W, Sander S, Trifonov EN. The ten helical twist angles of B-DNA. Nucleic Acids Res 1982; 10:1097-1104.
46. Koo HS, Drak J, Rice JA et al. Determination of the extent of DNA bending by an adenine-thymine tract. Biochemistry 1990; 29:4227-4234.
47. Rivetti C, Walker C, Bustamante C. Polymer chain statistics and conformational analysis of DNA molecules with bends or sections of different flexibility. J Mol Biol 1998; 280:41-59.

48. Ulanovsky L, Bodner M, Trifonov EN et al. Curved DNA: design, synthesis, and circularization. Proc Natl Acad Sci USA 1986; 83:862-866.
49. Tchernaenko V, Radlinska M, Drabik C et al. Topological measurement of an A-tract bend angle: comparison of the bent and straightened states. J Mol Biol 2003; 326:737-749.
50. Maroun RC, Olson WK. Base sequence effects in double-helical DNA. III. Average properties of curved DNA. Biopolymers 1988; 27:585-603.
51. Calladine CR, Drew HR, McCall MJ. The intrinsic curvature of DNA in solution. J Mol Biol 1988; 201:127-137.
52. Tjandra N, Bax A. Direct measurement of distances and angles in biomolecules by NMR in a dilute liquid crystalline medium. Science 1997; 278:1111-1114.
53. Tjandra N, Tate S-I, Ono A et al. The NMR structure of a DNA dodecamer in an aqueous dilute liquid crystalline phase. J Am Chem Soc 2000; 122:6190-6200.
54. Wu Z, Delaglio F, Tjandra N et al. Overall structure and sugar dynamics of a DNA dodecamer from homo- and heteronuclear dipolar couplings and ^{31}P chemical shift anisotropy. J Biomolec NMR 2003; 26:297-315.
55. MacDonald D, Herbert K, Zhang XL et al. Solution structure of an A-tract DNA bend. J Mol Biol 2001; 306:1081-1098.
56. Barbic A, Zimmer DP, Crothers DM. Structural origins of adenine-tract bending. Proc Natl Acad Sci USA 2003; 100:2369-2373.
57. DiGabriele AD, Sanderson MR, Steitz TA. Crystal lattice packing is important in determining the bend of a DNA dodecamer containing an adenine tract. Proc Natl Acad Sci USA 1989; 86:1816-1920.
58. Dlakic M, Park K, Griffith JD et al. The organic crystallizing agent 2-methyl-2,4-pentanediol reduces DNA curvature by means of structural changes in A-tracts. J Biol Chem 1996; 271:17911-17919.
59. Trifonov EN. Sequence-dependent deformational anisotropy of chromatin DNA. Nucleic Acids Res 1980; 8:4041-4053.
60. Olson WK, Marky NL, Jernigan RL et al. Influence of fluctuations on DNA curvature. A comparison of flexible and static wedge models of intrinsically bent DNA. J Mol Biol 1993; 232:530-554.
61. Geanacopoulos M, Vasmatzis G, Zhurkin VB et al. Gal repressosome contains an antiparallel DNA loop. Nat Struct Biol 2001; 8:432-436.
62. Mehta RA, Kahn JD. Designed hyperstable Lac repressor. DNA loop topologies suggest alternative loop geometries. J Mol Biol 1999; 294:67-77.
63. Young MA, Beveridge DL. Molecular dynamics simulations of an oligonucleotide duplex with adenine tracts phased by a full helix turn. J Mol Biol 1998; 281:675-687.
64. Beveridge DL, Dixit SB, Barreiro G et al. Molecular dynamics simulations of DNA curvature and flexibility: helix phasing and premelting. Biopolymers 2004; 73:380-403.
65. Ulyanov NB, James TL. Statistical analysis of DNA duplex structural features. Methods Enzymol 1995; 261:90-120.
66. Beutel BA, Gold L. In vitro evolution of intrinsically bent DNA. J Mol Biol 1992; 228:803-812.
67. Nagaich AK, Bhattacharyya D, Brahmachari SK et al. CA/TG sequence at the 5′end of oligo(A)-tracts strongly modulates DNA curvature. J Biol Chem 1994; 269:7824-7833.
68. Zhou N, Manogaran S, Zon G et al. Deoxyribose ring conformation of [d(GGTATACC)]$_2$: an analysis of vicinal proton-proton coupling constants from two-dimensional proton nuclear magnetic resonance. Biochemistry 1988; 27:6013-6020.
69. Ojha RP, Dhingra MM, Sarma MH et al. DNA bending and sequence-dependent backbone conformation NMR and computer experiments. Eur J Biochem 1999; 265:35-53.
70. Gorin AA, Ulyanov NB, Zhurkin VB. S-N transition of the sugar ring in B-form DNA. Mol Biol (Engl transl) 1990; 24:1036-1047.
71. Foloppe N, MacKerell AD. Intrinsic conformational properties of deoxyribonucleosides: implicated role for cytosine in the equilibrium among the A, B, and Z forms of DNA. Biophys J 1999; 76:3206-3218.
72 Berman HM, Olson WK. The many twists of DNA. In: Balaban M, ed. DNA50: The Secret of Life. London: Faircount LLC, 2003:104-124.
73. Kamath S, Sarma MH, Zhurkin VB et al. DNA bending and sugar switching. J Biomol Struct Dyn 2000; Conversation 11:317-325.
74. Hagerman P. Straightening out the bends in curved DNA. Biochim Biophys Acta 1992, 1131:125-132.
75. Haran TE, Cohen I, Spasic A et al. Dynamics of curved DNA molecules: prediction and experiment. J Am Chem Soc 2003; 125:11160-11161.

76. Haran TE, Cohen I, Spasic A et al. Characteristics of migration patterns of DNA oligomers in gels and the relationship to the question of intrinsic DNA bending. J Am Chem Soc 2004;126:2372-2377.
77. Tolstorukov M, Virnik K, Adhya S et al. Genome-wide A-tract distribution and DNA packaging in pro- and eukaryotes. J Biomol Struct Dyn 2003; 20: 869-870.
78. Bolshoy A, Nevo E. Ecologic genomics of DNA: upstream bending in prokaryotic promoters. Genome Res 2000; 10:1185-1193.
79. Luger K, Mader AW, Richmond RK et al. Crystal structure of the nucleosome core particle at 2.8 Å resolution. Nature 1997; 389:251-260.
80. Dickerson RE, Bansal M, Calladine CR et al. Definitions and nomenclature of nucleic acid structure parameters. J Mol Biol 1989; 205:787-791.
81. Lu XJ, Olson WK. 3DNA: a software package for the analysis, rebuilding and visualization of three-dimensional nucleic acid structures. Nucleic Acids Res 2003; 31:5108-5121.

Part II
Intrinsic DNA Curvature and Transcription

Chapter 3

Curved DNA and Prokaryotic Promoters:
A Mechanism for Activation of Transcription

Munehiko Asayama and Takashi Ohyama

Abstract

Intrinsically curved DNA structures often occur in or around origins of DNA replication, regions that regulate transcription, and DNA recombination loci, and are found in a wide variety of cellular and viral genomes from bacteria to man. In bacterial promoters, bent DNA structures are often located from immediately upstream of the -35 hexamer to around position -100 relative to the transcription start site (+1). They have a range of functions: facilitating RNA polymerase binding to the promoter, transition from closed to open promoter complexes, or transcription factor binding. To perform these functions, in some cases intrinsically curved structures function together with DNA bends that are induced by binding of RNA polymerase, transcription factors, or nucleoid-associated proteins. This chapter will describe how curved DNA structures are implicated in prokaryotic transcription.

Introduction

DNA can become bent either by an exterior force such as a protein binding, or by the nucleotide sequence per se. The former is called protein-induced DNA bending or simply DNA bending, and the latter is called bent DNA, curved DNA, or intrinsic DNA curvature. Initially, the formation of a stable bent configuration of DNA was proposed by Crick and Klug to explain the mechanism underlying DNA packaging into nucleosomes, which dates back to 1975.[1] Several different models soon followed.[2-6] Among them, the "wedge model" [5] and the "junction model"[3] are famous for predicting intrinsic DNA curvatures[6] (Chapter 2). The first naturally occurring curved DNA was discovered in 1982, by electrophoretic analyses of the minicircle fragments of mitochondrial (kinetoplast) DNA (k-DNA) from a parasite, *Leishmania tarentolae.*[7] A DNA fragment from the organism migrated unusually slowly in non-denaturing polyacrylamide gels. This fragment contained a unique nucleotide sequence, with regularly distributed runs of adenines or thymines, with a periodicity of one run per helical repeat. Although such regular runs were soon proved to be the cause of intrinsic DNA curvature, the molecular mechanism to form curved DNA structure is still a matter of controversy.[6,8-15]

Since the initial discovery, curved DNA structures have been identified in a wide variety of cellular and viral genomes from bacteria to man. Interestingly, curved DNA structures often occur in or around origins of DNA replication,[16-21] promoters and enhancers (Table 1; for eukaryotes see Table 1 in Chapter 5) and DNA recombination loci,[22,23] irrespective of the origin of the DNA, suggesting that DNA curvature is important in many basic genetic processes (for reviews, see refs. 24-27). In order to reveal its role, a great many studies have been carried out. Here, focusing on the roles in prokaryotic transcription, we describe the fruits of these studies.

DNA Conformation and Transcription, edited by Takashi Ohyama. ©2005 Eurekah.com and Springer Science+Business Media.

Table 1. Prokaryotic genes that have a curved DNA structure in or around the promoter

Promoter[a]	Origin	Locus	Reference(s)
argT	*Escherichia coli*	upstream	45
argU	*Escherichia coli*	upstream	31
bolAp1	*Escherichia coli*	core	33
caa	*Escherichia coli*	downstream	61,62
gal	*Escherichia coli*	upstream	54,123
glnAp2	*Escherichia coli*	upstream	59,96
glpF	*Escherichia coli*	core	31
helD	*Escherichia coli*	core	31
ilvIH	*Escherichia coli*	upstream	46
ilvP$_G$2	*Escherichia coli*	upstream	125
katEp	*Escherichia coli*	core	33
lac	*Escherichia coli*	core, upstream	30,83
leuV	*Escherichia coli*	upstream	37
ompF	*Escherichia coli*	upstream	55
proV	*Escherichia coli*	core	31
rrnB	*Escherichia coli*	upstream	39,57
rrnD	*Escherichia coli*	upstream	31
bla	*Escherichia coli* plasmid pUC19	upstream	40,41
*spo0FP*2	*Bacillus subtilis*	core	32
P$_{A3}$	*Bacillus subtilis* phage f29	upstream	58,89
Alu156	*Bacillus subtilis* phage SP82	upstream	42
Bal129	*Bacillus subtilis* phage SP82	upstream	74
plc	*Clostridium perfringens*	upstream	43
nifLA	*Klebsiella pneumoniae*	upstream	60
psbA2	*Microcystis aeruginosa*	upstream	47,48
rpoD1	*Microcystis aeruginosa*	upstream	44
Plkt	*Pasteurella haemolytica*	core ~ upstream	35
hisR	*Salmonella typhimurium*	upstream	38
skc	*Streptococcus equisimilis*	upstream	49
PctII	*Streptococcus pneumoniae* plasmid pLS1	upstream	50,118

[a] Gene or operon names are used except a few cases

The Shape of Curved DNA

Naturally occurring curved DNA can adopt various conformations, which are determined by runs of adenines or thymines (A-tracts or T-tracts) as described above. When the A- or T-tracts occur with a periodicity almost equal to that of the DNA helical repeat length of around 10.5 bp, DNA forms a planar curve (i.e., a two-dimensional curve). However, if the tract periodicity is other than this, then either a right- or a left-handed superhelical conformation is formed.[12] These three-dimensional (3D) structures are sometimes called "space curves" (Fig. 1). When the periodicity is less than 10.5 bp (e.g., 9 or 10 bp), DNA adopts a left-handed superhelical conformation, and when it is more than 10.5 bp (e.g., 11 or 12 bp), DNA adopts a right-handed superhelical conformation. As the tract periodicity departs further from 10.5 bp, the helical axis becomes almost straight; e.g., periodicities of 5 to 7 bp result in a nearly straight (actually, a zigzag) trajectory of the helical axis.

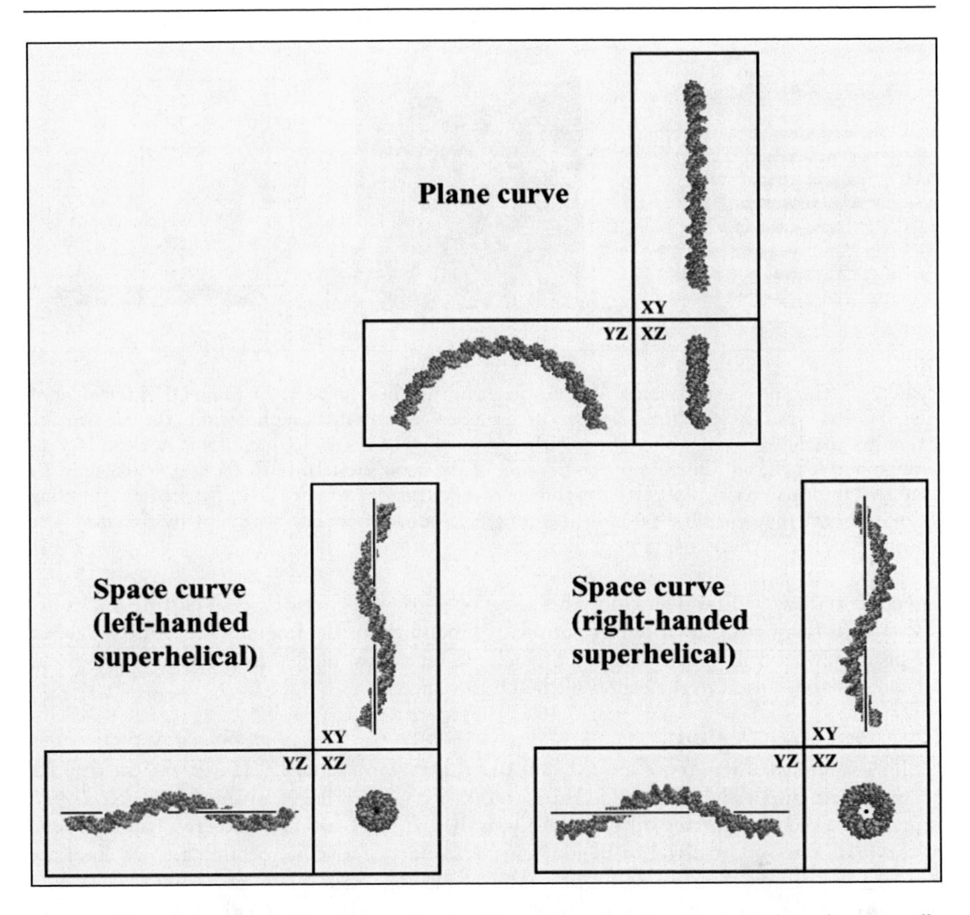

Figure 1. Three-dimensional views along each axis, of two-dimensionally curved and three-dimensionally curved DNA structures. The figure shows the structures formed by the nucleotide sequences $(A_5GTAC)_9$ (left), $(A_5GTACGA_5GTGCAC)_4$ (top) and $(A_5GTACGTC)_7$ (right). For DNA-drawing softwares, see Chapter 5 (Fig. 2).

Mapping Curved DNA Loci

Bacterial core promoters have conserved sequences at positions -35 (TTGACA) and -10 (TATAAT) relative to the transcription start site, which are recognized by RNA polymerase (RNAP) holoenzyme carrying a principal sigma factor such as sigma 70 (σ^{70}) (*Escherichia coli*) or SigA (σ^{A}) (*Bacillus subtilis*). This section describes where curved DNAs occur, and a representative technique for studying DNA curvature.

Circular Permutation Analysis

Circular permutation analysis, cyclization analysis, and electron microscopy are often used to find the position or magnitude of bends in DNA.[6,28,29] Circular permutation analysis, devised by Wu and Crothers in 1984, has now become routine. It can reveal the position, magnitude, and direction of a DNA bend compared to known standards. Here, we will describe only this assay (see Fig. 2). A DNA bend slows migration through a polyacrylamide gel. This phenomenon is more pronounced when the bend is in the center of a DNA fragment as opposed to near the ends. Thus to find the center of a bend, the DNA fragment of interest is first cloned

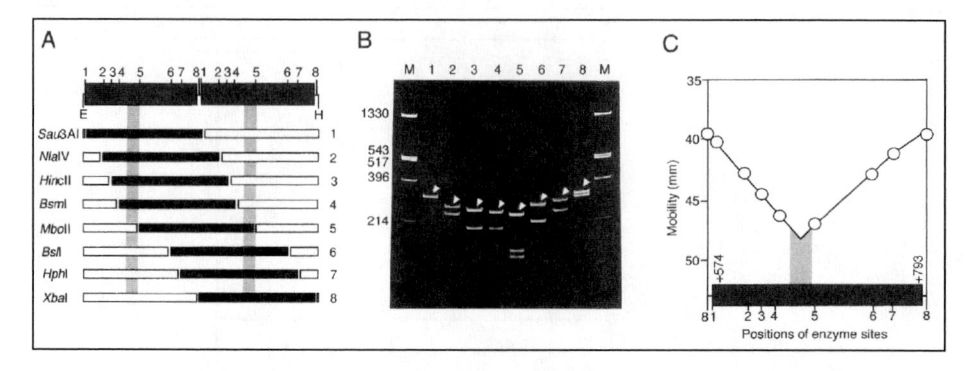

Figure 2. Circular permutation assay. A) DNA fragments used in the assay. First, the DNA fragment of interest is cloned as a tandem dimer. A variety of restriction enzymes that cut once within the fragment are used to cut the dimer, creating a set of circularly permuted DNA fragments, identical in nucleotide composition and length, but differing in the position of the bend site relative to the fragments' ends. B) Non-denaturing polyacrylamide gel electrophoresis of the circularly permuted DNA fragments. C) Plotting the mobility as a function of the position of cutting. Reproduced from ref. 44 with permission from The Japanese Biochemical Society ©1999.

as a tandem dimer, and is then cut with a series of restriction enzymes, each cutting once only within each fragment. The mobility (or relative mobility) of the fragments is plotted against the position of cutting. In the fragment which migrates most rapidly in the gel, the DNA cut was presumably closest to the center of the DNA bend.

Core Promoter Region

There are several reports of curved DNA in this region (Table 1). The *E. coli lac* core P1 promoter contains a slightly curved DNA around -31.[30] In addition, promoters of *E. coli glpF* (a gene involved in facilitated diffusion of glycerol), *helD* (helicase IV), and *proV* (the first gene of the *proU* operon, involved in high-affinity transport of glycine, betaine and proline) are involved in curved DNA.[31] The P2 promoter of the *B. subtilis spo0F* gene, which encodes a protein required to initiate sporulation, contains curved DNA centered around -10.[32] The curved region contains the recognition site of phospho-Spo0A, a transcriptional regulator essential for the initiation of sporulation.

In the following promoters, absence of the consensus hexamer sequences may be offset by the presence of curved DNA. In *E. coli*, expression of a number of genes during the stationary phase is controlled by σ^S (KatF). Interestingly, the σ^S-regulated promoters do not present good consensus sequences in -10 and -35 regions, but are probably located in intrinsically curved DNA.[33] Recently, a similar report has been presented on the *Campylobacter jejuni* promoters which are recognized by house-keeping sigma-factor, RpoD.[34] *C. jejuni* is a human gastrointestinal pathogen. Petersen et al showed that the promoters recognized by RpoD do not contain the conserved -35 motif, but instead show very strong periodic variation in AT-content, and also semi-conserved T-tracts with a periodicity of 10-11 nucleotides that could form curved DNA. In cattle, *Pasteurella haemolytica* causes necrosis of the lung and the production of a fibrinocellular exudate. In the *P. haemolytica* leukotoxin promoter, the conserved -35 motif is also absent, and instead four repeats of CA_6(C/T)A are present with approximately 10 bp periodicity. The repeats form a curved DNA structure, which influences transcription.[35]

Upstream of the Core Promoter

Based on computer modeling, an early study suggested that strong *E. coli* promoters harbor a bent DNA structure in their upstream sequences.[36] Indeed, most of the curved DNA structures that have been experimentally identified are upstream of core promoters. Roughly, they fall into two groups: one located immediately upstream of the -35 hexamer, and the other further upstream. The first group includes promoters of the following genes: *E. coli argU* (arginine tRNA-4),[31] the *E. coli leuV* tRNA operon,[37] a *Salmonella hisR* tRNA operon,[38] the *E. coli rrnB* rRNA operon,[39] the *E. coli rrnD* rRNA operon,[31] *bla* (β-lactamase gene) in *E. coli* plasmid pUC19,[40,41] the *Alu156* of *Bacillus subtilis* phage SP82,[42] *Clostridium perfringens plc* (encoding phospholipase C (α-toxin)),[43] and *Microcystis aeruginosa* (cyanobacterium) K-81 *rpoD1* (encoding principal sigma factor, σ^{A1}).[44]

The second group includes DNA curvatures of *E. coli argT,* and the *ilvIH* operon (encodes acetohydroxyacid synthase III), with centers of curvature between -90 and -95,[45] and at -120,[46] respectively. *M. aeruginosa* K-81, a photosynthetic freshwater cyanobacterium, can grow between 5 and 38°C, with optimal growth at 28 to 30°C. Its *psbA2* gene (encoding a core protein in photosystem II) contains a unique curved DNA upstream of the light-responsive promoter, from -180 to -140. Interestingly, the center of curvature depends on temperature: it was at bases -180, -160, and -140, at 50, 30, and 4°C, respectively.[47] In many other organisms, the light-responsive genes also have curved DNA in the upstream region, e.g., in *Synechocystis* PCC6803 (blue-green algae), *Cyanidium caldarium* (red algae), *Oryza sativa* (plant, rice) and *Nicotiana tabacum* (plant, tobacco).[48] This may suggest that the curved DNA is highly conserved among light responsive genes from cyanobacteria to higher plants. *Streptococcus equisimilis* H46A, a human serogroup C strain, is a potent producer of the plasminogen activator streptokinase. The streptokinase gene (*skc*) has an intrinsic DNA curvature located at -100.[49]

Some promoters have DNA curvatures in both regions. For example, the P_{ctII} promoter of plasmid pLS1, which can replicate in both Gram-positive and Gram-negative bacteria, carries two curvatures. One is in the proximal region (~-50 to -80), and the other is in the distal region (~-140 to -220).[50]

Transcription factor binding sites are often located in curved DNA, irrespective of the location of the curvature. Both *E. coli lac* and *gal* operons possess dual promoters, P1 and P2. In the *lac* operon, slight curves form around positions -30 and -100 relative to P1's transcription start point.[30] The catabolite gene activator protein (CAP; sometimes known as CRP, the cyclic AMP receptor protein), a positive regulator responding to carbon source limitation,[51] binds the region within this curvature.[30,51] In the *gal* operon, the curved structure is located between -60 and -90 relative to P1's transcription start point. This region is slightly upstream of the CAP binding site, which is centered at -41.5.[52-54] The *E. coli ompF* gene, coding for a major outer membrane protein, also carries curved DNA. The sequence causing the curve lies between -101 and -71, where two sets of periodically spaced A_4 tracts are present. The binding site of OmpR (an activator) overlaps the curved DNA.[55] In the *E. coli rrnB* P1 promoter, the curved DNA center overlaps one of three FIS (factor for inversion stimulation) binding sites.[56,57] The *B. subtilis* phage φ29 has a curved DNA that falls into this category.[58] The promoters P_{A2b} (the promoter of early genes) and P_{A3} (the promoter of late genes) partially overlap, and drive transcription in opposite directions. Two p4 binding sites, separated by 15 bp, are within a segment of curved DNA. When p4 binds, it increases the curvature.[58] In the *E. coli glnAp2* promoter (from the *glnALG* operon, encoding glutamine synthetase), computer simulation suggested that a 70 degree curve forms between the binding sites for the activator (nitrogen regulator I, NR_I, = NtrC) and for σ^N (σ^{54})-RNAP.[59] Similarly, in the *Klebsiella pneumoniae nifLA* promoter, a curved structure forms between two NtrC binding sites and the core promoter.[60]

Downstream of the Core Promoter

In this region, curved DNA is rare. One example is the slight curve in the operator region of the *E. coli caa* gene (colicin A).[61,62] However, in this region, protein-induced DNA bends may be more important than intrinsic curvature, as regulators of transcription: when LexA (the repressor of the SOS system) binds here, it bends the DNA substantially.[61,62]

Role of the Curved DNA

First, we briefly describe how researchers have tried to understand the function of DNA conformation. We then detail how curved DNA is implicated in the transcription cascade.

Short History

The first functional analysis of DNA curvature was performed in 1984 using a promoter of a tRNA operon of *Salmonella*.[38] A 3 bp deletion at position -70 disrupted a curved structure, and reduced in vivo transcription to 40%. This study suggested that curved DNA could control transcription. The *E. coli argT* promoter requires an upstream region for high in vivo activity. Deletion mutants were used to study this promoter,[45] which required an upstream curved DNA region for high activity. Techniques using synthetic bent DNA are sometimes very useful, and have suggested that a DNA curvature close to the -35 hexamer is important.[63] Furthermore, linker scanning mutations were used to study the DNA curvature just upstream of the -35 sequence of *E. coli rrnB* operon. They revealed that the angular orientation of the DNA curvature determines promoter activation.[39]

Insertion of a short DNA segment(s) into sites in or around the curved DNA region is also useful. In 1989, McAllister and Achberger investigated the function of curved DNA upstream of the *Alu 156* promoter of *B. subtilis* phage SP82.[42] By introducing short DNA fragments (6 to 29 bp) between the core promoter and the curved DNA, they changed the rotational phase between them. These changes correlated with the changes in promoter function in vivo. The most efficient mutant promoters contained insertions of 11 and 21 base pairs, and the least efficient promoters contained insertions of 15 and 25 base pairs. In vitro these mutations influenced the efficiency of RNA polymerase binding to the promoter. These findings demonstrate that the rotational phase between core promoter and the curved DNA is significant. The same methodology was used to study the promoter of *C. perfringens plc* gene and it was shown that upstream curved DNA stimulates transcription both in vivo and in vitro.[43]

In 1994, using deletion mutants, Pérez-Martín and Espinosa showed that curved DNA increases transcription from the P_{ctII} promoter of pLS1 in vivo and in vitro, apparently independently of any activator protein.[50] Furthermore, an upstream curved DNA was replaced by the target sequence of IHF (integration host factor) or that of CopG (both are DNA-bending proteins), which activated transcription in the presence of these proteins but did not in their absence or deficiency. This study indicated that the curved conformation of DNA increases the number of contacts between the RNA polymerase and the promoter DNA, and that this increase is important in transcription initiation.

In order to alter shapes of DNA curvatures per se with minimal changes to sequence, short DNA segments were inserted into the center region of curved DNA in the pUC19 β-lactamase gene promoter,[40] and into the cyanobacterium *M. aeruginosa rpoD1* gene promoter.[44] The resulting promoters were less active than the wild-type promoters, indicating that activity depended on the gross geometry. These promoters have right-handed curved DNA. Such DNAs are often located just upstream of promoters,[64] but promoter DNA itself wraps around RNAP left-handedly (Fig. 3).[65-67] Thus it is tempting to speculate that if RNAP changed the writhe of the helical axis from right-handed to left-handed, it might deform the DNA and lead to local unwinding (and formation of an open promoter complex).[64,65] To test this hypothesis, right-handed curved DNA, left-handed curved DNA, two-dimensionally curved DNA and straight DNA segments were synthesized and tested for their effect on transcription in vivo. Right-handed curved DNAs clearly facilitated formation of the open promoter complex and activated transcription.[41] Curved DNA can also change its shape depending on temperature.[68] Thus, temperature can be used to study the relationship between DNA conformation and function.[69,70] Recently, a mechanism of transcriptional regulation has been proposed that depends on a temperature-induced conformational change (Chapter 4).

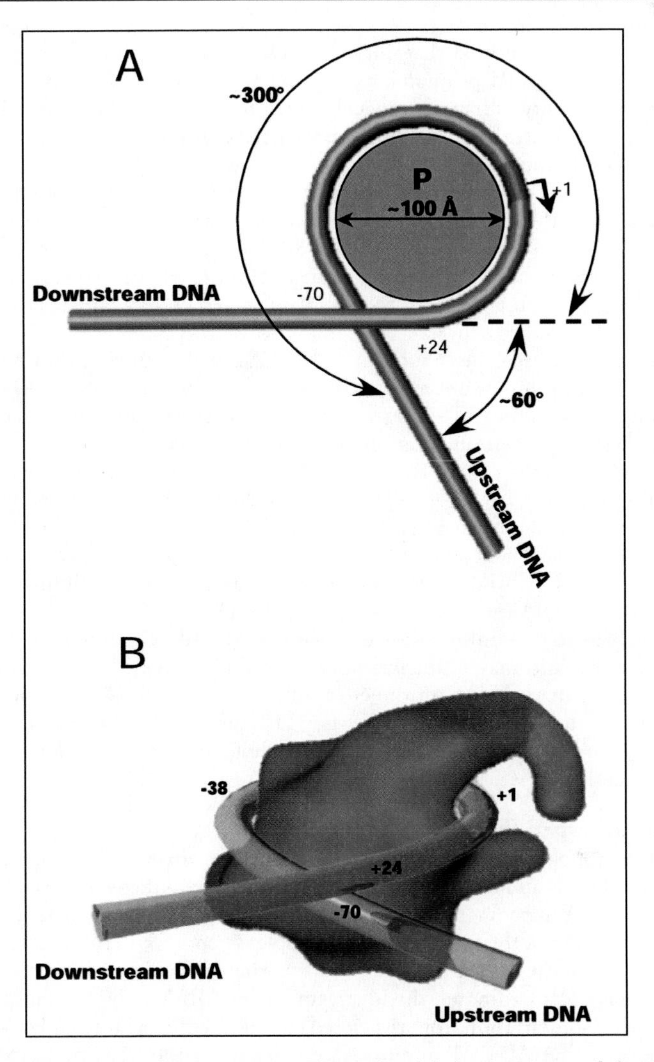

Figure 3. Schematic view showing the RNA polymerase and the trajectory of the DNA (A) and three-dimensional representation of how promoter DNA wraps around the polymerase (B). "P" indicates RNAP. Reproduced from ref. 67 with permission from the Nature publishing group ©1999.

Mechanistic Roles

Transcription initiation involves several steps.[71,72] Briefly, an RNAP binds to promoter (P), to yield RNAP-promoter closed complex (RP_C) with an equilibrium constant K_B. Next, RNAP melts approximately 14 bp of promoter DNA surrounding the transcription start site, with a rate constant k_2, to yield an RNAP-promoter open complex (RP_O). Subsequently, RNAP begins to synthesize RNA as an RNAP-promoter initial transcribing complex (RP_{ITC}). There are several abortive cycles of RNA synthesis, which yield RNA of 2-8 nucleotides long, with a rate constant k_i. When by chance a 9 nucleotide strand is synthesized, RNAP releases the promoter DNA (promoter clearance) and synthesizes RNA as an RNAP-DNA elongation complex (RD_E). Bent DNA is implicated in the steps described below.

RNAP Binding

Several studies have indicated that curved DNA is important for binding RNAP to the promoter.[70,73-75] In the *gal* P1 promoter, a point mutation from G•C to T•A at position -19, which abolishes P2 activity, enhances contact between the *E. coli* RNAP and the DNA between -49 and -54, and activates transcription from the P1 promoter, even in the absence of cAMP-CAP.[73] This mutation generates a run of six consecutive thymines (5′-TTTTTT-3′ is formed from 5′-TTTGTT-3′), which presumably influences the helical trajectory of the promoter, and helps RNAP binding. Hybrid *E. coli* λ phage promoters (λp_R) have been created, carrying curved DNA from the *B. subtilis* phage SP82 promoter, *Alu156*, or *Bal129*, immediately upstream of the -35 region. These promoters bound *E. coli* RNAP more efficiently than did the original promoter. Interestingly, the upstream curved DNA wrapped around the RNAP in a nucleosomal-DNA-like fashion.[75]

The interaction between the α-subunit of RNAP and the curved DNA next to the core promoter may be important for initiating transcription. Many genes have an AT-rich upstream (UP) element upstream of the -35 hexamer, and this was originally identified in the *E. coli rrnB* P1 promoter. This element stimulates transcription through contact with the C-terminal domain of the α-subunit (α-CTD) of RNAP.[76-79] The consensus sequence of the UP element is $N_2A_3(A/T)(A/T)T(A/T)T_4N_2A_4N_3$.[80] This region behaved slightly abnormally during electrophoresis, suggesting that it may be curved. The curved conformation may influence interactions between the UP element and α-CTD.[57] A study using the *E. coli lacUV5* promoter revealed that in the RNAP-promoter open complex, α-CTD makes alternative nonspecific interactions with the DNA minor grooves at positions -43, -53, -63, -73, -83, and -93.[81] Thus around the UP element, the conformation of DNA may assist the DNA to interact with α-CTD efficiently, and as a result, may influence transcription. However, it is difficult to distinguish the effects due to the shape of DNA from effects due to an AT-rich DNA sequence per se.[78,82] None of the above studies paid attention to the 3D architecture of DNA. Interestingly, an in vitro experiment indicated that in the step of binding, *E. coli* RNAP favors DNA with a right-handed superhelical writhe.[41]

Formation of the Open Promoter Complex

After the RNAP holoenzyme binds to the promoter, approximately 14 bp of the promoter (between positions -11 and +3) melts.[72] An early study hypothesized that DNA curvature allows upstream DNA to contact the promoter-bound RNAP, and that these interactions help the DNA for formation of the open promoter complex.[42] Indeed, in some cases, curved DNA enhances formation of the open complex.[41,54,83] For example, an in vitro experiment indicated that in the *E. coli gal* P1 promoter, the upstream curved DNA activates transcription by enhancing the rate of isomerization from the closed complex to the open complex (k_2) at P1, both in the absence of the cAMP-CAP and in its presence (the study used mutants where the P2 promoter was silenced).[54]

DNA architectures are not necessarily optimally oriented towards the surface of the RNAP. The RNAP must therefore distort the DNA.[84] Actually, the promoter DNA becomes bent when RNAP binds to it, and it wraps around the enzyme with a left-handed superhelical conformation (Fig. 3).[65-67,85,86] Energy invested in DNA bending could open the double helix.[87] These considerations suggest that DNA curvatures can enhance stress, depending on the direction of the curvature, which explains well why many bacterial promoters have a right-handed curved DNA immediately upstream of the -35 region.[64] This right-handed curvature presumably increases stress when the DNA wraps around RNAP, reversing its superhelical handedness. Indeed, when a synthetic right-handed curved DNA segment was placed immediately upstream of the -35 region of pUC19 β-lactamase promoter, it facilitated DNA melting at positions -11, -9, and +3 and activated transcription, compared to left-handed curved, two-dimensionally curved, and zigzag (straight) DNA segments.[41] Positions -11 and +3 correspond to the limits of RNAP-induced promoter melting.[72] Based on this putative mechanism, modulating the shape of DNA can also produce negative effects. A T_5 tract introduced just upstream of the -35 hexamer reduced promoter activity.[88] As another example, curved DNA inhibited transcription from the *B. subtilis* P_{A2b} promoter, by reducing the ability of RNAP to form transcriptionally active open complexes (in this case, binding of RNAP to the promoter was also impaired).[89] However, it is still not clear why DNA curvatures which are preset to modulate promoter melting, are usually located upstream of the core promoter.

Promoter Clearance

In the last stage of transcription initiation, promoter escape or promoter clearance occurs. "Escape" deals directly with issues affecting the downstream movement of a polymerase molecule. On the other hand, "clearance" implies that the polymerase moves far enough downstream to make the core promoter available to a second polymerase.[90,91] These phenomena are also rate-limiting steps in transcription initiation. In the *B. subtilis* page ϕ29 A2c promoter, preventing RNAP escape represses transcription.[92] Are DNA curvatures involved in promoter escape or promoter clearance? The following study suggests they are. In a study using two promoters isolated from the *B. subtilis* bacteriophage SP82, curved DNA with a strong affinity for RNAP reduced transcription, while curved DNA with a weaker affinity stimulated transcription.[74] DNA curvatures may collaborate with H-NS to reduce promoter clearance - it is known that they can repress transcription (Chapter 4). Images from scanning force microscopy (SFM) showed that in an *E. coli rrnB* P1 promoter with an upstream DNA curvature, H-NS trapped RNA polymerase in the open initiation complex. The SFM images suggested that H-NS-mediated trapping of RNAP could prevent promoter clearance.[93] To the best of our knowledge, a positive influence of DNA curvature has not been reported.

Transcription Factor Binding

If a curved DNA overlaps with some *cis*-DNA element, the curve's role may be to recruit *trans*-acting factors. In the *M. aeruginosa psbA* promoter, the curved region (approximately -180 to -140) is bound by a protein factor, and mutants having altered curvature had decreased basal transcription.[48] *E. coli* OmpR and *B. subtilis* Spo0A are response regulators in bacterial two-component regulatory systems.[94] Although OmpR can stimulate transcription of *ompF* (at low osmotic strength), and of *ompC* (at high osmotic strength), the OmpR binding site in the *ompC* promoter seems not to form a curved DNA. Thus, the presence or absence of curved DNA may be distinguished by OmpR in the differential activation of *ompF* and *ompC* promoters.[55] The *B. subtilis spo0F* gene, which decides the cell fate in sporulation initiation, has a tandem promoter, P1 and P2. P1 is recognized by σ^A-RNAP during exponential cell growth, and P2 is recognized by σ^H-RNAP during initiation of sporulation. For Spo0A binding at P1, the center of DNA curvature is close to the 0A-boxes. Increased Spo0A binding to the 0A-boxes represses transcription from the upstream P1, and simultaneously induces transcription from the downstream P2.[32] Curved DNA may facilitate Spo0A binding. However, no report has shown conclusively that curved DNA conformation per se is involved in factor binding.

Curved DNA as a Framework for Interaction between Activator and RNAP

Intrinsically curved DNA is sometimes located between activator binding sites and promoters. Several examples indicate that these structures enable activator and RNAP to interact. In the σ^N-dependent *glnAp2* or *glnHp2* promoter of *E. coli*, the DNA between the NR_I (NtrC)-binding site and the σ^N-recognition region must be either intrinsically curved, or curved by binding of the integration host factor, IHF. The *glnAp2* has an intrinsic DNA curvature in the relevant region. Although *glnHp2* does not carry curved DNA, it has an IHF-binding site in the relevant region. The DNA bend allows the activator to contact the σ^N-RNAP-promoter complex. This activates transcription by catalyzing the isomerization of the closed σ^N-RNAP-promoter complex to form an open complex.[59,95,96]

However, when supercoiled DNA templates are used, transcription can be initiated even in the absence of such a bend,[96] meaning that the spatial alignment between NR_I and RNAP on the templates may be significant. Similarly, in the *K. pneumoniae nifLA* promoter, an activator protein NtrC may interact with the σ^N-RNAP holoenzyme bound to the promoter with the help of an intervening curved DNA.[60] As another example, in the *E. coli rrnB* P1 promoter, the rotational phase between the FIS binding site and the promoter is important in activating transcription.[97] The curvature between these regions may also help FIS and RNAP to spatially position themselves correctly.

Other Roles

Curved DNA can also play a role in packaging of genomes. In eukaryotes, this role is played in nucleosome or chromatin formation (see Chapter 13). In *E. coli*, some nucleoid-associated proteins, such as H-NS (histone-like nucleoid-structuring protein), CbpA (curved DNA-binding protein A), Hfq (host factor for phage Q_β), and IciA (inhibitor of chromosome initiation A) preferentially bind to curved DNA.[98-100] Of these, H-NS can condense DNA both in vitro and in vivo[101,102] and regulates expression of a number of genes,[26,103] which are described in Chapter 4.

Curved DNA may also modulate the geometry of promoters in collaboration with structure-specific transcriptional regulators. The *B. subtilis* LrpC protein, which belongs to the LrpC/AsnC family of transcriptional regulators, forms stable complexes with curved DNAs. Interestingly, LrpC proteins wrap DNA to form nucleosome-like structures (but containing positively supercoiled DNA), and it is speculated that these could regulate transcription.[104] As another function, temperature-dependent conformational changes of bent DNA, as observed for *M. aeruginosa psbA2* gene,[47] can regulate transcription of several genes (Chapter 4).

Intrinsic DNA Curvature and Transcription-Factor Induced DNA Bend

A great many DNA binding proteins, including transcription factors, from both eukaryotes and prokaryotes, bend DNA on binding.[26,105-107] *E. coli* CAP, FIS and IHF, and *B. subtilis* ϕ29 p4 protein are well known to induce DNA bends (for a review, see ref. 26). CAP activates transcription through protein-protein interaction with RNAP.[108,109] The CAP-induced DNA bend might be required to create the appropriate geometry for this interaction.[110] IHF and FIS are nucleoid-associated proteins. IHF is distributed uniformly within the nucleoid, while FIS accumulates at specific loci.[111] They help formation of DNA micro-loops and/or nucleoprotein complexes.[112-114] RNAP itself also induces DNA bends.[115,116] In addition, there are many reports of repressors causing DNA to bend.[116-121]

For transcriptional activation, it is thought that induced DNA bends promote polymerase binding, protein-protein interactions, wrapping of DNA around the RNAP or multi-component DNA-protein complexes, local unwinding of DNA, and in some cases dissociation of nucleoprotein complexes.[106,122] Intrinsic DNA bends can also generate most of these effects. Indeed, CAP-induced DNA bends were functionally substituted by replacing the CAP binding site of the *gal* promoter or the *lac* promoter with synthetic bent DNA sequences.[83,123] Also, the IHF binding site was functionally replaced by intrinsically bent DNA sequences.[124] Conversely, as described above, intrinsic DNA curvature can be functionally replaced by protein-induced DNA bends.[50] These studies shed light on the significance of DNA bend per se in transcription initiation. The same is true for transcriptional repression. The p4 bends its target site on binding, excluding *B. subtilis* σ^A-RNA polymerase from the P_{A2b} promoter, and directing it to the P_{A3} promoter. As a result, transcription from the ϕ29 P_{A2b} promoter is repressed. Interestingly, the p4-induced DNA bend can be functionally substituted by intrinsically bent DNA.[89]

As described above, intrinsic DNA bends and protein-induced DNA bends can play the same role. In some cases, they collaborate. In *E. coli*, the *ilv*P_G2 promoter drives transcription of the *ilvGMEDA* operon whose products are required for the biosynthesis of L-leucine, L-isoleucine and L-valine. The promoter is activated both by intrinsic bend around position -50, and by IHF-induced DNA bend around -90.[125] The DNA geometry and localized destabilization of the DNA helix, generated by these two bends, might facilitate RNAP-promoter interaction and DNA unwinding. Proteins that bend intrinsically curved DNA structures toward different direction can also facilitate DNA unwinding as discussed for open promoter complex formation. This phenomenon is another example of the collaboration. The general importance of bending induced by transcription factors is described in Chapter 11.

Conclusion

Intrinsically curved DNA structures can modulate transcription initiation. Although such structures have various functions in the transcription initiation cascade, it appears that they mainly work by helping RNAP to bind to the promoter, and by facilitating formation of the open promoter complex. Bacteria are able to control promoter activity by carefully tuning the three-dimensional architecture of the promoter.

Acknowledgements

The authors would like to acknowledge the contribution of Yoshiro Fukue. Our own studies reported in this chapter were partly supported by Grants-in-Aid for Scientific Research from the Ministry of Education, Science, Sports and Culture of Japan.

References

1. Crick FH, Klug A. Kinky helix. Nature 1975; 255:530-533.
2. Sobell HM, Tsai CC, Gilbert SG et al. Organization of DNA in chromatin. Proc Natl Acad Sci USA 1976; 73:3068-3072.
3. Selsing E, Wells RD, Alden CJ et al. Bent DNA: visualization of a base-paired and stacked A-B conformational junction. J Biol Chem 1979; 254:5417-5422.
4. Zhurkin VB, Lysov YP, Ivanov VI. Anisotropic flexibility of DNA and the nucleosomal structure. Nucleic Acids Res 1979; 6:1081-1096.
5. Trifonov EN, Sussman JL. The pitch of chromatin DNA is reflected in its nucleotide sequence. Proc Natl Acad Sci USA 1980; 77:3816-3820.
6. Wu HM, Crothers DM. The locus of sequence-directed and protein-induced DNA bending. Nature 1984; 308:509-513.
7. Marini JC, Levene SD, Crothers DM et al. Bent helical structure in kinetoplast DNA. Proc Natl Acad Sci USA 1982; 79:7664-7668.
8. Hagerman PJ. Sequence-directed curvature of DNA. Nature 1986; 321:449-450.
9. Diekmann S. Sequence specificity of curved DNA. FEBS Lett 1986; 195:53-56.
10. Ulanovsky LE, Trifonov EN. Estimation of wedge components in curved DNA. Nature 1987; 326:720-722.
11. Koo HS, Crothers DM. Calibration of DNA curvature and a unified description of sequence-directed bending. Proc Natl Acad Sci USA 1988; 85:1763-1767.
12. Calladine CR, Drew HR, McCall MJ. The intrinsic curvature of DNA in solution. J Mol Biol 1988; 201:127-137.
13. Olson WK, Marky NL, Jernigan RL et al. Influence of fluctuations on DNA curvature. A comparison of flexible and static wedge models of intrinsically bent DNA. J Mol Biol 1993; 232:530-554.
14. Mack DR, Chiu TK, Dickerson RE. Intrinsic bending and deformability at the T-A step of CCTTTAAAGG: a comparative analysis of T-A and A-T steps within A-tracts. J Mol Biol 2001; 312:1037-1049.
15. Barbic A, Zimmer DP, Crothers DM. Structural origins of adenine-tract bending. Proc Natl Acad Sci USA 2003; 100:2369-2373.
16. Zahn K, Blattner FR. Direct evidence for DNA bending at the λ replication origin. Science 1987; 236:416-422.
17. Hertz GZ, Young MR, Mertz JE. The A+T-rich sequence of the simian virus 40 origin is essential for replication and is involved in bending of the viral DNA. J Virol 1987; 61:2322-2325.
18. Williams JS, Eckdahl TT, Anderson JN. Bent DNA functions as a replication enhancer in *Saccharomyces cerevisiae.* Mol Cell Biol 1988; 8:2763-2769.
19. Du C, Sanzgiri RP, Shaiu WL et al. Modular structural elements in the replication origin region of *Tetrahymena* rDNA. Nucleic Acids Res 1995; 23:1766-1774.
20. Polaczek P, Kwan K, Liberies DA et al. Role of architectural elements in combinatorial regulation of initiation of DNA replication in *Escherichia coli.* Mol Microbiol 1997; 26:261-275.
21. Kusakabe T, Sugimoto Y, Hirota Y et al. Isolation of replicational cue elements from a library of bent DNAs of *Aspergillus oryzae.* Mol Biol Rep 2000; 27:13-19.

22. Milot E, Belmaaza A, Wallenburg JC et al. Chromosomal illegitimate recombination in mammalian cells is associated with intrinsically bent DNA elements. EMBO J 1992; 11:5063-5070.
23. Mazin A, Milot E, Devoret R et al. KIN17, a mouse nuclear protein, binds to bent DNA fragments that are found at illegitimate recombination junctions in mammalian cells. Mol Gen Genet 1994; 244:435-438.
24. Travers A. DNA structure. Curves with a function. Nature 1989; 341:184-185.
25. Hagerman PJ. Sequence-directed curvature of DNA. Annu Rev Biochem 1990; 59:755-781.
26. Pérez-Martín J, Rojo F, de Lorenzo V. Promoters responsive to DNA bending: a common theme in prokaryotic gene expression. Microbiol Rev 1994; 58:268-290.
27. Ohyama T. Intrinsic DNA bends: an organizer of local chromatin structure for transcription. Bioessays 2001; 23:708-715.
28. Levene SD, Crothers DM. Ring closure probabilities for DNA fragments by Monte Carlo simulation. J Mol Biol 1986; 189:61-72.
29. Laundon CH, Griffith JD. Curved helix segments can uniquely orient the topology of supertwisted DNA. Cell 1988; 52:545-549.
30. Liu-Johnson HN, Gartenberg MR, Crothers DM. The DNA binding domain and bending angle of *E. coli* CAP protein. Cell 1986; 47:995-1005.
31. Tanaka K, Muramatsu S, Yamada H et al. Systematic characterization of curved DNA segments randomly cloned from *Escherichia coli* and their functional significance. Mol Gen Genet 1991; 226:367-376.
32. Asayama M, Yamamoto A, Kobayashi Y. Dimer form of phosphorylated Spo0A, a transcriptional regulator, stimulates the *spo0F* transcription at the initiation of sporulation in *Bacillus subtilis.* J Mol Biol 1995; 250:11-23.
33. Espinosa-Urgel M, Tormo A. σ^S-dependent promoters in *Escherichia coli* are located in DNA regions with intrinsic curvature. Nucleic Acids Res 1993; 21:3667-3670.
34. Petersen L, Larsen TS, Ussery DW et al. RpoD promoters in *Campylobacter jejuni* exhibit a strong periodic signal instead of a -35 box. J Mol Biol 2003; 326:1361-1372.
35. Highlander SK, Weinstock GM. Static DNA bending and protein interactions within the *Pasteurella haemolytica* leukotoxin promoter region: development of an activation model for leukotoxin transcriptional control. DNA Cell Biol 1994; 13:171-181.
36. Plaskon RR, Wartell RM. Sequence distributions associated with DNA curvature are found upstream of strong *E. coli* promoters. Nucleic Acids Res 1987; 15:785-796.
37. Bauer BF, Kar EG, Elford RM et al. Sequence determinants for promoter strength in the *leuV* operon of *Escherichia coli.* Gene 1988; 63:123-134.
38. Bossi L, Smith DM. Conformational change in the DNA associated with an unusual promoter mutation in a tRNA operon of *Salmonella.* Cell 1984; 39:643-652.
39. Zacharias M, Theissen G, Bradaczek C et al. Analysis of sequence elements important for the synthesis and control of ribosomal RNA in *E. coli.* Biochimie 1991; 73:699-712.
40. Ohyama T, Nagumo M, Hirota Y et al. Alteration of the curved helical structure located in the upstream region of the β-lactamase promoter of plasmid pUC19 and its effect on transcription. Nucleic Acids Res 1992; 20:1617-1622.
41. Hirota Y, Ohyama T. Adjacent upstream superhelical writhe influences an *Escherichia coli* promoter as measured by in vivo strength and in vitro open complex formation. J Mol Biol 1995; 254:566-578.
42. McAllister CF, Achberger EC. Rotational orientation of upstream curved DNA affects promoter function in *Bacillus subtilis.* J Biol Chem 1989; 264:10451-10456.
43. Matsushita C, Matsushita O, Katayama S et al. An upstream activating sequence containing curved DNA involved in activation of the *Clostridium perfringens plc* promoter. Microbiology 1996; 142:2561-2566.
44. Asayama M, Hayasaka Y, Kabasawa M, Shirai M, Ohyama A. An intrinsic DNA curvature found in the cyanobacterium *Microcystis aeruginosa* K-81 affects the promoter activity of *rpoD1* encoding a principal σ factor. J Biochem (Tokyo) 1999; 125:460-468.
45. Hsu LM, Giannini JK, Leung TW et al. Upstream sequence activation of *Escherichia coli argT* promoter in vivo and in vitro. Biochemistry 1991; 30:813-822.
46. Wang Q, Albert FG, Fitzgerald DJ et al. Sequence determinants of DNA bending in the *ilvIH* promoter and regulatory region of *Escherichia coli.* Nucleic Acids Res 1994; 22:5753-5760.
47. Agrawal GK, Asayama M, Shirai M. A novel bend of DNA CIT: changeable bending-center sites of an intrinsic curvature under temperature conditions. FEMS Microbiol Lett 1997; 147:139-145.
48. Asayama M, Kato H, Shibato J et al. The curved DNA structure in the 5'-upstream region of the light-responsive genes: its universality, binding factor and function for cyanobacterial *psbA* transcription. Nucleic Acids Res 2002; 30:4658-4666.

49. Groß S, Gase K, Malke H. Localization of the sequence-determined DNA bending center upstream of the streptokinase gene *skc*. Arch Microbiol 1996; 166:116-121.
50. Pérez-Martín J, Espinosa M. Correlation between DNA bending and transcriptional activation at a plasmid promoter. J Mol Biol 1994; 241:7-17.
51. Kolb A, Busby S, Buc H et al. Transcriptional regulation by cAMP and its receptor protein. Annu Rev Biochem 1993; 62:749-795.
52. Taniguchi T, O'Neill M, de Crombrugghe B. Interaction site of *Escherichia coli* cyclic AMP receptor protein on DNA of galactose operon promoters. Proc Natl Acad Sci USA 1979; 76:5090-5094.
53. Taniguchi T, de Crombrugghe B. Interactions of RNA polymerase and the cyclic AMP receptor protein on DNA of the *E. coli* galactose operon. Nucleic Acids Res 1983; 11:5165-5180.
54. Lavigne M, Herbert M, Kolb A et al. Upstream curved sequences influence the initiation of transcription at the *Escherichia coli* galactose operon. J Mol Biol 1992; 224:293-306.
55. Mizuno T. Static bend of DNA helix at the activator recognition site of the *ompF* promoter in *Escherichia coli*. Gene 1987; 54:57-64.
56. Ross W, Thompson JF, Newlands JT et al. *E. coli* Fis protein activates ribosomal RNA transcription in vitro and in vivo. EMBO J 1990; 9:3733-3742.
57. Gaal T, Rao L, Estrem ST et al. Localization of the intrinsically bent DNA region upstream of the *E. coli rrnB* P1 promoter. Nucleic Acids Res 1994; 22:2344-2350.
58. Rojo F, Zaballos A, Salas M. Bend induced by the phage ϕ29 transcriptional activator in the viral late promoter is required for activation. J Mol Biol 1990; 211:713-725.
59. Carmona M, Magasanik B. Activation of transcription at σ^{54}-dependent promoters on linear templates requires intrinsic or induced bending of the DNA. J Mol Biol 1996; 261:348-356.
60. Cheema AK, Choudhury NR, Das HK. A- and T-tract-mediated intrinsic curvature in native DNA between the binding site of the upstream activator NtrC and the nifLA promoter of *Klebsiella pneumoniae* facilitates transcription. J Bacteriol 1999; 181:5296-5302.
61. Lloubès R, Granger-Schnarr M, Lazdunski C et al. LexA repressor induces operator-dependent DNA bending. J Mol Biol 1988; 204:1049-1054.
62. Lloubès R, Lazdunski C, Granger-Schnarr M et al. DNA sequence determinants of LexA-induced DNA bending. Nucleic Acids Res 1993; 21:2363-2367.
63. Collis CM, Molloy PL, Both GW et al. Influence of the sequence-dependent flexure of DNA on transcription in *E. coli*. Nucleic Acids Res 1989; 17:9447-9468.
64. Travers AA. Why bend DNA? Cell 1990; 60:177-180.
65. Amouyal M, Buc H. Topological unwinding of strong and weak promoters by RNA polymerase. A comparison between the *lac* wild-type and the UV5 sites of *Escherichia coli*. J Mol Biol 1987; 195:795-808.
66. Coulombe B, Burton ZF. DNA bending and wrapping around RNA polymerase: a "revolutionary" model describing transcriptional mechanisms. Microbiol Mol Biol Rev 1999; 63:457-478.
67. Rivetti C, Guthold M, Bustamante C. Wrapping of DNA around the *E. coli* RNA polymerase open promoter complex. EMBO J 1999; 18:4464-4475.
68. Diekmann S. Temperature and salt dependence of the gel migration anomaly of curved DNA fragments. Nucleic Acids Res 1987; 15:247-265.
69. Ohyama T. Bent DNA in the human adenovirus type 2 E1A enhancer is an architectural element for transcription stimulation. J Biol Chem 1996; 271:27823-27828.
70. Katayama S, Matsushita O, Jung CM et al. Promoter upstream bent DNA activates the transcription of the *Clostridium perfringens* phospholipase C gene in a low temperature-dependent manner. EMBO J 1999; 18:3442-3450.
71. Gussin GN. Kinetic analysis of RNA polymerase-promoter interactions. Methods Enzymol 1996;273:45-59.
72. Ebright RH. RNA polymerase-DNA interaction: structures of intermediate, open, and elongation complexes. Cold Spring Harb Symp Quant Biol 1998; 63:11-20.
73. Busby S, Spassky A, Chan B. RNA polymerase makes important contacts upstream from base pair -49 at the *Escherichia coli* galactose operon *P1* promoter. Gene 1987; 53:145-152.
74. McAllister CF, Achberger EC. Effect of polyadenine-containing curved DNA on promoter utilization in *Bacillus subtilis*. J Biol Chem 1988; 263:11743-11749.
75. Nickerson CA, Achberger EC. Role of curved DNA in binding of *Escherichia coli* RNA polymerase to promoters. J Bacteriol 1995; 177:5756-5761.
76. Ross W, Gosink KK, Salomon J et al. A third recognition element in bacterial promoters: DNA binding by the α subunit of RNA polymerase. Science 1993;262:1407-1413.
77. Estrem ST, Ross W, Gaal T et al. Bacterial promoter architecture: subsite structure of UP elements and interactions with the carboxy-terminal domain of the RNA polymerase α subunit. Genes Dev 1999; 13:2134-2147.

78. Gourse RL, Ross W, Gaal T. UPs and downs in bacterial transcription initiation: the role of the α subunit of RNA polymerase in promoter recognition. Mol Microbiol 2000; 37:687-695.
79. Ross W, Ernst A, Gourse RL. Fine structure of *E. coli* RNA polymerase-promoter interactions: α subunit binding to the UP element minor groove. Genes Dev 2001; 15:491-506.
80. Estrem ST, Gaal T, Ross W et al. Identification of an UP element consensus sequence for bacterial promoters. Proc Natl Acad Sci USA 1998; 95:9761-9766.
81. Naryshkin N, Revyakin A, Kim Y et al. Structural organization of the RNA polymerase-promoter open complex. Cell 2000; 101:601-611.
82. Aiyar SE, Gourse RL, Ross W. Upstream A-tracts increase bacterial promoter activity through interactions with the RNA polymerase α subunit. Proc Natl Acad Sci USA 1998; 95:14652-14657.
83. Gartenberg MR, Crothers DM. Synthetic DNA bending sequences increase the rate of in vitro transcription initiation at the *Escherichia coli lac* promoter. J Mol Biol 1991; 219:217-230.
84. Borowiec JA, Gralla JD. High-resolution analysis of lac transcription complexes inside cells. Biochemistry 1986; 25:5051-5057.
85. Mekler V, Kortkhonjia E, Mukhopadhyay J et al. Structural organization of bacterial RNA polymerase holoenzyme and the RNA polymerase-promoter open complex. Cell 2002; 108:599-614.
86. Murakami KS, Masuda S, Campbell EA et al. Structural basis of transcription initiation: an RNA polymerase holoenzyme-DNA complex. Science 2002; 296:1285-1290.
87. Ramstein J, Lavery R. Energetic coupling between DNA bending and base pair opening. Proc Natl Acad Sci USA 1988; 85:7231-7235.
88. Lozinski T, Adrych-Rozek K, Markiewicz WT et al. Effect of DNA bending in various regions of a consensus-like *Escherichia coli* promoter on its strength in vivo and structure of the open complex in vitro. Nucleic Acids Res 1991; 19:2947-2953.
89. Rojo F, Salas M. A DNA curvature can substitute phage ϕ29 regulatory protein p4 when acting as a transcriptional repressor. EMBO J 1991; 10:3429-3438.
90. Gralla JD. Promoter recognition and mRNA initiation by *Escherichia coli* $E\sigma^{70}$. Methods Enzymol 1990; 185:37-54.
91. Hsu LM. Promoter clearance and escape in prokaryotes. Biochim Biophys Acta 2002; 1577:191-207.
92. Salas M. Control mechanisms of bacteriophage ϕ29 DNA expression. Int Microbiol 1998; 1:307-310.
93. Dame RT, Wyman C, Wurm R et al. Structural basis for H-NS-mediated trapping of RNA polymerase in the open initiation complex at the *rrnB* P1. J Biol Chem 2002; 277:2146-2150.
94. Stock JB, Stock AM, Mottonen JM. Signal transduction in bacteria. Nature 1990; 344:395-400.
95. Claverie-Martin F, Magasanik B. Positive and negative effects of DNA bending on activation of transcription from a distant site. J Mol Biol 1992; 227:996-1008.
96. Carmona M, Claverie-Martin F, Magasanik B. DNA bending and the initiation of transcription at σ^{54}-dependent bacterial promoters. Proc Natl Acad Sci USA 1997; 94:9568-9572.
97. Newlands JT, Josaitis CA, Ross W et al. Both fis-dependent and factor-independent upstream activation of the *rrnB* P1 promoter are face of the helix dependent. Nucleic Acids Res 1992; 20:719-726.
98. Yamada H, Muramatsu S, Mizuno T. An *Escherichia coli* protein that preferentially binds to sharply curved DNA. J Biochem (Tokyo) 1990; 108:420-425.
99. Ueguchi C, Kakeda M, Yamada H et al. An analogue of the DnaJ molecular chaperone in *Escherichia coli.* Proc Natl Acad Sci USA 1994; 91:1054-1058.
100. Azam TA, Ishihama A. Twelve species of the nucleoid-associated protein from *Escherichia coli.* Sequence recognition specificity and DNA binding affinity. J Biol Chem 1999; 274:33105-33113.
101. Spassky A, Rimsky S, Garreau H et al. H1a, an *E. coli* DNA-binding protein which accumulates in stationary phase, strongly compacts DNA in vitro. Nucleic Acids Res 1984; 12:5321-5340.
102. Spurio R, Durrenberger M, Falconi M et al. Lethal overproduction of the *Escherichia coli* nucleoid protein H-NS: ultramicroscopic and molecular autopsy. Mol Gen Genet 1992; 231:201-211.
103. Atlung T, Ingmer H. H-NS: a modulator of environmentally regulated gene expression. Mol Microbiol 1997; 24:7-17.
104. Beloin C, Jeusset J, Révet B et al. Contribution of DNA conformation and topology in right-handed DNA wrapping by the *Bacillus subtilis* LrpC protein. J Biol Chem 2003; 278:5333-5342.
105. Horikoshi M, Bertuccioli C, Takada R et al. Transcription factor TFIID induces DNA bending upon binding to the TATA element. Proc Natl Acad Sci USA 1992; 89:1060-1064.
106. van der Vliet PC, Verrijzer CP. Bending of DNA by transcription factors. Bioessays 1993; 15:25-32.
107. Werner MH, Gronenborn AM, Clore GM. Intercalation, DNA kinking, and the control of transcription. Science 1996; 271:778-784.
108. Zhou Y, Zhang X, Ebright RH. Identification of the activating region of catabolite gene activator protein (CAP): isolation and characterization of mutants of CAP specifically defective in transcription activation. Proc Natl Acad Sci USA 1993; 90:6081-6085.

109. Heyduk T, Lee JC, Ebright YW et al. CAP interacts with RNA polymerase in solution in the absence of promoter DNA. Nature 1993; 364:548-549.
110. Kapanidis AN, Ebright YW, Ludescher RD et al. Mean DNA bend angle and distribution of DNA bend angles in the CAP-DNA complex in solution. J Mol Biol 2001; 312:453-468.
111. Azam TA, Hiraga S, Ishihama A. Two types of localization of the DNA-binding proteins within the *Escherichia coli* nucleoid. Genes Cells 2000; 5:613-626.
112. Goosen N, van de Putte P. The regulation of transcription initiation by integration host factor. Mol Microbiol 1995; 16:1-7.
113. Travers A, Muskhelishvili G. DNA microloops and microdomains: a general mechanism for transcription activation by torsional transmission. J Mol Biol 1998; 279:1027-1043.
114. Travers A, Schneider R, Muskhelishvili G. DNA supercoiling and transcription in *Escherichia coli*: the FIS connection. Biochimie 2001; 83:213-217.
115. Kuhnke G, Fritz HJ, Ehring R. Unusual properties of promoter-up mutations in the *Escherichia coli* galactose operon and evidence suggesting RNA polymerase-induced DNA bending. EMBO J 1987; 6:507-513.
116. Kuhnke G, Theres C, Fritz HJ et al. RNA polymerase and gal repressor bind simultaneously and with DNA bending to the control region of the *Escherichia coli* galactose operon. EMBO J 1989; 8:1247-1255.
117. Pérez-Martín J, del Solar GH, Lurz R et al. Induced bending of plasmid pLS1 DNA by the plasmid-encoded protein RepA. J Biol Chem 1989; 264:21334-21339.
118. Pérez-Martín J, Espinosa M. The RepA repressor can act as a transcriptional activator by inducing DNA bends. EMBO J 1991; 10:1375-1382.
119. Ansari AZ, Chael ML, O'Halloran TV. Allosteric underwinding of DNA is a critical step in positive control of transcription by Hg-MerR. Nature 1992; 355:87-89.
120. Wang L, Winans SC. High angle and ligand-induced low angle DNA bends incited by OccR lie in the same plane with OccR bound to the interior angle. J Mol Biol 1995;253:32-38.
121. Tian G, Lim D, Carey J et al. Binding of the arginine repressor of *Escherichia coli* K-12 to its operator sites. J Mol Biol 1992; 226:387-397.
122. Zinkel SS, Crothers DM. Catabolite activator protein-induced DNA bending in transcription initiation. J Mol Biol 1991; 219:201-215.
123. Bracco L, Kotlarz D, Kolb A et al. Synthetic curved DNA sequences can act as transcriptional activators in *Escherichia coli*. EMBO J 1989; 8:4289-4296.
124. Pérez-Martín J, Timmis KN, de Lorenzo V. Co-regulation by bent DNA. Functional substitutions of the integration host factor site at σ^{54}-dependent promoter Pu of the upper-TOL operon by intrinsically curved sequences. J Biol Chem 1994; 269:22657-22662.
125. Pagel JM, Winkelman JW, Adams CW et al. DNA topology-mediated regulation of transcription initiation from the tandem promoters of the *ilvGMEDA* operon of *Escherichia coli*. J Mol Biol 1992; 224:919-935.

CHAPTER 4

Repression of Transcription by Curved DNA and Nucleoid Protein H-NS:

A Mode of Bacterial Gene Regulation

Cynthia L. Pon, Stefano Stella and Claudio O. Gualerzi

Abstract

Nucleoid-associated protein H-NS has emerged as one of the most intriguing and versatile global regulators of enterobacterial gene expression acting primarily yet not exclusively at the transcriptional level where it generally acts as a repressor. H-NS is also believed to contribute to the architectural organization of the nucleoid by causing DNA compaction, although the evidence for such a role is not overwhelming. H-NS binds preferentially to DNA elements displaying intrinsic curvatures and can induce DNA bending. These functions are determined by its quaternary tetrameric structure. In turn, the existence of an intrinsic DNA curvature separating two or more H-NS binding sites seems to be characteristic of the H-NS-sensitive promoters and a prerequisite for the transcriptional repressor activity of this protein. In some cases, like that of the *virF* promoter, the temperature-sensitivity of the DNA curvature represents a key element in the thermo-regulation of pathogenicity gene expression.

Introduction

Histone-like nucleoid-structuring protein (H-NS), was discovered in *Escherichia coli* approximately two decades ago;[1,2] the claim of an earlier discovery of this protein is in fact devoid of any scientific foundation. Indeed, the H1 protein described in 1971-72[3,4] was a <10 kDa thermostable (at 100°C) transcriptional enhancer composed of 67-70 amino acids eluting as a monomer from size-exclusion chromatography (SEC).[3] Instead, it is well established that H-NS, which is almost invariably a strong transcriptional repressor in vivo and under all in vitro conditions (for reviews see refs. 5-7), is inactivated by a brief exposure to 55°C[8] and its monomeric mass is 15.6 kDa, being constituted by 136 amino acids.[9] Furthermore, H-NS elutes from SEC as a mixture of dimers and tetramers even at μM concentrations[10,11] and its amino acid composition[9] is distinctly different from that reported for H1.[4]

After elucidation of the H-NS primary sequence,[9] the monocistronic gene (*hns*) encoding this protein was isolated and characterized.[12,13] However, due to a mistake in the orientation of the Kohara-Isono blot used in the physical mapping of the gene,[12] its chromosomal position was erroneously reported to be 6.1 min instead of 27 min where it is actually located.[14] Further biochemical characterization demonstrated that H-NS binds double stranded (ds)DNA better than single-stranded DNA and RNA[15] and displays a marked preference for bent DNA.[16-18] An architectural role in the organization of "bacterial chromatin", was inferred from its abundance (20,000 copies / cell),[1,19] from its biochemical properties and from its nucleoid localization,[20] but evidence also began to accumulate that H-NS controls, mainly at the

DNA Conformation and Transcription, edited by Takashi Ohyama.

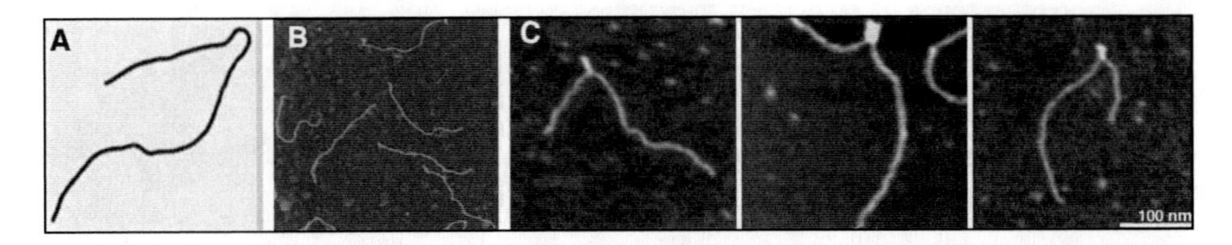

Figure 1. Specific recognition of curved DNA by H-NS. A) Curvature prediction and B) scanning force microscopy of a 1.1 kb linear fragment of naked DNA containing an in phase triple repeat of 5A and 6A tracts inserted at one third of its length; C) specific complexes formed by H-NS at the position of the curvature of the same DNA molecule at 1 protein dimer / 10 bp. Reprinted from: Dame RT, Wyman C, Goosen N. Biochimie 2001; 83:231-234 with permission from Elsevier.

transcriptional level, the expression of several genes since different *hns* mutations were shown to cause extremely pleiotropic phenotypes (for a review see ref.14). More recently H-NS has emerged not only as one of the most important and intriguing global transcriptional regulators, but also as a participant in other processes such as translation and the control of RNA and protein stability.

Genes encoding proteins homologous to H-NS were in the meantime discovered in other enterobacteriaceae[13, 21] but no H-NS-like protein was detected in Gram positive bacteria (*Bacillus stearothermophilus* and *B. subtilis*). Thus, unlike with the ubiquitous nucleoid-associated protein HU, H-NS was considered to be restricted to enteric bacteria. However, recent genomic data has shown that several Gram negatives from ecologically different habitats, contain H-NS-related proteins sharing a similar two-module structural organization.[22] However, it is not clear whether the roles of these proteins and of enterobacterial H-NS are the same. Finally, a gene (*stpA*) orthologous to *hns* was also identified in *E. coli* [23] and its product (StpA) was shown to share at least some properties with H-NS.

In the following sections we describe in more detail some structural and functional facets of H-NS with emphasis on its regulation of transcription in combination with bent DNA.

Relationship between Function and Three-Dimensional and Quaternary Structure of H-NS

As mentioned above, the preference for curved DNA is a major characteristic of H-NS. This binding property is clearly demonstrated by biochemical evidence and by electron micrographs showing that H-NS clusters precisely in the position of an intrinsic bend within a linear segment of DNA (Fig. 1). The functional significance of this preference for curved DNA[16-18] is suggested by the frequent occurrence of an intrinsic DNA bend flanked by two or more extended H-NS binding sites in the promoter regions susceptible to H-NS repression. (e.g., refs. 8, 24, 25) Among different types of curved DNA, H-NS seems to prefer AT-rich planar curvatures.[26,27] These findings, along with the fact that H-NS binds also to non-AT curvatures lacking a defined sequence-specificity,[28] suggest that this protein prefers a specific geometric structure of the duplex, probably corresponding to a narrow minor groove. However, little is known of the molecular nature of the H-NS-DNA interaction from both the protein and the nucleic acid side and nothing is known of the structural consequences locally induced by such an interaction. Thus, although H-NS binding results in the stabilization or induction of DNA bends from which transcriptional repression normally ensues, the microscopic consequences of the binding from which these macroscopic effects ultimately stem remain mysterious.

Structural and biochemical data show that H-NS consists of two domains whose three-dimensional (3D) structures have been elucidated[29-35] (Fig. 2). However, considerable controversy surrounds the actual structure of the N-terminal domain for which three different analyses have yielded three completely different structures,[33-35] two of which are shown in Figure 2.

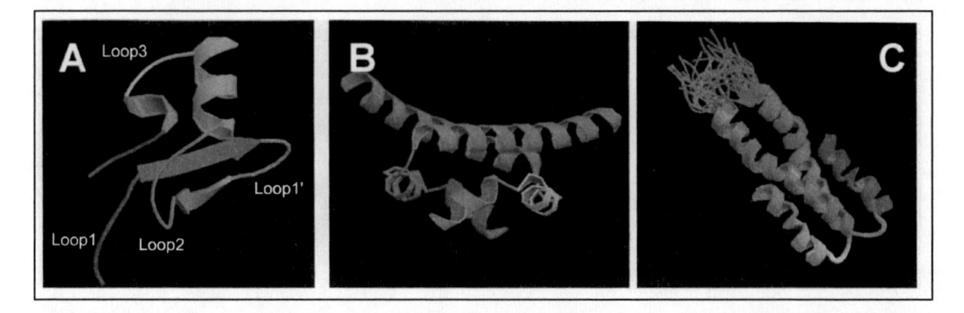

Figure 2. 3D structures determined in solution by NMR spectroscopy of (A) the C-domain[29,30] (1HNR) and (B) N-domain[35] (1NI8) of *E. coli* H-NS. The alternative 3D structure of the N-domain of *Salmonella typhimurium* H-NS[34] (1LR1) is shown in (C). The structures were obtained from the Protein Data Bank. The loops indicated in panel A are numbered according to Schroeder and Wagner.[7]

The C-terminal domain, which is separated from the N-domain by a linker, is considered to be the DNA binding domain. The interaction of this domain plus the linker (residues 60-137) with a 14 bp synthetic DNA (CAAAATATATTTTG) was investigated by nuclear magnetic resonance (NMR) spectroscopy; loops L1 and L2, which are spatially close to each other and display marked positive surface charges, were identified as the structural elements involved in this interaction. The N-terminal domain, which is ~45 Å away[36] from the C-domain, consists primarily of α-helices and its function is to promote protein dimerization. However, in addition to the controversy concerning the 3D structure of this domain, there is disagreement concerning the quaternary structure of H-NS which consists of a heterodisperse aggregate of trimers[32,33] or of monomers, dimers and tetramers in dynamic equilibrium.[9-11,37]

In spite of the commonplace attribution of the DNA-binding and the protein dimerization functions to the C- and the N-domain, respectively, not all genetic and biochemical data can be easily fit within this schematic model. Therefore, the relationship between structure and function of H-NS may be much more complicated than it superficially appears.

Indeed, probably because of the contribution of loop L1 to the H-NS-DNA interaction, the DNA affinity of H-NS depends substantially on the linker,[30] which is otherwise considered responsible for tetramerization of the protein.[38,39] Furthermore, also the N-terminal 46 residues, which should constitute the heart of the dimerization domain of H-NS[10,33-35] may contribute to the interaction with DNA.[10,35,38] This agrees with genetic data suggesting that deletion of the first 20 N-terminal residues abolishes the capacity of H-NS to repress transcription of at least some genes.[10] On the other hand, both recognition of bent DNA and bending non-curved DNA by H-NS are severely impaired by mutations of P115, (within the DNA binding domain), which affect H-NS oligomerization without influencing its basal DNA binding capacity. Similar properties are also conferred by other mutations in both N-domain (residues 1-20) and C-domain (W108).[10,11,35,38,39] Although these phenotypes could arise from defects of protein-protein and/or protein-DNA interaction, overall these results indicate the role played by H-NS oligomerization in the selective binding to bent DNA and induction of DNA curvature. The images of H-NS-plasmid DNA complexes obtained by atomic force microscopy (AFM)[40] (see also Fig. 3) are fully compatible with this premise.

Indeed, at one H-NS dimer / 12 bp of nicked circular plasmid, H-NS was found to cause two types of DNA condensation. In the first type of complex, large tracts of dsDNA are held together by H-NS bridges while the rest of the DNA forms double stranded loops. The bridges were interpreted as the result of tetramerization of H-NS dimers bound to two separate double stranded helices. The second type of complex is characterized by globular foci of H-NS incorporating large amounts of DNA while the rest of the plasmid remains partly naked and partly subject to the lateral condensation characteristic of the first type of complex. The DNA contour length is reduced ~3% in the first type of complex, possibly due to interwinding of the

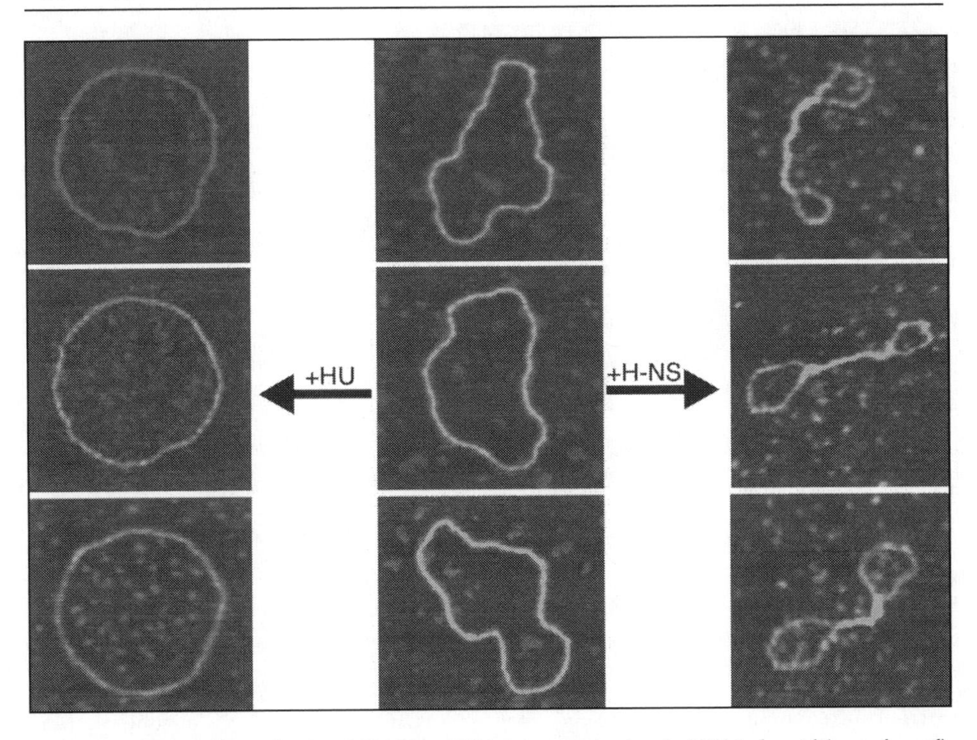

Figure 3. Opposite effects of HU and H-NS on DNA structure. A relaxed pUC19 plasmid (central panel) is opened by addition of 1 dimer HU / 9 bp (left panel) and compacted by 1 dimer H-NS / 12 bp (right panel). Reprinted by permission of Federation of the European Biochemical Societies from: Dame RT, Goosen N. FEBS Lett 2002; 529:151-156.

two helices, and by ~25% in the second. When two-times more protein is added, unstable rod-like structures whose compaction may be as high as 50% may appear.[40]

Although it is difficult to judge whether and how the complexes observed by electron microscopy (EM) and AFM[40] could be related to complexes surmised from other types of experimental evidence, the occurrence of at least two types of H-NS-DNA interactions leading to the formation of different types of nucleoprotein complexes is also indicated by fluorescence spectroscopy[15,41] and by H-NS footprinting analysis.[27,42] One type of complex could involve a small number of H-NS molecules nucleated around select positions of the chromosome and would depend on the H-NS capacity to form tetramers or small oligomers inducing duplex-duplex bridging. Another, less specific type of complex, would rely on the basic DNA binding capacity of H-NS and would engage a much larger number of H-NS molecules linearly polymerized along the duplex, eventually resulting in larger aggregates with strongly condensed DNA.[27,40,42,43] The moderate compaction of the nucleoids clearly observed in cells overproducing a mutant H-NS (ΔG112-P115) having wild-type (wt) affinity for non-curved DNA but with strongly reduced affinity for curved DNA[11,44] (Fig. 4) could be due to this second type of H-NS-DNA interaction. Likewise, the repression of the *bgl* operon by a transcriptional silencing mode could involve this type of H-NS-DNA interaction and could account for the suppression of the *bgl* phenotype apparently sustained by the formation of heterodimers between H-NS mutants lacking the DNA-binding domain and wt StpA.[45,46] Indeed, formation of these chimeric dimers containing only one DNA binding site is incapable of suppressing other, obviously more stringent *hns* phenotypes (e.g., see refs. 10, 47 and references therein).

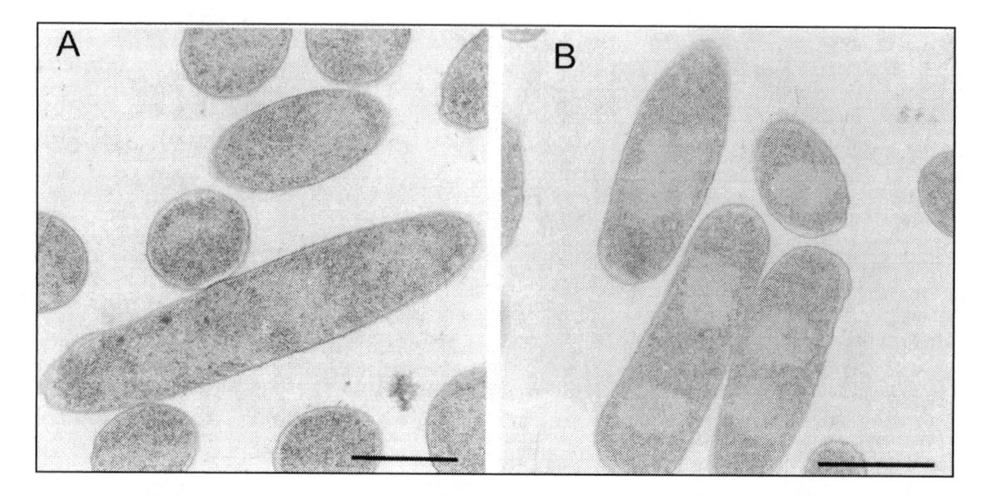

Figure 4. Appearance of the nucleoids of *E. coli* K12ΔH1ΔTrp subjected to cryo-fixation before (A) and after (B) overexpression of H-NS (ΔGly112-Pro115), a mutant protein with intact basal DNA-binding capacity but unable to recognize bent DNA and to bend DNA.[11] The black bars correspond to 1 μm. Reprinted from: Spurio R, Duerrenberger M, Falconi M et al. Mol Gen Genet 1992; 231:201-211 with permission from Springer-Verlag.

The observation that different H-NS-DNA complexes are formed as a function of the variation of environmental parameters[40] matches the finding that H-NS oligomerization equilibria are sensitive both in vitro[37] and in vivo (Stella et al, in preparation) to variations of physical parameters (protein concentration, temperature and ionic strength) corresponding to the environmental cues to which H-NS responds in vivo. These findings suggest that changes of the intracellular milieu may modulate H-NS function through the formation of different types of complexes.

Architectural Role of H-NS

DNA compaction is believed to involve DNA supercoiling and looping. In light of their capacity to affect these parameters, proteins HU and H-NS have long been considered responsible for condensing chromosomal DNA (~1.5 mm in *E. coli*) inside the nucleoid. However, this viewpoint has been challenged by the observation that the HU/H-NS ratio varies considerably (2.5-fold) in the cell as a function of the growth phase,[19] by data showing that none of the nucleoid-associated proteins contributes to chromosome looping[48] and by recent AFM observations suggesting that HU and H-NS have antagonistic effects on chromosomal architecture,[49] with HU stretching out and extending a circular duplex and H-NS causing instead its compaction (Fig. 3).[49]

Thus, although H-NS is commonly credited with the dual roles of transcriptional repressor and of architectural protein of the nucleoid, the evidence for the latter role is somewhat weak and indirect. The nucleoid localization of H-NS in both cryo-fixed[20] and viable cells[50] (Fig. 5) is compatible with but does not constitute proof in its favor. The early evidence that H-NS causes DNA compaction[2] has recently been challenged by Amit et al[51] who suggested that the increased sedimentation velocity acquired by a plasmid upon H-NS binding is due to the binding of a large number of protein molecules. Furthermore, the finding that H-NS polymerization (one dimer/15-20 bp) increased the bending rigidity of the double helix and increased the end-to-end distance of a DNA fragment clashes with the notion that H-NS causes DNA

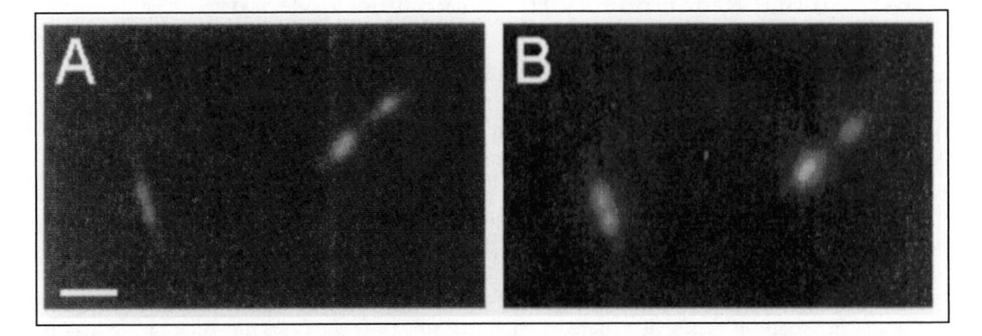

Figure 5. In vivo fluorescence localization of H-NS-GFP (green fluorescent protein) fusion protein (A) and DAPI-stained nucleoids (B) in exponential phase *E. coli* JM109 cells 30 min after IPTG induction of p*hns-gfp* – ASV (Alanine, Serine, Valine). The bar represents 1 μm. Reprinted from: Giangrossi M, Exley RM, Le Hegarat F et al. FEMS Microbiol Lett 2001; 202:171-176 with permission from Elsevier.

compaction.[2] Thus, the most convincing evidence for an architectural role comes from the H-NS-induced DNA condensation seen by EM (Fig. 3)[24,40] and from the observation that nucleoids are clearly condensed upon overproduction of H-NS (ΔG112-P115), a non-lethal event which does not cause generalized transcriptional repression (Fig. 4A,B).[44] The altered DNA-binding properties of this H-NS mutant (normal capacity to bind non-curved DNA but reduced capacity to bind curved DNA) support our previous interpretation that it is the non-specific coating of DNA by H-NS which ultimately causes clumping and compaction of the duplex.

DNA Bending and H-NS Activity

Beyond their immediate informational significance, the nucleotide sequences characterizing the individual genomes determine intrinsic structural properties and spatial structures of the DNA which may represent an additional source of genetic information exploited by cells to regulate life processes. Sequence-directed DNA curvature represents one of the most important and widespread of these structural features. These curvatures, present in a substantial fraction of the promoters (Chapter 3), show conservative patterns of distribution in the genomes of mesophilic bacteria, being preferentially located 40-200 bp upstream from the nearest transcriptional start.[52] Thus, DNA curvatures represent a well known example of the functional importance which the local architecture of the genome might acquire[53-55] since they can affect a large number of cellular phenomena, transcription being one of the most prominent.[56,57] Indeed, bent DNA has been shown to affect bacterial transcription sometimes facilitating the binding to promoters of RNA polymerase and/or of nucleoid-associated proteins (e.g., IHF, FIS, HU, H-NS, Lrp, Crp) functioning as activators and/or repressors, depending on the genetic system. Whereas in some cases bent DNA may play a direct role in transcriptional regulation, in other cases it plays an indirect role insofar as the binding of DNA-binding proteins stabilizes or enhances the curvature of a pre-existing DNA loop giving rise to nucleoprotein complexes which block downstream transcription. In fact, these nucleoid-associated proteins are not only able to recognize and bind curved DNA regions displaying more or less stringent consensus sequences, but are often able to induce DNA bending.[54-58] Concerning H-NS in particular the presence of intrinsically curved DNA regions is a common occurrence in H-NS-sensitive promoters, and the molecular basis of its regulatory activity often resides in its preferential interaction with intrinsically curved DNA and in its ability to induce DNA bending.[11,59,60]

Transcriptional Regulation by H-NS and Thermosensing through Curved DNA

H-NS affects the expression of a large number of enterobacterial genes, some coding for housekeeping functions, and many implicated in cellular responses to environmental changes, including the virulence factors whose expression is triggered by the passage from the external to the intestinal environment.[6,58] Thus, a variety of phenotypes is associated with *hns* mutations: increase in pH resistance;[61,62] loss of mobility;[63,64] serine-[65] and cold-susceptibility[66] (the latter phenotype likely related to the fact that H-NS is a cold-shock protein).[67,68] However, H-NS does not regulate all its target genes through the same mechanism. In fact, although H-NS acts as a transcriptional repressor in the majority of the cases, it can also act as a translational repressor[69] or affect gene expression at different post-transcriptional levels.[70] Nevertheless, to focus on the specific scope of this review, here we shall describe only the cases in which H-NS acts as a transcriptional repressor, with emphasis on the cases in which the repression mechanism can be related to the presence and participation of a curved fragment of DNA. However, it should be borne in mind that even in those cases in which H-NS clearly acts as a transcriptional repressor on a promoter containing a curved segment of DNA, the repression may involve different mechanisms. In fact, depending on the genetic system, the repression has been attributed to the capacity of H-NS to increase DNA compaction, to alter its topology or to induce the formation of more or less complex nucleoprotein particles as a result of its preferential interaction with bent DNA.[71-73] Furthermore, while for some genes the actual molecular mechanism causing repression has been attributed to promoter occlusion (Fig. 6A)[25,71,74] in other cases it has been demonstrated that it is due to RNA polymerase entrapment in the promoter caused by H-NS-mediated looping of DNA (Fig. 6B).[60,75]

As mentioned above, intrinsic DNA curvatures may be the primary actors in determining transcriptional regulation insofar as DNA curvatures are sensitive to changes in environmental parameters such as temperature, magnesium and polyamine concentrations.[76] Furthermore, the fact that these parameters differentially affect the curvature depending on the DNA sequence indicates that at least some DNA curvatures could be ideal and specific sensors implicated in gearing gene expression to environmental changes. It should not be ignored, however, that the same (or similar) physical parameters which can influence DNA curvature can also affect, both in vitro[37] and in vivo (Stella et al, manuscript in preparation), the oligomerization equilibria of H-NS which, in turn, can influence its activity on the DNA.

H-NS is involved in many regulatory circuits controlling the expression of virulence genes[6,58,71,77-85] and different mechanisms may account for this function. In fact, while in some cases transcription is prevented by direct binding of H-NS to target genes,[71] in others, the effect is indirectly caused by the influence of H-NS on the expression of specific transcriptional regulators or by its inhibition of DNA modifications involved in gene activation.[84]

A particular type of H-NS-dependent transcriptional regulation, which has recently emerged as a widespread strategy for controlling the expression of pathogenicity, is that mediated by environmental-dependent changes of DNA curvature. Some detailed information concerning this type of regulation is available in the case of *Shigella virF*, the gene whose product triggers the pathogenicity cascade in the etiological agent of human bacillary dysentery.[86] Like in other pathogenic enterobacteriaceae the transcription of the virulence genes of *Shigella* is strictly dependent on temperature, being repressed in non-intestinal environments characterized by lower temperature and osmolarity than in the intestine. Thus, the entry of the bacterium into the warm (37°C) host milieu represents a central cue triggering the expression of virulence factors and, ultimately, of the virulence phenotype.[87,88] The primary event following the temperature upshift is the synthesis of VirF, a transcriptional activator encoded by *virF*, a gene transcriptionally repressed below the threshold temperature of ~32°C.[81,89] In turn, the increased intracellular concentration of VirF triggers the activation of *virB*, a second regulator gene[89,90] whose product activates several invasion operons.[58,85] The transcriptional repression of *virF* below 32°C is mediated by H-NS and depends on the presence of an intrinsically bent

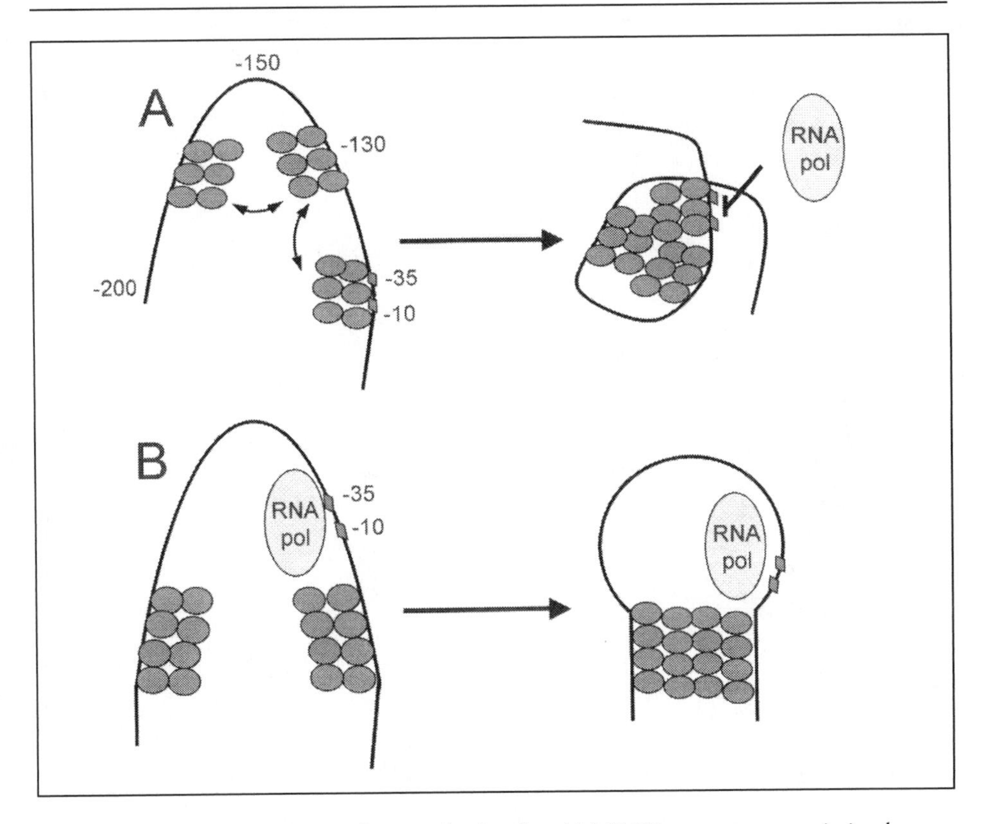

Figure 6. Schematic representation of two mechanisms by which H-NS may cause transcriptional repression. A) Promoter occlusion as postulated in the case of *hns* autorepression.[74] B) RNA polymerase entrapping as shown for the P1 promoter of *rrnB*.[75] Further details can be found in the text.

region whose center at 4°C is located at 137 bp upstream from the transcriptional start. This bend is flanked by two rather extended H-NS sites, one of which overlaps the core elements of the promoter. Both extent and localization of this curvature were found to be temperature-sensitive. In fact, gel electrophoretic analyses between 20°C and 60°C revealed that with increasing temperature there is a reduction of DNA curvature (which is maximum at 4°C) so that the bend seems to collapse when the temperature approaches the transcription-permissive conditions (32°C) thereby allowing the transition of *virF* promoter from the repressed to the derepressed state.[81,83] Furthermore, mutations affecting the curvature's amplitude or the relative orientation of the two H-NS sites severely affected in vivo and in vitro H-NS binding to *virF* and the thermoregulation of its expression.[83] In addition, it was shown that when the temperature increases from 4° to 60°C, the center of the DNA bend within the *virF* promoter slides downstream by almost eight helical turns; it is also noteworthy that the sliding rate is not linear with temperature but undergoes the largest increase within the narrow range (28°C-32°C) which corresponds to the transition from transcriptional repression to derepression. Within this range, the center of bending moves downstream by one helical turn going from -116 to -106 (Fig. 7).[83] Sliding of a bending center has also been reported in other cases such as in the upstream region of the light-responsive promoters of the cyanobacterial *psbA* gene.[91,92] Although the biological significance of these slidings remains largely unclear, the observation that the bending center within the *virF* promoter region "moves" towards the boundary of a FIS binding site (site III) and that the movement of the bending center is slow up to 28°C while becoming rapid when the transition temperature is approached suggests that

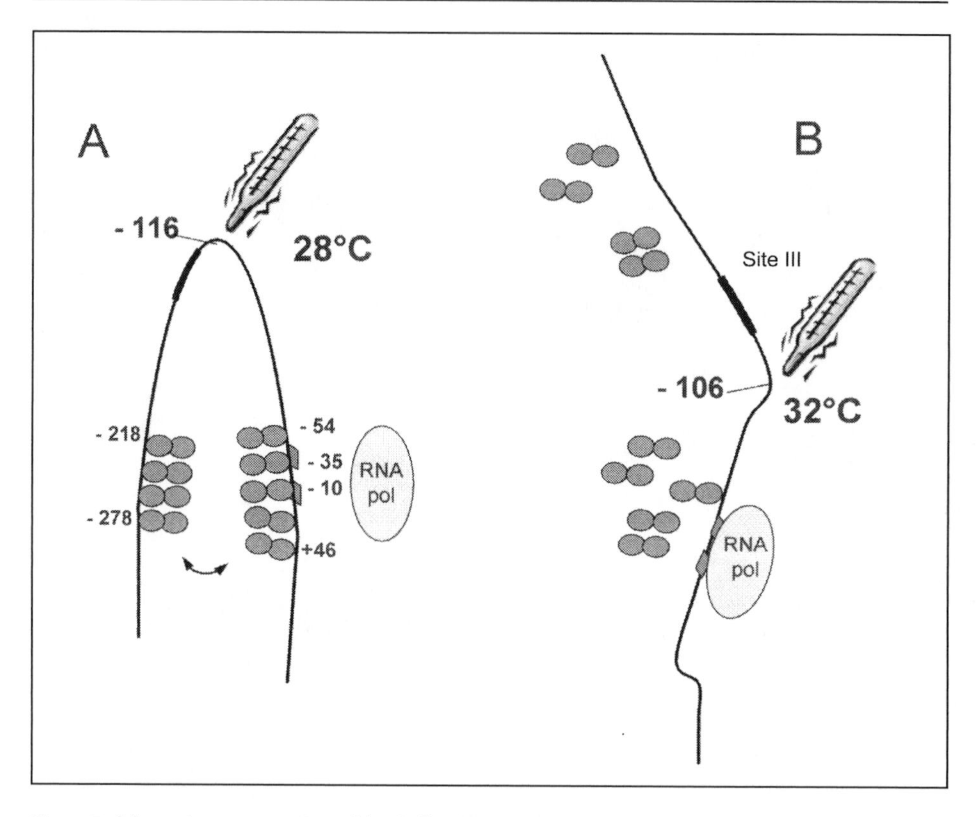

Figure 7. Schematic representation of the *S. flexneri virF* promoter in the (A) repressed configuration and (B) derepressed configuration. As described in the text, the H-NS- mediated repression is switched on and off by the presence of a thermosensor represented by an intrinsic DNA curvature.[81,83] The portion of DNA indicated by a thicker line corresponds to the FIS site III which spans from approximately -119 to -145.[82] The center of the bend was determined to be at -136, -118, -106 and -91 at 4°, 28° 32° and 40°, respectively.[83] Further details can be found in the text.

this shift might be related to the ability of FIS to partially relieve H-NS-induced *virF* repression at the transition temperature.[82] If it could be demonstrated that the sliding of the *virF* bending center renders this promoter region more accessible to FIS binding, thereby facilitating this protein in antagonizing H-NS-mediated repression, it would be tempting to interpret the bend-movement in *virF* as an example of a (general?) strategy to optimize gene control in multifactorial regulatory systems by the gradual environmentally-guided unmasking of specific DNA targets.

Taken together, the available data indicate that an intrinsic DNA bend within the promoter region of *virF* behaves as a thermal sensor and the geometric characteristics of this DNA curvature play an essential role in the thermoregulation of *Shigella* virulence expression (Fig. 7).[81,83]

Other hypotheses were also presented in the past to explain the H-NS-dependent thermoregulation of *virF* expression. They were formulated in terms of changes in H-NS structure and modifications of DNA supercoiling (reviewed in refs. 31, 93). However, they have been disproven eventually. In fact, transcriptional repression by H-NS requires a supercoiled DNA target but the role of DNA supercoiling is only allowing H-NS to build a repressing complex and not acting as the temperature sensor.[81] Furthermore, recent in vitro[37] and in vivo (Stella et al, manuscript in preparation) data indicate that H-NS oligomerization does change with temperature but not within the temperature interval in which the *virF* promoter undergoes the inactive/active transition.

In addition to the just described case of the *virF* thermoregulation, other cases are known in which a nucleic acid acts as a thermosensor in the control of virulence expression. For instance, a more stable binding of the YmoA repressor to the *virF* promoter of *Yersinia enterocolitica*, determined by a stronger DNA curvature induced by low temperature, results in silencing the virulence genes encoded by the Yops plasmid.[94,95] Furthermore, several genetic systems are known in which an increased DNA curvature induced by low temperature favors transcription instead of repressing it. For instance, like in the case of *Shigella virF*, the promoter of *Clostridium perfringens plcC* gene displays an increased bent structure at lower temperatures but in this case the expression is more efficient at low temperature because of the acquisition of a suitable template conformation.[96] Similarly, an increased DNA curvature at low temperature favors the binding of IHF and stimulates transcription from the P_L promoter of bacteriophage λ.[97]

A temperature-mediated change in DNA flexibility has also been implicated in the repression at low temperature of *E.coli hly*, an operon which is either plasmid- or chromosome-encoded, depending on the bacterial strain.[79] This operon, whose thermoregulated expression is under the control of H-NS and DNA curvature, encodes the α-hemolysin Hly, a toxin produced by several uropathogenic strains of *E. coli*. In the case of the *hly* operon carried by pHly152, a regulatory sequence (the 650 bp-long *hlyR*) is located > 1.5 kb upstream of the three promoters of *hlyC*, the first gene of the operon, being separated from the latter by an IS2 insertion element. Deletion of *hlyR* results in constitutive repression of hemolysin expression, a phenotype presumably connected with the loss of *ops*, a transcriptional antitermination element located in *hlyR*. Several lines of evidence indicate that the thermoregulated expression of *hly* operon is regulated by a nucleoprotein complex formed by H-NS and Hha, a protein homologous to YmoA, the above-mentioned temperature-dependent modulator of *Yersinia* virulence factors.[98] In fact, while *hha* mutants display only a partial derepression of temperature- and osmolarity-mediated expression of hemolysin, *hha-hns* double mutants were defective in both thermo- and osmo-regulation. Furthermore, while Hha displays a strong affinity for H-NS, this protein alone does not show any DNA binding preference, unlike its partner H-NS which preferentially binds to two sites in the regulatory region of the *hly* operon. The two sites, the first partially overlapping the core elements of the two most upstream promoters of *hlyC* and the second located ~2 kb upstream, are separated by an intrinsic DNA curvature predicted in silico and observed by AFM. Deletion analysis demonstrated that the upstream site is important for thermoregulation of the operon, and temperature was found to influence the affinity of H-NS for a DNA fragment containing both sites. The higher affinity of H-NS for its sites at low temperature correlates well with its more efficient transcriptional repression observed in vitro and with the constitutive hemolytic phenotype of the *hns* mutants observed in vivo. A current model explaining the H-NS/Hha-mediated temperature-dependent repression of hemolysin expression is that Hha generates hetero-oligomeric complexes with H-NS which are better suited than H-NS homo-oligomers in the temperature dependent repression of the *hly* operon. The H-NS binding specificity ensures that these complexes interact specifically with the DNA targets at the two sides of the bend, simultaneously occluding the transcription antiterminator sequence *ops* and at least two *hlyC* promoters.

Conclusion

H-NS has emerged as a major actor in the transcriptional regulation of gene expression, especially for genes involved in the cellular response to environmental changes and in virulence. While a general feature of the genes subject to H-NS control is the presence of an intrinsic DNA curvature, more work is necessary to establish whether a unitary mechanism underlies the H-NS functions and to clarify the molecular basis of the specificity governing H-NS-DNA interaction. Also to be clarified are the mechanism and the structural consequences of the binding.

Acknowledgements

This work was supported by MIUR grants PRIN 2001 (to CLP and COG) and PRIN 2002 (to COG).

References

1. Lammi M, Paci M, Pon C et al. Proteins from the prokaryotic nucleoid. Biochemical and ^{1}H-NMR studies on three bacterial histone-like proteins. In: Hubscher H, Spadari S, eds. Proteins Involved in DNA Replication. New York: Plenum Publishing Co., 1984:467-477.
2. Spassky A, Rimsky S, Garreau H et al. H1a, an *E. coli* DNA-binding protein which accumulates in stationary phase, strongly compacts DNA in vitro. Nucleic Acids Res 1984; 12:5321-5340.
3. Jacquet M, Cukier-Kahn R, Pla J et al. A thermostable protein factor acting on in vitro DNA transcription. Biochem Biophys Res Commun 1971; 45:1597-1607.
4. Cukier-Kahn R, Jacquet M, Gros F. Two heat-resistant, low molecular weight proteins from *Escherichia coli* that stimulate DNA-directed RNA synthesis. Proc Natl Acad Sci USA 1972; 69:3643-3647.
5. Ussery DW, Hinton JC, Jordi BJ et al. The chromatin-associated protein H-NS. Biochimie 1994; 76:968-980.
6. Atlung T, Ingmer H. H-NS: a modulator of environmentally regulated gene expression. Mol Microbiol 1997; 24:7-17.
7. Schroeder O, Wagner R. The bacterial regulatory protein H-NS – a versatile modulator of nucleic acid structures. Biol Chem 2002; 383:945-960.
8. Tippner D, Afflerbach H, Bradaczek C et al. Evidence for a regulatory function of the histone-like *Escherichia coli* protein H-NS in ribosomal RNA synthesis. Mol Microbiol 1994;11:589-604.
9. Falconi M, Gualtieri MT, La Teana A et al. Proteins from the prokaryotic nucleoid: primary and quaternary structure of the 15-kD *Escherichia coli* DNA binding protein H-NS. Mol Microbiol 1988; 2:323-329.
10. Ueguchi C, Suzuki T, Yoshida T et al. Systematic mutational analysis revealing the functional domain organization of *Escherichia coli* nucleoid protein H-NS. J Mol Biol 1996; 263:149-162.
11. Spurio R, Falconi M, Brandi A et al. The oligomeric structure of nucleoid protein H-NS is necessary for recognition of intrinsically curved DNA and for DNA bending. EMBO J 1997; 16:1795-1805.
12. Pon CL, Calogero R, Gualerzi CO. Identification, cloning, nucleotide sequence and chromosomal map location of *hns*, the structural gene for *Escherichia coli* DNA-binding protein H-NS. Mol Gen Genet 1988; 212:199-202.
13. La Teana A, Falconi M, Scarlato V et al. Characterization of the structural genes for the DNA-binding protein H-NS in Enterobacteriaceae. FEBS Lett 1989; 244:34-38.
14. Higgins CF, Hinton JC, Hulton CS et al. Protein H1: a role for chromatin structure in the regulation of bacterial gene expression and virulence? Mol Microbiol 1990; 4:2007-2012.
15. Friedrich K, Gualerzi CO, Lammi M et al. Proteins from the prokaryotic nucleoid. Interaction of nucleic acids with the 15 kDa *Escherichia coli* histone-like protein H-NS. FEBS Lett 1989; 229:197-202.
16. Bracco L, Kotlarz D, Kolb A et al. Synthetic curved DNA sequences can act as transcriptional activators in *Escherichia coli*. EMBO J 1989; 8:4289-4296.
17. Yamada H, Muramatsu S, Mizuno T. An *Escherichia coli* protein that preferentially binds to sharply curved DNA. J Biochem (Tokyo) 1990; 108:420-425.
18. Yamada H, Yoshida T, Tanaka K et al. Molecular analysis of the *Escherichia coli hns* gene encoding a DNA binding protein, which preferentially recognizes curved DNA sequences. Mol Gen Genet 1991; 230:332-336.
19. Azam TA, Iwata A, Nishimura A et al. Growth phase-dependent variation in protein composition of the *Escherichia coli* nucleoid. J Bacteriol 1999; 181:6361-6370.
20. Dürrenberger M, La Teana A, Citro G et al. *Escherichia coli* DNA-binding protein H-NS is localized in the nucleoid. Res Microbiol 1991; 142:373-380.
21. Marsh M, Hillyard DR. Nucleotide sequence of *hns* encoding the DNA-binding protein H-NS of *Salmonella typhimurium*. Nucleic Acids Res 1990; 18:3397.
22. Tendeng C, Krin E, Soutourina OA et al. A novel H-NS-like protein from an antarctic psychrophilic bacterium reveals a crucial role for the N-terminal domain in thermal stability. J Biol Chem 2003; 278:18754-18760.
23. Zhang A, Belfort M. Nucleotide sequence of a newly-identified *Escherichia coli* gene, *stpA*, encoding an H-NS-like protein. Nucleic Acids Res 1992; 20:6735.

24. Dame RT, Wyman C, Goosen N. Structural basis for preferential binding of H-NS to curved DNA. Biochimie 2001; 83:231-234.
25. Falconi M, Higgins NP, Spurio R et al. Expression of the gene encoding the major bacterial nucleoid protein H-NS is subject to transcriptional auto-repression. Mol Microbiol 1993; 10:273-282.
26. Zuber F, Kotlarz D, Rimsky S et al. Modulated expression of promoters containing upstream curved DNA sequences by the *Escherichia coli* nucleoid protein H-NS. Mol Microbiol 1994; 12:231-240.
27. Rimsky S, Zuber F, Buckle M et al. A molecular mechanism for the repression of transcription by the H-NS protein. Mol Microbiol 2001; 42:1311-1323.
28. Jordi BJ, Fielder AE, Burns CM. DNA binding is not sufficient for H-NS-mediated repression of *proU* expression. J Biol Chem 1997; 272:12083-12090.
29. Shindo H, Iwaki T, Ieda R et al. Solution structure of the DNA binding domain of a nucleoid-associated protein, H-NS, from *Escherichia coli*. FEBS Lett 1995; 360:125-131.
30. Shindo H, Ohnuki A, Ginba H et al. Identification of the DNA binding surface of H-NS protein from *Escherichia coli* by heteronuclear NMR spectroscopy. FEBS Lett 1999; 455:63-69.
31. Dorman CJ, Hinton JC, Free A. Domain organization and oligomerization among H-NS-like nucleoid-associated proteins in bacteria. Trends Microbiol 1999; 7:124-128.
32. Smyth CP, Lundback T, Renzoni D et al. Oligomerization of the chromatin-structuring protein H-NS. Mol Microbiol 2000; 36:962-972.
33. Renzoni D, Esposito D, Pfuhl M et al. Structural characterization of the N-terminal oligomerization domain of the bacterial chromatin-structuring protein, H-NS. J Mol Biol 2001; 306:1127-1137.
34. Esposito D, Petrovic A, Harris R et al. H-NS oligomerization domain structure reveals the mechanism for high order self-association of the intact protein. J Mol Biol 2002; 324:841-850.
35. Bloch V, Yang Y, Margeat E et al. The H-NS dimerization domain defines a new fold contributing to DNA recognition. Nat Struct Biol 2003; 10:212-218.
36. Schroeder O, Tippner D, Wagner R. Toward the three-dimensional structure of the *Escherichia coli* DNA-binding protein H-NS: a CD and fluorescence study. Biochem Biophys Res Commun 2001; 282:219-227.
37. Ceschini S, Lupidi G, Pon CL et al. Multimeric self-assembly equilibria involving the histone-like protein H-NS. J Biol Chem 2000; 275:729-734.
38. Williams RM, Rimsky S, Buc H. Probing the structure, function, and interactions of the *Escherichia coli* H-NS and StpA proteins by using dominant negative derivatives. J Bacteriol 1996; 178:4335-4343.
39. Ueguchi C, Seto C, Suzuki T et al. Clarification of the dimerization domain and its functional significance for the *Escherichia coli* nucleoid protein H-NS. J Mol Biol 1997; 274:145-151.
40. Dame RT, Wyman C, Goosen N. H-NS mediated compaction of DNA visualised by atomic force microscopy. Nucleic Acid Res 2000; 28:3304-3510.
41. Tippner D, Wagner R. Fluorescence analysis of the *Escherichia coli* transcription regulator H-NS reveals two distinguishable complexes dependent on binding to specific or nonspecific DNA sites. J Biol Chem 1995; 270:22243-22247.
42. Badaut C, Williams R, Arluison V et al. The degree of oligomerization of the H-NS nucleoid structuring protein is related to specific binding to DNA. J Biol Chem 2002; 277:41657-41666.
43. Rimsky S, Spassky A. Sequence determinants for H1 binding on *Escherichia coli lac* and *gal* promoters. Biochemistry 1990; 29:3765-3771.
44. Spurio R, Duerrenberger M, Falconi M et al. Lethal overproduction of the *Escherichia coli* nucleoid protein H-NS: ultramicroscopic and molecular autopsy. Mol Gen Genet 1992; 231:201-211.
45. Johansson J, Eriksson S, Sonden B et al. Heteromeric interactions among nucleoid-associated bacterial proteins: localization of StpA-stabilizing regions in H-NS of *Escherichia coli*. J Bacteriol 2001; 183:2343-2347.
46. Free A, Porter ME, Deighan P et al. Requirement for the molecular adapter function of StpA at the *Escherichia coli bgl* promoter depends upon the level of truncated H-NS protein. Mol Microbiol 2001; 42:903-917.
47. Ohta T, Ueguchi C, Mizuno T. *rpoS* function is essential for *bgl* silencing caused by C-terminally truncated H-NS in *Escherichia coli*. J Bacteriol 1999; 181:6278-6283.
48. Brunetti R, Prosseda G, Beghetto E et al. The looped domain organization of the nucleoid in histone-like protein defective *Escherichia coli* strains. Biochimie 2001; 83:873-882
49. Dame RT, Goosen N. HU: promoting or counteracting DNA compaction? FEBS Lett 2002; 529:151-156.
50. Giangrossi M, Exley RM, Le Hegarat F et al. Different in vivo localization of the *Escherichia coli* proteins CspD and CspA. FEMS Microbiol Lett 2001; 202:171-176.
51. Amit R, Oppenheim AB, Stavans J. Increased bending rigidity of single DNA molecules by H-NS, a temperature and osmolarity sensor. Biophys J 2003; 84:2467-2473.

52. Bolshoy A, Nevo E. Ecologic genomics of DNA: upstream bending in prokaryotic promoters. Genome Res 2000; 10:1185-1193.
53. Ohyama T. Intrinsic DNA bends: an organizer of local chromatin structure for transcription. Bioessays 2001; 23:708-715.
54. Travers AA. DNA conformation and protein binding. Annu Rev Biochem 1989; 58:427-452.
55. Travers AA. Reading the minor groove. Nat Struct Biol 1995; 2:615-618.
56. Perez-Martin J, de Lorenzo V. Clues and consequences of DNA bending in transcription. Annu Rev Microbiol 1997; 51: 593-628.
57. McLeod SM, Johnson RC. Control of transcription by nucleoid proteins. Curr Opin Microbiol 2001; 4:152-159.
58. Prosseda G, Falconi M, Nicoletti M et al. Histone-like proteins and the *Shigella* invasivity regulon. Res Microbiol 2002; 153:461-468.
59. Afflerbach H, Schroeder O, Wagner R. Conformational changes of the upstream DNA mediated by H-NS and FIS regulate *E. coli rrnB* P1 promoter activity. J Mol Biol 1999; 286:339-353.
60. Schroeder O, Wagner R. The bacterial DNA-binding protein H-NS represses ribosomal RNA transcription by trapping RNA polymerase in the initiation complex. J Mol Biol 2000; 298:737-748.
61. Hommais F, Krin E, Laurent-Winter C et al. Large-scale monitoring of pleiotropic regulation of gene expression by the prokaryotic nucleoid-associated protein, H-NS. Mol Microbiol 2001; 40:20-36.
62. De Biase D, Tramonti A, Bossa F et al. The response to stationary-phase stress conditions in *Escherichia coli*: role and regulation of the glutamic acid decarboxylase system. Mol Microbiol 1999; 32:1198-1211.
63. Soutourina O, Kolb A, Krin E et al. Multiple control of flagellum biosynthesis in *Escherichia coli*: role of H-NS protein and the cyclic AMP-catabolite activator protein complex in transcription of the *flhDC* master operon. J Bacteriol 1999; 181:7500-7508.
64. Soutourina OA, Krin E, Laurent-Winter C et al. Regulation of bacterial motility in response to low pH in *Escherichia coli*: the role of H-NS protein. Microbiology 2002; 148:1543-1551.
65. Lejeune P, Bertin P, Walon C et al. A locus involved in kanamycin, chloramphenicol and L-serine resistance is located in the *bglY-galU* region of the *Escherichia coli* K12 chromosome. Mol Gen Genet 1989; 218:361-363.
66. Dersch P, Kneip S, Bremer E. The nucleoid-associated DNA-binding protein H-NS is required for the efficient adaptation of *Escherichia coli* K-12 to a cold environment. Mol Gen Genet 1994; 245:255-259.
67. La Teana A, Brandi A, Falconi M et al. Identification of a cold shock transcriptional enhancer of the *Escherichia coli* gene encoding nucleoid protein H-NS. Proc Natl Acad Sci USA 1991; 88:10907-10911.
68. Tendeng C, Badaut C, Drin E et al. Isolation and characterization of *vicH*, encoding a new pleiotropic regulator in *Vibrio cholerae*. J Bacteriol 2000; 182:2026-2032.
69. Yamashino T, Ueguchi C, Mizuno T. Quantitative control of the stationary phase-specific sigma factor, σ^S, in *Escherichia coli*: involvement of the nucleoid protein H-NS. EMBO J 1995; 14:594-602.
70. Deighan P, Free A, Dorman CJ. A role for the *Escherichia coli* H-NS-like protein StpA in OmpF porin expression through modulation of *micF* RNA stability. Mol Microbiol 2000; 38:126-139.
71. Goransson M, Sonden B, Nilsson P et al. Transcriptional silencing and thermoregulation of gene expression in *Escherichia coli*. Nature 1990; 344:682-685.
72. Ueguchi C, Mizuno T. The *Escherichia coli* nucleoid protein H-NS functions directly as a transcriptional repressor. EMBO J 1993; 12:1039-1046.
73. Tupper AE, Owen-Hughes TA, Ussery DW et al. The chromatin-associated protein H-NS alters DNA topology in vitro. EMBO J 1994; 13:258-268.
74. Falconi M, Brandi A, La Teana et al. Antagonistic involvement of FIS and H-NS proteins in the transcriptional control of *hns* expression. Mol Microbiol 1996; 19:965-975.
75. Dame RT, Wyman C, Wurm R et al. Structural basis for H-NS mediated trapping of RNA polymerase in the open initiation complex at the *rrnB* P1. J Biol Chem 2002; 277:2146-2150.
76. Ussery DW, Higgins CF, Bolshoy A. Environmental influences on DNA curvature. J Biomol Struct Dyn 1999; 16:811-823.
77. Maurelli AT, Sansonetti PJ. Identification of a chromosomal gene controlling temperature regulated expression of *Shigella virulence*. Proc Natl Acad Sci USA 1988; 85:2820-2824.
78. Beloin C, Dorman CJ. An extended role of the nucleoid structuring protein H-NS in the virulence gene regulatory cascade of *Shigella flexneri*. Mol Microbiol 2003; 47:825-838.
79. Madrid C, Nieto JM, Paytubi S et al. Temperature-and H-NS-dependent regulation of a plasmid-encoded virulence operon expressing *Escherichia coli* haemolysin. J Bacteriol 2002; 184:5058-5066.

80. Tobe T, Yoshikawa M, Mizuno T et al. Transcriptional control of the invasion regulatory gene *virB* of *Shigella flexneri*: activation by VirF and repression by H-NS. J Bacteriol 1993; 175:6142-6149.
81. Falconi M, Colonna B, Prosseda G et al. Thermoregulation of *Shigella* and *Escherichia coli* EIEC pathogenicity. A temperature-dependent structural transition of DNA modulates accessibility of *virF* promoter to transcriptional repressor H-NS. EMBO J 1998; 17:7033-7043.
82. Falconi M, Prosseda G, Giangrossi M et al. Involvement of FIS on the H-NS mediated regulation of the *virF* gene of *Shigella* and *Escherichia coli* EIEC. Mol Microbiol 2001; 42:439-452.
83. Prosseda G, Falconi M, Giangrossi M et al. The *virF* promoter in *Shigella*: more than just a curved DNA stretch. Mol Microbiol 2004; 51:523-537.
84. White-Ziegler CA, Angus Hill ML, Braaten BA et al. Thermoregulation of *Escherichia coli pap* transcription: H-NS is a temperature-dependent DNA methylation blocking factor. Mol Microbiol 1998; 28:1121-1137.
85. Dorman CJ, Porter ME. The *Shigella* virulence gene regulatory cascade: a paradigm of bacterial gene control mechanisms. Mol Microbiol 1998; 29:677-684.
86. Sansonetti PJ. Rupture, invasion and inflammatory destruction of the intestinal barrier by *Shigella*, making sense of prokaryote-eukaryote cross-talks. FEMS Microbiol Rev 2001; 25:3-14.
87. Hurme R, Rhen M. Temperature sensing in bacterial gene regulation-what it all boils down to. Mol Microbiol 1998; 30:1-6.
88. Konkel ME, Tilly K. Temperature-regulated expression of bacterial virulence genes. Microbes Infect 2000; 2:157-166.
89. Colonna B, Casalino M, Fradiani PA et al. H-NS regulation of virulence gene expression in enteroinvasive *Escherichia coli* harboring the virulence plasmid integrated into the host chromosome. J Bacteriol 1995; 177:4703-4712.
90. Durand JMB, Dagberg B, Uhlin BE et al. Transfer RNA modification, temperature and DNA superhelicity have a common target in the regulatory network of the virulence of *Shigella flexneri*: the expression of the *virF* gene. Mol Microbiol 2000; 35:924-935.
91. Agrawal GK, Asayama M, Shirai M. A novel bend of DNA CIT: changeable bending-center sites of an intrinsic curvature under temperature conditions. FEMS Microbiol Lett 1997; 147:139-145.
92. Asayama M, Kato H, Shibato J et al. The curved DNA structure in the 5'- upstream region of the light-responsive genes : its universality, binding factor and function for cyanobacterial *psbA* transcription. Nucleic Acid Res 2002; 30:4658-4666.
93. Dorman CJ, Deighan P. Regulation of gene expression by histone-like proteins in bacteria. Curr Opin Genet Dev 2003; 13:179-184.
94. Rohde JR, Luan X, Rohde H et al. The *Yersinia enterocolitica* pYV virulence plasmid contains multiple intrinsic DNA bends which melt at 37°C. J Bacteriol 1999; 181:4198-4204.
95. Drlica K, Perl-Rosenthal NR. DNA switches for thermal control of gene expression. Trends Microbiol 1999; 7:425-426.
96. Katayama S, Matsushita O, Jung C et al. Promoter upstream bent DNA activates the transcription of the *Clostridium perfringens* phospholipase C gene in a low temperature-dependent manner. EMBO J 1999; 18:3442-3450.
97. Giladi H, Goldenberg D, Koby S et al. Enhanced activity of the bacteriophage λP_L promoter at low temperature. Proc Natl Acad Sci USA 1995; 92:2184-2188.
98. Cornelis GR, Sluiters C, Delor I et al. *ymoA*, a *Yersinia enterocolitica* chromosomal gene modulating the expression of virulence functions. Mol Microbiol 1991; 5:1023-1034.

CHAPTER 5

Curved DNA and Transcription in Eukaryotes

Takashi Ohyama

Abstract

Intrinsically curved DNA structures are often found in or around transcriptional control regions of eukaryotic genes, and curved DNA may be common to all class I gene promoters. Although not all class II gene promoters contain curved DNA structures, both TATA-box-containing and TATA-box-less promoters often contain such structures. Furthermore, several studies have suggested that the TATA box itself adopts a curved DNA conformation. Curved DNA structures are likely to function in transcription in several ways. These include acting as a conformational signal for transcription factor binding; juxtaposition of the basal machinery with effector domains on upstream-bound factors; regulation of transcription in association with transcription-factor-induced bending of DNA; and organization of local chromatin structure to increase the accessibility of *cis*-DNA elements. This chapter presents a concise overview of studies of these functions.

Introduction

Eukaryotic cells use three different RNA polymerases, referred to as pols I, II, and III, to transcribe their genes. Pol I produces rRNAs, pol II produces mRNA molecules and most small nuclear (sn) RNAs, and pol III produces small RNA molecules such as tRNA, 5S rRNA and U6 snRNA. DNA templates that are transcribed by pols I, II and III are often called class I, II and III genes, respectively. They have their own (class-specific) promoters. Curved DNA is often found in the transcriptional control regions,[1] with many reports having identified curved DNA in class II gene promoters. Both TATA-box-containing and TATA-box-less promoters often contain curved DNA. However, curved DNA is not necessarily confined to class II gene promoters. Class I gene promoters also often contain it. The role of curved DNA in eukaryotic transcription is being gradually revealed.

Here, I review what has been clarified in the field of eukaryotic transcription, and what is still unknown. Several studies have indicated that curved DNA is implicated in the packaging promoter DNA into chromatin, and this is described in Chapter 13.

Intrinsic DNA Bends in Control Regions of Transcription

Class I Gene Promoters

A large number of ribosomes are required in each cell generation. To meet the requirement, genomes have multiple copies of the rRNA gene (rDNA). For example, the human genome has about 200 copies of the gene. They are organized in tandem arrays in eukaryotes, in which the transcribed regions are separated by an intergenic spacer (Fig. 1). Three of the four rRNAs

DNA Conformation and Transcription, edited by Takashi Ohyama. ©2005 Eurekah.com and Springer Science+Business Media.

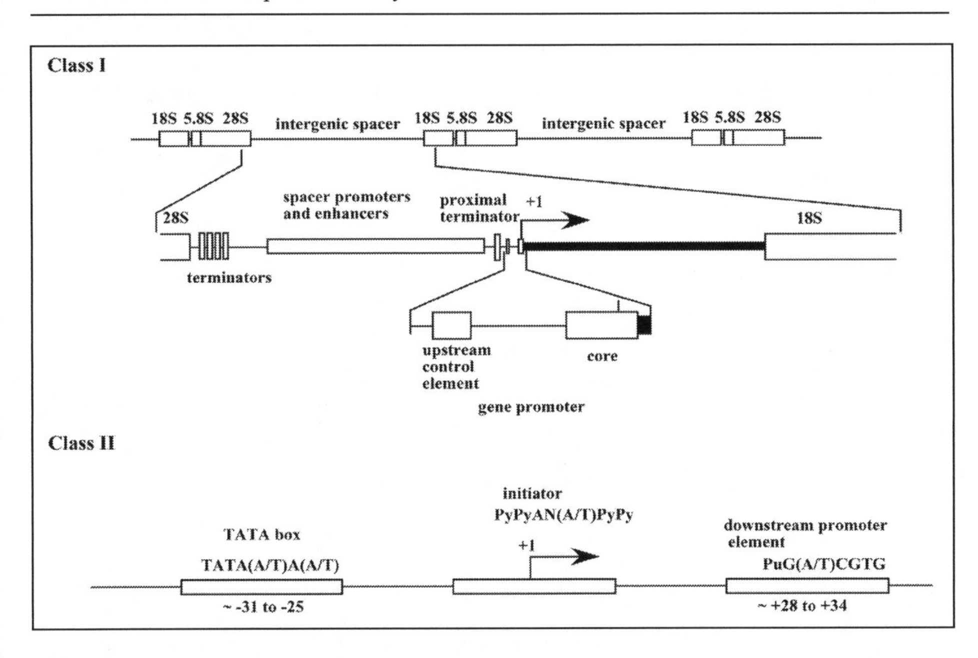

Figure 1. Schematic representation of class I and II gene promoters.

(18S, 5.8S and 28S molecules) are generated from a single precursor rRNA. The nucleotide sequences of the rRNA gene promoters are highly divergent across species. As a result, in this class of genes, transcription is restricted by the cognateness between gene and transcription machinery. For example, the human rRNA gene is not transcribed by the mouse pol I machinery, or vice versa.[2]

More than a decade ago, by employing a circular permutation polyacrylamide gel electrophoresis assay (Chapter 3), Schroth et al identified two intrinsically curved DNA structures near the transcription initiation site of the *Physarum polycephalum* rRNA gene (Table 1).[3] The center of one curved DNA was 160 nucleotides upstream (-160) of the transcription start site (+1) and the center of the other was about 150 nucleotides downstream (+150). Subsequently, a curved DNA was also experimentally identified upstream of the promoter (center position: -260 - -280) for the *Arabidopsis thaliana* rRNA gene.[4] Marilley and Pasero used several curvature prediction models to analyze rRNA gene promoters from mammals (*Rattus norvegicus*), amphibians (*Xenopus laevis, Xenopus borealis*), insects (*Drosophila melanogaster*), echinoderms (*Paracentrotus lividus*), protozoans (*Tetrahymena pyriformis, Dictyostelium discoideum*), dicotyledons (*Arabidopsis thaliana, Pisum sativum*), monocotyledons (*Triticum aestivum, Zea mays*), and myxomycetes (*Physarum polycephalum*). Each promoter had at least one curved DNA structure, although the position and extent of DNA curvature differed.[5] Computational modeling also suggested that the promoter of the mouse rDNA overlaps with a curved DNA structure.[6] In conclusion, a curved DNA structure may be a common feature of class I gene promoters. To the best of my knowledge, however, no study has succeeded in clarifying the function of this curved DNA.

Class II Gene Promoters and Enhancers

The TATA box, the initiator element (INR), and the downstream promoter element (DPE), are generally regarded as the core elements of promoters of class II genes (Fig. 1). The TATA box, with the consensus TATA(A/T)A(A/T), is located approximately 25-31 bp upstream of the transcription start site. The INR is centered at the transcription initiation site and has a

Table 1. Curved DNA-containing promoters, enhancers and regulatory sequences[a]

Gene	Class	Origin	Locus	Reference
rRNA	I	human	promoter	5
rRNA	I	*Rattus norvegicus*	promoter	5
rRNA	I	mouse	promoter	6
rRNA	I	*Xenopus laevis*	promoter	5
rRNA	I	*Xenopus borealis*	promoter	5
rRNA	I	*Drosophila melanogaster*	promoter	5
rRNA	I	*Paracentrotus lividus*	promoter	5
rRNA	I	*Tetrahymena pyriformis*	promoter	5
rRNA	I	*Dictyostelium discoideum*	promoter	5
rRNA	I	*Arabidopsis thaliana*	promoter	5
rRNA	I	*Arabidopsis thaliana*	around position -270	4
rRNA	I	*Pisum sativum*	promoter	5
rRNA	I	*Triticum aestivum*	promoter	5
rRNA	I	*Zea mays*	promoter	5
rRNA	I	*Physarum polycephalum*	both sides of transcription start site	3
rRNA	I	*Physarum polycephalum*	promoter	5
β-actin	II	human	promoter	20
cdc2	II	human	promoter	16
c-*myc*	II	human	promoter	42
E2F1	II	human	E2F binding site	34
erythropoietin receptor	II	human	promoter	42
ε-globin	II	human	promoter	42
Gγ-globin	II	human	promoter	38
Aγ-globin	II	human	promoter	38
ψβ-globin	II	human	promoter	38
δ-globin	II	human	promoter	38
β-globin	II	human	promoter	42
IFNβ	II	human	enhancer	24
β^{major}-globin	II	mouse	promoter	42
A2 vitellogenin	II	*Xenopus*	upstream regulatory region	18
BhC4-1	II	*Bradysia hygida*	promoter	39
AaH I′ toxin	II	*Androctonus australis*	promoter	15
DNA polymerase δ	II	*Plasmodium falciparum*	promoter	37
E1A	II	human adenovirus type 2	enhancer	23
E1A	II	human adenovirus type 5	enhancer	41
E2	II	adenovirus	E2F binding site	34
E6-E7	II	human papillomavirus type 16	E2 protein binding site	21
IE94	II	simian cytomegalovirus	upstream of the enhancer	25
rbcS-3A	II	pea	light responsive elements	17
rbcS-3.6	II	pea	light responsive elements	17
MF α 1	II	*Saccharomyces cerevisiae*	upstream activation site	22
STE3	II	*Saccharomyces cerevisiae*	upstream activation site	22
GAL1-10	II	yeast	promoter	19
GAL80	II	yeast	promoter	19
AKY2	II	*Saccharomyces cerevisiae*	promoter	40

[a] Reports on the TATA box conformation are not involved.

loose consensus sequence, PyPyAN(T/A)PyPy, where Py is a pyrimidine and N is any nucleotide. Many promoters have either the TATA box (including TATA-like sequences) or some kind of INR, and some have both. Many *Drosophila* TATA-less promoters, and most likely many mammalian TATA-less promoters, seem to contain a DPE.[7] The DPE, with the consensus PuG(A/T)CGTG, where Pu is a purine, is centered approximately 30 bp downstream of the initiation site. There is also a class of pol II promoters comprised of G/C rich sequences, and containing multiple binding sites for transcription factor Sp1 instead of the elements described above. Such promoters are often associated with housekeeping genes.

As shown in Table 1, many class II gene promoters have a curved DNA structure. Although the table does not list the reports on the conformation of TATA box per se, several groups have discussed this issue. Using models of Bolshoy et al,[8] Calladine et al[9] and Satchwell et al,[10] Schätz and Langowski calculated helix trajectories of 504 vertebrate promoters, which suggested that the regions containing the TATA box (either canonical or non-canonical) all form curved DNA structures.[11] They also predicted that the upstream half of the TATA box would be curved towards the major groove, and the remainder would be straight, or even curved towards the minor groove. de Souza and Ornstein subjected the TATA box of the adenovirus major late promoter (whose sequence is TATAAAA) to molecular dynamics simulations. Their results suggested that the direction of the curvature is toward the major groove.[12] However, based on DNA cyclization kinetics, Davis et al reported that the curvature is toward the minor groove.[13] Thus the issue of the TATA box conformation is not yet settled. What about the other promoter elements? Based on the consensus sequence for the INR, AAA or TTT could be formed within the box. These tracts change the direction of the helical path of DNA. However, the DNA conformation of this region has not yet been reported. DPE itself cannot possess DNA curvature as judged from its consensus sequence. To the best of my knowledge, there are no reports describing the DNA conformation of this region, though the main reason may simply be lack of studies. Concerning the G/C rich promoters, Sp1-induced DNA bending is well known.[14] However, intrinsic DNA curvature has not been reported.

The DNA elements for binding transcription factors are sometimes located in curved DNA structures. Delabre et al reported that in the scorpion (*Androctonus australis*) AaH I' toxin gene promoter, the region containing the putative TATA box and the CCAAT box is located in a large curved DNA structure.[15] In the human *cdc2* promoter,[16] the binding sites for c-Myb, ATF, Sp1, E2F and CBF are located in curved DNA. In the pea *rbcS-3A* and *rbcS-3.6* genes, both of which code for the small subunit of ribulose-1,5-bisphosphate carboxylase/oxygenase, each light responsive element overlaps with a curved DNA structure.[17] The *Xenopus* A2 vitellogenin gene presents another example of a curved DNA structure that occurs between *cis*-DNA elements: curved DNA was found between a DNA element located around position -100, which could activate transcription in a cell-specific manner, and the estrogen-responsive element located around -300.[18] Also, in the yeast *GAL80* gene, a single curved DNA structure lies between the UAS_{GAL80} (an upstream *GAL80*-specific activation sequence) and the more gene-proximal promoter elements.[19] The *GAL1* gene has two curved DNA structures between UAS_G (an upstream activation sequence) and the TATA box.[19] The curvatures found near *GAL1* and *GAL80* may be involved in nucleosome formation (Chapter 13). *Cis*-DNA elements themselves form curved DNA structures in the following cases. The human β-actin promoter has a β-actin-specific conserved sequence (conserved among humans, rats and chickens) between the TATA and CCAAT boxes, which forms a curved DNA structure.[20] In human papillomavirus type 16 (HPV-16), the E2 protein binding site in the upstream regulatory region forms a curved DNA structure.[21] At present, however, we cannot find any common features shared by these regions, except that a curved DNA structure is present.

Curved DNA structures have also been found in enhancers, or adjacent to enhancers. In the *Saccharomyces cerevisiae* genome, the UAS of the mating pheromone α factor 1 (*MF α 1*) gene, and the UAS of the *STE3* gene, adopt curved DNA structures.[22] The enhancer of the human adenovirus type 2 (Ad2) *E1A* gene has a sharply curved DNA structure.[23] As shown in Figure

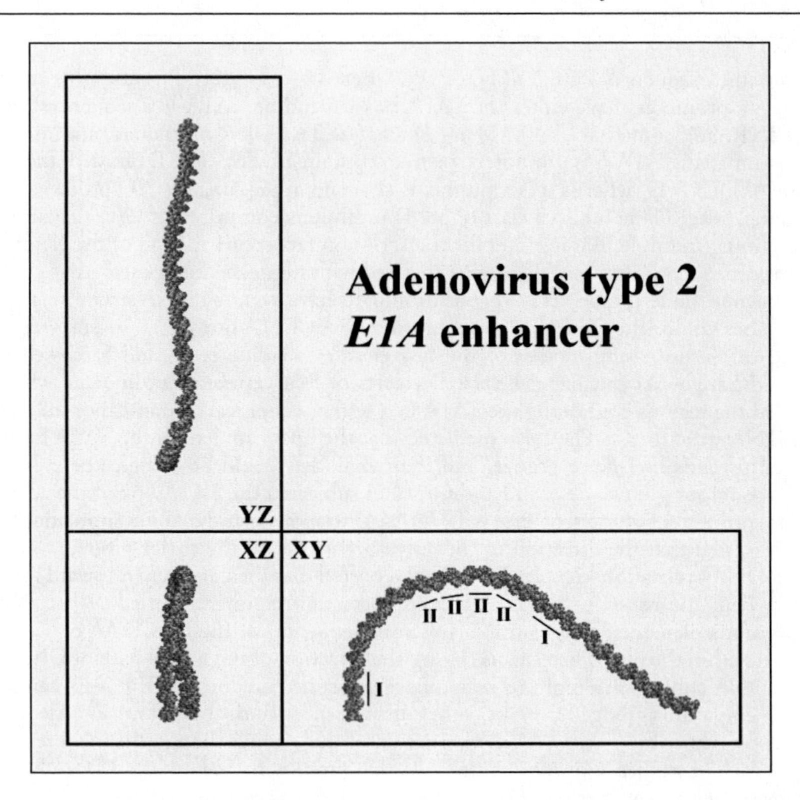

Figure 2. Computer modeling of the curved DNA found in the human adenovirus type 2 *E1A* enhancer. The upstream half of the enhancer, where most of the enhancer elements are contained, adopts a curved DNA conformation.[23] The three-dimensional structure of the enhancer was drawn by a combination of DIAMOD[43] and RASMOL.[44] The modeling algorithm was based on Bolshoy et al[8] with the twist angles from Kabsch et al.[45] The repeated sequence elements required for enhancer function are indicated by I and II.

2, most of the enhancer elements are within the curved DNA. In the enhancer of the human interferon-β (*IFNβ*) gene, positive regulatory domains (PRDs) II and IV also form curved DNA structures: PRDII is curved towards the minor groove and PRDIV towards the major groove.[24] The simian cytomegalovirus *IE94* gene has a wide curved DNA region (about 1 kb) upstream of the enhancer.[25]

Class III Gene Promoters

The important promoter elements of the 5S rRNA genes and tRNA genes are located downstream of the transcription start site. On the other hand, the promoters of human and *Saccharomyces cerevisiae* U6 snRNA genes are similar to class II gene promoters in the sense that they adopt a TATA box at the "correct" site (-32 to -25 and -30 to -23, respectively).[26] The TATA box is likely to have a curved conformation as described above. Therefore, they may be counted as promoters that contain curved DNA. To the best of my knowledge, there are no reports of curved DNA structures in or around the promoters of 5S rRNA genes and tRNA genes, but presumably this is simply due to lack of studies.

Role of Intrinsic DNA Curvature in Transcription

Does curved DNA actually function in eukaryotic transcription? The sequence and conformation of DNA are two sides of the same coin. Thus, we must always consider the possibility that it is the DNA sequence itself that is important, and the curved conformation may simply be a by-product. It is usually very difficult to experimentally prove significance of DNA conformation per se without impairing the DNA sequence. In the following two cases, however, this difficulty was overcome. Using a constrained minicircle strategy,[27] Parvin et al showed that the bending of the TATA box enhances its affinity for the TATA box binding protein (TBP).[28] Moreover, they suggested that the direction of bending is important: a bend towards the major groove showed 100-fold higher affinity than an unbent structure, and 300-fold higher affinity than a bend towards the minor groove. They observed similar discrimination with the holo-TFIID transcription complex. However, this study did not describe the effect of the intrinsically curved DNA conformation of the TATA box, but reported the effect of "constrained" bending of the box. If the TATA box intrinsically curves toward major groove, as suggested by Schätz and Langowski[11] and de Souza and Ornstein,[12] the role of the conformation may be to facilitate TBP binding. In the adenovirus *E1A* gene enhancer, the bent conformation per se activated transcription.[29] Run-off transcription assays using the wild-type DNA template, with a temperature-dependent conformational change of DNA curvature, showed that a moderately curved enhancer was superior to a highly curved enhancer in stimulating transcription. In the study, curved DNA of the phage λ origin of replication could substitute for the *E1A* enhancer, and stimulated transcription of the *E1A* gene to some extent, when it had an appropriate DNA conformation.

Synthetic curved DNA fragments sometimes help the analysis of the structure-function relationships. An early study reported that $(CAAAAATGCC)_{19}$ and $(CAAAAATGCC)_{35}$ activated transcription driven by the promoter element CGTATTTATTTG by two- and three-fold, respectively.[30] Using synthetic curved DNA segments, Kim et al also showed that transcription could be activated by curved DNA.[31] As described in Chapter 3, curved DNA with a right-handed superhelical writhe activates transcription in *Escherichia coli*. However, a recent study showed that it was left-handed curved DNA that activated transcription in eukaryotes.[32] In the study, synthetic DNA fragments with different geometries were constructed by changing the spacing between T_5 tracts, which was based on the principle described in Chapter 3. When left-handed curved DNA was linked to the herpes simplex virus thymidine kinase promoter at a specific rotational orientation and distance, the curved DNA activated transcription. Neither planar curving, nor right-handedly curved DNA, nor straight DNA had this effect (Chapter 13). Thus, in studying the role of DNA curvatures, it seems important to consider the shape of DNA curvature. It has been suggested that many eukaryotic promoters can adopt superhelical conformations.[33]

How does curved DNA function in transcription? The most likely role is that curved DNA structures function as recognition signals for transcription factors (Fig. 3). In the study of the human β-actin promoter described above, a protein in the HeLa cell extracts bound to the curved DNA.[20] The above report by Parvin et al[28] seems to support this role as well. However, the TATA box may curve toward the minor groove as suggested by Davis et al.[13] Their study leads us to another important concept: the binding of a transcription factor can change the direction of DNA curvature. They used the TATA box of the adenovirus major late promoter and showed, by cyclization kinetics, that the direction of bend induced by TBP (toward the major groove of the TATA box) was opposite to the direction of bend of the TATA box per se (toward the minor groove).[13] They speculated that the intrinsic bend of the TATA box could repress transcription complex assembly in the absence of TBP. A similar phenomenon has been reported for the interaction between the E2F transcription factor and the E2F site.[34] In the human *E2F1* promoter, a consensus E2F site adopts a curved DNA structure oriented toward the major groove relative to the center of the E2F site (magnitude of the bend; ~40°). When E2F binds, the bend is reversed and oriented toward the minor groove (net magnitude of the resulting bend; ~25°). Furthermore, in the human *IFNβ* enhancer, the directions of the two intrinsic bends are also counteracted or reversed by the binding of NF-κB, ATF-2/c-Jun, and HMGI(Y).[24] Alteration of DNA architecture may facilitate protein-protein interactions between the transcription factors that constitute the enhanceosome.

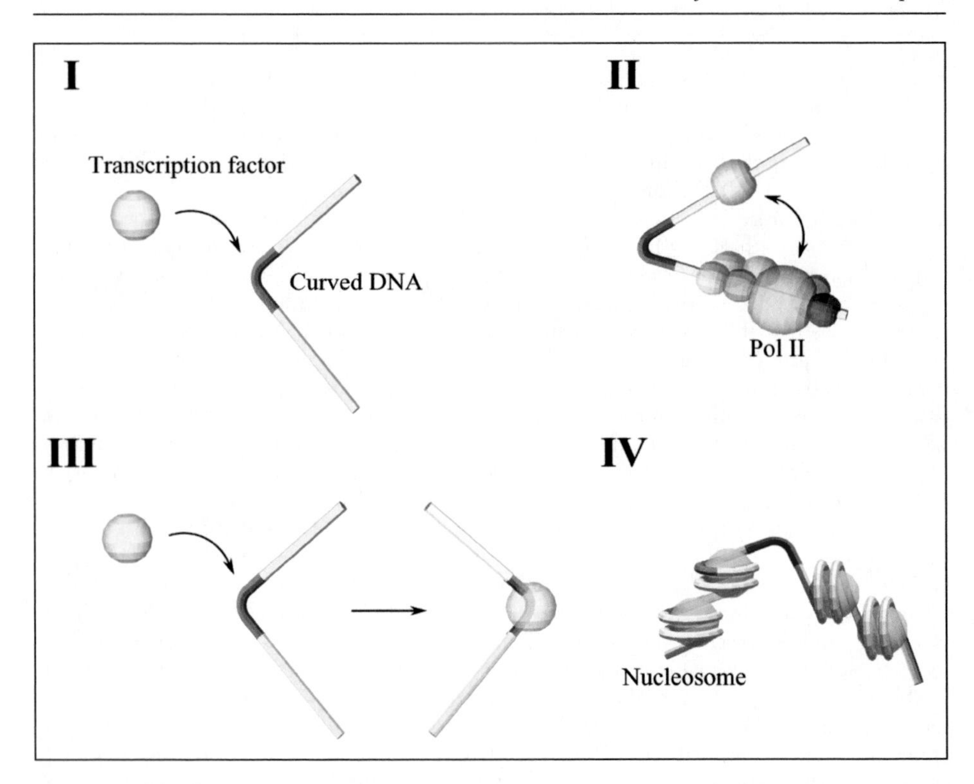

Figure 3. Possible roles of curved DNA in eukaryotic transcription. I) Transcription factors may recognize the curved shape of DNA. II) Curved DNA can juxtapose two distant sites on a sequence. For example, factors bound upstream can be brought close to the basal machinery. III) Transcription factor binding may induce a change in the curved conformation, which in turn regulates transcription. IV) Curved DNA could regulate nucleosome positioning.

As another possibility, curved DNA-mediated nucleosome positioning may be used to juxtapose two distant sites, and to allow interaction between transcription factors. The regulatory region of the *Xenopus* A2 vitellogenin gene may be such an example. In this case, curved DNA is located between the estrogen-responsive element and the cell type specific activator sequence.[18] These elements may be juxtaposed by nucleosome positioning directed by the curved DNA. In fact, in the corresponding region of the *Xenopus* vitellogenin B1 gene, the DNA sequence strongly directs the positioning of a nucleosome. This nucleosome creates a static loop. As the result, the distal estrogen receptor binding site is brought close to the proximal promoter elements, which presumably allows interaction between transcription factors, and facilitates transcription.[35] Without the help of a nucleosome, curved DNA conformation could play the same role. An experiment using an artificial promoter showed that a curved DNA juxtaposed bound factors, and activated transcription.[36] Thus, there are several lines of direct or indirect evidence for the function of curved DNAs. In some cases, they may play a single role in transcription, but in others they may play multiple roles simultaneously.

Some of the putative roles of DNA curvature described above seem to be the same, in essence, as those for prokaryotic transcription. The important question we must consider next is the relationship between curved DNA, chromatin structure and transcription. This issue is described in Chapter 13.

Conclusion

The presence of a curved DNA structure may be a common feature shared by class I gene promoters. In transcription of some class II genes, curved DNA conformations per se may also play an important role. Curved DNA sequences seem to be used in several ways; some roles are presumably common between prokaryotes and eukaryotes, while others are used in transcription in the chromatin environment, and are thus specific to eukaryotes.

Acknowledgements

The author would like to acknowledge the contributions of Yoshiro Fukue, Jun-ichi Nishikawa, and Junko Ohyama. The studies reported from my laboratory were supported in part by Grants-in-Aid for Scientific Research from the Ministry of Education, Science, Sports and Culture of Japan.

References

1. Ohyama T. Intrinsic DNA bends: an organizer of local chromatin structure for transcription. Bioessays 2001; 23:708-715.
2. Learned RM, Cordes S, Tjian R. Purification and characterization of a transcription factor that confers promoter specificity to human RNA polymerase I. Mol Cell Biol 1985; 5:1358-1369.
3. Schroth GP, Siino JS, Cooney CA et al. Intrinsically bent DNA flanks both sides of an RNA polymerase I transcription start site. Both regions display novel electrophoretic mobility. J Biol Chem 1992; 267:9958-9964.
4. Kneidl C, Dinkl E, Grummt F. An intrinsically bent region upstream of the transcription start site of the rRNA genes of *Arabidopsis thaliana* interacts with an HMG-related protein. Plant Mol Biol 1995; 27:705-713.
5. Marilley M, Pasero P. Common DNA structural features exhibited by eukaryotic ribosomal gene promoters. Nucleic Acids Res 1996; 24:2204-2211.
6. Längst G, Schätz T, Langowski J et al. Structural analysis of mouse rDNA: coincidence between nuclease hypersensitive sites, DNA curvature and regulatory elements in the intergenic spacer. Nucleic Acids Res 1997; 25:511-517.
7. Burke TW, Kadonaga JT. *Drosophila* TFIID binds to a conserved downstream basal promoter element that is present in many TATA-box-deficient promoters. Genes Dev 1996; 10:711-724.
8. Bolshoy A, McNamara P, Harrington RE et al. Curved DNA without A-A: experimental estimation of all 16 DNA wedge angles. Proc Natl Acad Sci USA 1991; 88:2312-2316.
9. Calladine CR, Drew HR, McCall MJ. The intrinsic curvature of DNA in solution. J Mol Biol 1988; 201:127-137.
10. Satchwell SC, Drew HR, Travers AA. Sequence periodicities in chicken nucleosome core DNA. J Mol Biol 1986; 191:659-675.
11. Schätz T, Langowski J. Curvature and sequence analysis of eukaryotic promoters. J Biomol Struct Dyn 1997; 15:265-275.
12. de Souza ON, Ornstein RL. Inherent DNA curvature and flexibility correlate with TATA box functionality. Biopolymers 1998; 46:403-415.
13. Davis NA, Majee SS, Kahn JD. TATA box DNA deformation with and without the TATA box-binding protein. J Mol Biol 1999; 291:249-265.
14. Sjottem E, Andersen C, Johansen T. Structural and functional analyses of DNA bending induced by Sp1 family transcription factors. J Mol Biol 1997; 267:490-504.
15. Delabre ML, Pasero P, Marilley M et al. Promoter structure and intron-exon organization of a scorpion α-toxin gene. Biochemistry 1995; 34:6729-6736.
16. Nair TM. Evidence for intrinsic DNA bends within the human *cdc2* promoter. FEBS Lett 1998; 422:94-98.
17. Cacchione S, Savino M, Tufillaro A. Different superstructural features of the light responsive elements of the pea genes *rbcS-3A* and *rbcS-3.6*. FEBS Lett 1991; 289:244-248.
18. Döbbeling U, Roß K, Klein-Hitpaß L et al. A cell-specific activator in the *Xenopus* A2 vitellogenin gene: promoter elements functioning with rat liver nuclear extracts. EMBO J 1988; 7:2495-2501.
19. Bash RC, Vargason JM, Cornejo S et al. Intrinsically bent DNA in the promoter regions of the yeast *GAL1-10* and *GAL80* genes. J Biol Chem 2001; 276:861-866.
20. Kawamoto T, Makino K, Orita S et al. DNA bending and binding factors of the human β-actin promoter. Nucleic Acids Res 1989; 17:523-537.

21. Bedrosian CL, Bastia D. The DNA-binding domain of HPV-16 E2 protein interaction with the viral enhancer: protein-induced DNA bending and role of the nonconserved core sequence in binding site affinity. Virology 1990; 174:557-575.
22. Inokuchi K, Nakayama A, Hishinuma F. Sequence-directed bends of DNA helix axis at the upstream activation sites of α-cell-specific genes in yeast. Nucleic Acids Res 1988; 16:6693-6711.
23. Ohyama T, Hashimoto S. Upstream half of adenovirus type 2 enhancer adopts a curved DNA conformation. Nucleic Acids Res 1989; 17:3845-3853.
24. Falvo JV, Thanos D, Maniatis T. Reversal of intrinsic DNA bends in the IFNβ gene enhancer by transcription factors and the architectural protein HMG I(Y). Cell 1995; 83:1101-1111.
25. Chang YN, Jeang KT, Chiou CJ et al. Identification of a large bent DNA domain and binding sites for serum response factor adjacent to the NFI repeat cluster and enhancer region in the major IE94 promoter from simian cytomegalovirus. J Virol 1993; 67:516-529.
26. Schramm L, Hernandez N. Recruitment of RNA polymerase III to its target promoters. Genes Dev 2002;16:2593-2620.
27. Kahn JD, Crothers DM. Protein-induced bending and DNA cyclization. Proc Natl Acad Sci USA 1992; 89:6343-6347.
28. Parvin JD, McCormick RJ, Sharp PA et al. Pre-bending of a promoter sequence enhances affinity for the TATA-binding factor. Nature 1995; 373:724-727.
29. Ohyama T. Bent DNA in the human adenovirus type 2 E1A enhancer is an architectural element for transcription stimulation. J Biol Chem 1996; 271:27823-27828.
30. Delic J, Onclercq R, Moisan-Coppey M. Inhibition and enhancement of eukaryotic gene expression by potential non-B DNA sequences. Biochem Biophys Res Commun 1991; 181:818-826.
31. Kim J, Klooster S, Shapiro DJ. Intrinsically bent DNA in a eukaryotic transcription factor recognition sequence potentiates transcription activation. J Biol Chem 1995; 270:1282-1288.
32. Nishikawa J, Amano M, Fukue Y et al. Left-handedly curved DNA regulates accessibility to *cis*-DNA elements in chromatin. Nucleic Acids Res 2003; 31:6651-6662.
33. Miyano M, Kawashima T, Ohyama T. A common feature shared by bent DNA structures locating in the eukaryotic promoter region. Mol Biol Rep 2001; 28:53-61.
34. Cress WD, Nevins JR. A role for a bent DNA structure in E2F-mediated transcription activation. Mol Cell Biol 1996; 16:2119-2127.
35. Schild C, Claret FX, Wahli W et al. A nucleosome-dependent static loop potentiates estrogen-regulated transcription from the *Xenopus* vitellogenin B1 promoter in vitro. EMBO J 1993; 12:423-433.
36. Kim J, de Haan G, Shapiro D. DNA bending between activator sequences increases transcriptional synergy. Biochem Biophys Res Commun 1996; 226:638-644.
37. Porter ME. The DNA polymerase δ promoter from *Plasmodium falciparum* contains an unusually long 5' untranslated region and intrinsic DNA curvature. Mol Biochem Parasitol 2001; 114:249-255.
38. Wada-Kiyama Y, Kiyama R. An intrachromosomal repeating unit based on DNA bending. Mol Cell Biol 1996; 16:5664-5673.
39. Fiorini A, Basso LR Jr, Paçó-Larson ML et al. Mapping of intrinsic bent DNA sites in the upstream region of DNA puff *BhC4-1* amplified gene. J Cell Biochem 2001; 83:1-13.
40. Angermayr M, Oechsner U, Gregor K et al. Transcription initiation in vivo without classical transactivators: DNA kinks flanking the core promoter of the housekeeping yeast adenylate kinase gene, *AKY2*, position nucleosomes and constitutively activate transcription. Nucleic Acids Res 2002; 30:4199-4207.
41. Eckdahl TT, Anderson JN. Bent DNA is a conserved structure in an adenovirus control region. Nucleic Acids Res 1988; 16:2346.
42. Wada-Kiyama Y, Kiyama R. Periodicity of DNA bend sites in human ε-globin gene region. J Biol Chem 1994; 269:22238-22244.
43. Dlakic M, Harrington RE. DIAMOD: display and modeling of DNA bending. Bioinformatics 1998; 14:326-331.
44. Sayle RA, Milner-White EJ. RASMOL: biomolecular graphics for all. Trends Biochem Sci 1995; 20:374-376.
45. Kabsch W, Sander C, Trifonov EN. The ten helical twist angles of B-DNA. Nucleic Acids Res 1982; 10:1097-1104.

Chapter 6

Putative Roles of kin17, a Mammalian Protein Binding Curved DNA, in Transcription

Jaime F. Angulo, Philippe Mauffrey, Ghislaine Pinon-Lataillade, Laurent Miccoli and Denis S.F. Biard

Abstract

In bacteria, RecA protein is indispensable for recombination, mutagenesis and for the induction of SOS genes. Curiously, anti-RecA antibodies recognize kin17, a human nuclear Zn-finger protein of 45 kDa that preferentially binds to curved DNA and participates in a general response to diverse genotoxics. *KIN17* gene is conserved from yeast to man and codes for a protein involved in DNA replication. Recent observations suggest that kin17 protein may also participate in RNA metabolism. Taken together all these data indicate the participation of kin17 protein in a pathway that harmonizes transcription, replication and repair in order to circumvent the topological constraints caused by unusually complex lesions like multiply damaged sites.

Introduction

DNA conformation undergoes important and dynamic changes during transcription and replication. The molecular characterization of these changes is essential in predicting the progress of the cell cycle and when evaluating the consequences of the damage to mammalian cells caused by endogenous or environmental genotoxic agents. Therefore it is useful to identify proteins that recognize the particular structures formed during the metabolic processing of DNA. Several transcription factors and other nuclear proteins have been identified and shown to specifically recognize DNA deformations like cyclobutane pyrimidine dimers, platinum-DNA cross-links, cruciform DNA and curved regions.[1-6] We have characterized a human protein that preferentially binds to curved DNA and is involved in DNA and RNA metabolism.

Identification of kin17 Protein

The analysis of human inherited disorders like xeroderma pigmentosum (xp) allowed the identification of genes coding for proteins that specifically bind to pyrimidine dimers and other types of DNA lesions.[7-10] Some of these proteins belong to the machinery of DNA transcription or replication. These data provide a first image of the drastic effects that the modification of these proteins may have for the whole organism. In parallel, other human genes of DNA metabolism have been identified thanks to the genetic analysis of bacteria. In *Escherichia coli* bacteria, RecA protein is essential for homologous recombination, mutagenesis

DNA Conformation and Transcription, edited by Takashi Ohyama. ©2005 Eurekah.com and Springer Science+Business Media.

and the induction of the SOS response.[11,12] The strong conservation of RecA protein among prokaryotes[13,14] stimulated the search for analogous proteins in mammalian cells. Numerous eukaryotic proteins that are structural and/or functional homologues of RecA protein have been identified.[15-25] All of them seem to evolve from a common ancestral protein but only a small subset of these RecA-like proteins may form helical filaments on DNA. Indeed, the functional divergence of RecA-like proteins is enormous, the members of this family being involved in different DNA transactions.[26] Anti-RecA antibodies cross-react with nuclear proteins from plant or mammalian cells.[27-33] Interestingly, the cross-reacting material is localized in structures directly involved in DNA transactions during mitosis or meiosis (e.g., the synaptonemal complex, the stroma or meiotic nodules).[29,30,33] In mammalian cells and tissues, the cross-reacting material is nuclear and their levels are low in quiescent cells as compared to proliferating cells. Genotoxic agents like mitomycin C (MMC) induced a striking increase of these antigens inside the nucleus. [27] Considering the cross-reactivity with anti-RecA antibodies we proposed the generic name of kin proteins (from immunological kinship to RecA protein). The screening of expression libraries using antibodies raised against the *E. coli* RecA protein allowed us the isolation of *recA* genes from *Streptococcus pneumoniae,*[34] *Streptomyces ambofaciens*[35] and Gram$^+$ bacteria.[36] Therefore, we used this approach to isolate 17 cDNA fragments from a mouse embryo cDNA library. The clone number 17 gave the strongest cross-reactivity and was used to isolate the full-length cDNA. The encoded protein has a primary structure different to that of RecA and is different from all the proteins reported in databases. The lack of information on the biological role of this gene led us to call it *KIN17* gene.[37,38]

Molecular Characterization of kin17 Protein

Modular Nature of kin17 Protein

The mouse kin17 protein, has a calculated molecular weight of 44,726 and an isolelectric point of 9.3. The primary structure contains the following domains that are located in hydrophilic helix regions of the protein.[37,39] (Fig. 1A):

1. a Zn-finger domain of 23 residues of the type $Cx_2Cx_{12}Hx_5H$ between residues 28 and 50 which mediated DNA recognition.[38,39]
2. an active bipartite nuclear localization signal of 31 residues (236 and 266)[38] that accounts for the nuclear localization of kin17 protein in mammalian tissues.[40-44]
3. a KOW motif (residues 330 to 363) supposed to participate in transcription elongation.[45,46]

Orthologs of *KIN17* gene have been reported from yeast to humans, however the family of encoded proteins do not present striking homologies with other proteins.[42] The evolutionary conservation of kin17 protein structure is significant since there are several mammalian nuclear proteins involved in important DNA processes, like tumour suppressor p53[47] and DNA-damage sensor protein poly(ADP-ribose) polymerase[48] which lack a yeast counterpart, indicating that some important DNA transactions are particular to mammals. A central fragment of 39 residues which is highly conserved between the human and mouse kin17 proteins (residues 163 to 201, Fig. 1A) has a sequence similarity with the fragment 308-346, within the C-terminal domain of RecA protein (residues 270-352),[32] as shown in Figure 1B. The core of RecA protein (residues 31-269) is followed by a smaller C-terminal domain that appears as a lobe on the surface of the RecA filament.[49] This lobe shifts after binding of nucleotides.[50-54] A major antigenic determinant is located between residues 260 and 330 of RecA protein.[55] Although the last 24 residues of the C-terminus are disordered in the proposed crystal structure of RecA, genetic and biochemical data indicate a role in modulating DNA binding, probably in regulating the access of double stranded (ds) DNA to the presynaptic filament.[56,57] Accordingly, the deletion of a C-terminal fragment of RecA protein reduces the efficiency of recombination,[58] probably because it nucleates to double-stranded DNA much more frequently than does the wild-type protein[57] resulting in increased sensitivity to MMC in vivo.[59-61] In vitro this deletion enhanced binding to duplex DNA.[56,61]

```
                                                          I
Homo sapiens          MGKSDFLTPKAIANRIKSKGLQKLRWYCQMCQKQCRDENGFKCHCMSESHQRQLLLASEN
Mus musculus          MGKSDFLSPKAIANRIKSKGLQNVRWYCQMCQKQCRDENGFKCHCMSESHQRQLLLASEN
Drosophila m.         MGRAEVGTPKYLANKMKSKGLQKLRWYCQMCEKQCRDENGFKCHTMSESHQRQLLLFADN
Anopheles g.          -GKAEVGTPKYLANKMKAKGLQKLRWYCQMCEKQCRDENGFKCHTMSESHQRQILLFADN
Caenorhabditis e.     MGKHEKGSSKDLANRTKSKGLQKLKFFCQMCQKQCRDANGFKCHLTSEAHQRQLLLFAEN
                       *: :   :.* :**: *:****::::****:***** ******   **:****:**  ::*

Homo sapiens          PQQFMDYFSEEFRNDFLELLRRRFGTKRVHNNIVYNEYISHREHIHMNATQWETLTDFTK
Mus musculus          PQQFMDYFSEEFRNDFLELLRRRFGTKRVHNNIVYNEYISHREHIHMNATQWETLTDFTK
Drosophila m.         PGKFLHSFSKEFSDGYMELLRRRFGTKRTSANKIYQEYIAHKEHIHMNATRWLTLSDYVK
Anopheles g.          AGRFIDGFSSEFLTGYLQILRRQFGTKRVAANKVYQEYIADRHHLHMNATKWHSLSDFVK
Caenorhabditis e.     SNSYLRQFSNDFEKNFMQLLRTSYGTKRVRANEVYNAFIKDKGHVHMNSTVWHSLTGFVQ
                      .  ::  **.:*  .::::**  :****.  * :*: :*  .: *:***:* *  :*:.:.:
                                                                    II
Homo sapiens          WLGREGLCKVDETPKGWYIQYIDRDPETIRRQLELEKKKQDLDDEEKTAKFIEEQVRRG
Mus musculus          WLGREGLCKVDETPKGWYIQYIDRDPETIRRQLELEKKKQDLDDEEKTAKFIEEQVRRG
Drosophila m.         WLGRTGQVIADETEKGWFVTYIDRSPEAMERQAKADRKEKMEKDDEERMADFIEQQIKNA
Anopheles g.          YLGRNGHCVADETDKGWFITYIDRDPETLAMQEKMAKKQKMDKDDAERLAEFIEEQVRRG
Caenorhabditis e.     YLGSSGKCKIDEGDKGWYIAYIDQ--EALIRKEEDQRKQQQEKDDEERHMQIMDGMVQRG
                      :**  *    **  ***:: ***:  *::  : :  :*:: : ** *:   .:::  ::..

Homo sapiens          L--EGKEQEVPTFTELSRENDEEKVTFNLSKGACSSSGATSSKSSTLGPSALKT------
Mus musculus          L--EGKEQETPVFTELSRENEEEKVTFNLNKGAGGSAGATTSKSSSLGPSALKL------
Drosophila m.         K--AKDGEEDEGQEKFTELKREENEPLKLDIRLEKK---FQPDTVLGKSALAKR------
Anopheles g.          K--TEEEPCTSG---YSELKRE-NEEDTIKIELKLG---SKQQQSTPSAVISKR------
Caenorhabditis e.     KELAGDDEHEYEATELIRDTPDQKIQLDLNLGILDRKLDVKSGVASAKISIFDMPKVKKE
                          .          . . : :    :.                  .
                              III
Homo sapiens          ------IGSSASVKRKESSQSSTQSKEKKKKKSALDEIMEIEEEKK-RTARTDYWLQPEI
Mus musculus          ------LGSAASGKRKESSQSSAQP--AKKKKSALDEIMELEEEKK-RTARTDAWLQPGI
Drosophila m.         ------PAPEAEEKVFKK---PKSVAGDSQTRSVLDEIIKQEESKKERANRKDYWLHKGI
Anopheles g.          ------PFDALDDGKKEKKIKAATSNGETKKLSALDELIQEEEQKKEKNNRKDYWLAEGI
Caenorhabditis e.     DPDEPGPSQPSRKSGKKRSRSRSPAAKKFSKKSALDEIKEMEERKKERKNRKDYWMREGI
                                  :              .. *.***: : ** **  :  *.* *:    *
                                                                   IV
Homo sapiens          IVKIITKKLGEKYHKKKAIVKEVIDKYTAVVKMIDSGDKLKLDQTHLETVIPAPGKRILV
Mus musculus          VVKIITKKLGEKYHKKKGVVKEVIDRYTAVVKMTDSGDRLKLDQTHLETVIPAPGKRVLV
Drosophila m.         VVKFISKSMGEKFFKQKAVVLDVIDRYQGKIKFLETGEKLKVDQAHLETVIPALDKPVMV
Anopheles g.          VVKLISRSLGEKYYKEKGVVVEVIEKYRAKIKLLETGEKLKVDQAHLETVIPAVGKQILV
Caenorhabditis e.     VVKVITKSLGSEYYKAKGVVRKVVDDYTAQVKLDD-GTVVKLDQEHVETVIPSLGRQMMI
                      :**.*::.:*.::.* *.:* .*:: * .  :*: : *  :*:** *:*****: .: :::

Homo sapiens          LNGGYRGNEGTLESINEKTFSATIVIETGPLKGRRVEGIQYEDISKLA--
Mus musculus          LNGGYRGNEGTLESINEKAFSATIVIETGPLKGRRVEGIQYEDISKLA--
Drosophila m.         VNGAYRGSEALLRKLDERRYSVSVEILHGPLKGRIVDNVQYEDISKLHGA
Anopheles g.          LNGGYRGCTAVLKAINTERYSVTIEIASGPLKGRLVSNVAYEDISKL---
Caenorhabditis e.     VNGAYRGQEATLESIDEKRFSLRLKIASGPTRGRQID-VPYEDASKLA--
                      :**.***   . *. :: . :*  : *  ** :** :. : *** ***
```

Figure 1A. Conservation of the modular structure of kin17 protein. Multiple sequence alignment of five metazoan kin17 proteins. The Zinc finger motif that mediates the interaction with DNA is enclosed with a box and named motif I. The RecA homologous region corresponding to a DNA-binding domain is marked as II. The nuclear localization signal is marked as III and the KOW motif supposed to interact with RNA as IV. The kin17 protein sequences correspond to the following species from top to bottom: *Homo sapiens*, *Mus musculus*, *Drosophila melanogaster*, *Anopheles gambiae* and *Caenorhabditis elegans*. The character '*' indicates positions which have a single, fully conserved residue. ':' indicates that one of the following groups is fully conserved: STA; NEQK; NHQK; NDEQ; QHRK; MILV; MILF; HY; FYW and the character '.' indicates the conservation of one of the following groups: CSA, ATV, SAG, STNK, STPA, SGND, SNDEQK; NDEQHK; NEQHRK; FVLIM; HFY.

Molecular Bases of the Cross-Reactivity

The cross-reactivity between kin17 protein and anti-RecA antibodies is due to a major antigenic determinant located in the core of kin17 protein between amino acids 129 and 228.

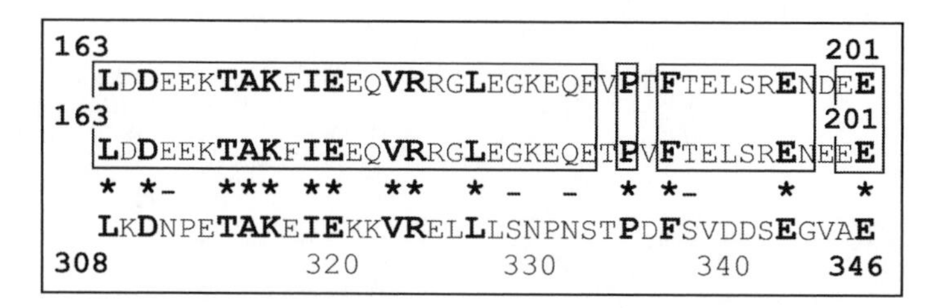

Figure 1B. Conservation of the modular structure of kin17 protein. Structure of the antigenic determinant common to kin17 and RecA proteins. The region of human and mouse kin17 proteins from the amino acid L^{163} to E^{201} is shown and compared with the C-terminal fragment of RecA protein between L^{308} and E^{346}. Identical amino acids in these three proteins are marked in bold letters and flagged with a star. A short line indicates similarities among the following groups of residues: NEQK; STA; NHQK and SAG. The identities concerning only human and mouse kin17 proteins are shown within boxes.

The deletion of this fragment avoids the cross-reactivity and dramatically affects the intranuclear localization. Indeed, the truncated kin17 protein forms large nucleoplasmic clusters that strongly interact with nuclear components.[62] This data indicates that the strong antigenic determinant is part of a functional domain.

Interaction between kin17 Protein and Curved DNA

The finger motif of kin17 protein binds Zn. The resultant Zn-finger mediates the interaction with double- and single-stranded (ss) DNA. More importantly, kin17 protein binds preferentially to pBR322 fragments carrying curved segments with an efficiency that seems to be correlated with the magnitude of DNA curvature.[39] Milot et al showed that chromosomal illegitimate recombination junctions in mammals are associated with the presence of curved DNA,[63] Curved DNA contains runs of adenines distributed regularly at one run per helical repeat and has a reduced electrophoretic mobility. The functional importance of curve DNA in several biological relevant processes will be discussed in the other chapters of this book. In the case of kin17 protein produced in bacteria, the relevance of the interaction with curved DNA was further assessed in vitro by using fragments found at illegitimate recombination. Indeed, Stary and Sarasin isolated several hot spots of illegitimate recombination from an HeLa derived-cell line carrying a single copy of an SV40 shuttle vector. In these cells, the overexpression of SV40 T-antigen produce heterogenous circular DNA molecules carrying the integrated vector and the boundary cellular DNA associated with palindromes, A + T-rich DNA segments, alternating purine/pyrimidine sequences and Alu family repeats.[64] Other sites of integration of polyomavirus in the mouse genome were tested in parallel. They confirmed a preferential binding of kin17 protein to curved DNA leading to the suggestion that this protein may be involved in the illegitimate recombination process.[65,66] The binding to curve DNA is mediated by a domain located in the core of kin17 protein between residues 71 and 281.[39] In vitro, the base composition of DNA may modify its interaction with the kin17 protein (Tran et al, 2004, submitted). The binding of kin17 protein to DNA has been also observed in cultured human cells.[67] As described below, the binding of kin17 protein to curved DNA was further confirmed in vivo by overexpressing the mouse kin17 protein in *E. coli*.[65,66]

Evidence for the Participation of kin17 Protein in Transcription

Expression of kin17 Protein in E. coli *and Trans Species Complementation*

Devoret and colleagues showed that kin17 protein produced in bacteria under the control of a *Tac* promoter is able to regulate gene expression by binding to curved DNA regions.[66] They used H-NS-deficient strains of *E. coli*. H-NS protein is a major component of the nucleoid which binds to curved DNA and regulates the expression of at least 36 genes. Among these

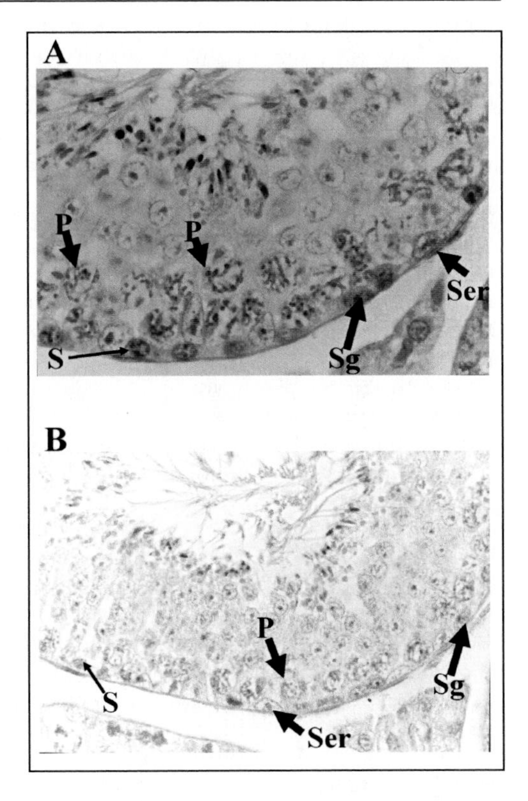

Figure 2. In vivo detection of kin17 protein in adult mouse testis sections. Kin17 protein was detected using the monoclonal antibody 58 (mAb k58) and revealed with 3,3'-diaminobenzidine tetrahydrochloride (DAB) brown staining as reported.[88,97] The sections were counterstained with Mayer's hemalun solution. A) Immunohistochemical detection of mouse kin17 protein in a seminiferous tubule of mouse testis. The brown staining corresponding to the presence of kin17 protein is clearly observed and marked with arrows in the following cells: spermatogonia (Sg), preleptotene (S) and pachytene (P) spermatocytes. Sertoli cells (Ser) are also stained as compared with the control. Magnification: x 1,000. B) Photomicrograph of a paraffin section of mouse seminiferous tubule. The monoclonal antibody against kin17 protein was omitted. The Mayer's hemalun staining reveal cells nuclei in blue. Magnification: x 800.

H-NS upregulates the *flhDC* gene, encoding flagellin, the major structural component of flagellae responsible for cell motility. The expression of kin17 protein in H-NS-deficient bacteria increases the synthesis of flagellin and leads to the recovery of motility. The same H-NS protein downregulates in vivo the *bgl* operon as kin17 protein does. However, not all H-NS functions can be recovered by the expression of kin17 protein. This is not surprising because even in two closely related species like *E. coli* and *Salmonella* only partial complementation of H-NS are observed. Whatever the complementation may be, these data indicate that kin17 protein may be involved in the regulation of gene expression in mammalian cells.[66] In mouse or human cells, the up- or downregulation of kin17 protein induced important changes in the pattern of gene expression leading to a complex pleiotrophic phenotype suggesting a role in transcription.[67-70] Furthermore, in vivo, kin17 protein is detected in all the cell types that compose the male mouse germinal tissue (Fig. 2). It is distributed in intranuclear clusters and interacts directly with RNA and DNA. In agreement with this data, Rappsilver et al have identified kin17 protein as part of the human spliceosome by large-scale proteomic analysis.[71] Indeed, under native conditions, a fraction of kin17 protein was co-purified with the fraction containing polyA$^+$ RNAs pointing to a direct role in RNA processing.[72]

Preferential Expression during Proliferation

In cultured cells, *KIN17* gene expression reaches its highest level during cell proliferation and falls in confluent cells in a way similar to that of the group of late-growth related genes. In proliferating cells, kin17 protein is located in nuclear foci.[68] This particular intranuclear distribution indicates its participation in a nuclear network required during cell growth.[40,41,73] However, the physiologic level of kin17 protein is tightly regulated since its ectopic overexpression triggers a strong deformation of nuclear morphology and is lethal for mammalian cells.[68,69] The overexpression of deletion mutants indicates that the C-terminal end of kin17 protein may interact with the nuclear matrix and is essential for the formation of the large intranuclear clusters. Indeed, a fraction of the endogenous nuclear kin17 protein is strongly anchored in the nuclear matrix.[74] Interestingly, one of the partners of kin17 protein in the nuclear networks is the viral T-antigen (T-Ag).[75] In certain cell types T-Ag is also associated with the nuclear matrix in the heterogeneous nuclear riboucleoprotein (hnRNP) network including peri- and interchromatin fibrils which are the centres of pre-mRNA splicing.[76]

Physical Interaction with SV40 T Antigen

The physical contact with kin17 protein takes place through the T2 region (amino acids 168 to 383) located in the NH_2-terminal region of T antigen.[75] The binding of T-Ag to the viral origin of replication enhances this interaction. The residues involved lie in the major DNA-binding domain of T-Ag which interacts with p53 and DNA polymerase α. The kin17 protein inhibits T-Ag-dependent DNA replication and suggests that these two proteins form part of a nuclear complex in vivo.[75] This tallies with the fact that nuclear kin17 protein molecules are in equilibrium between a fraction dispersed through the nucleoplasm and fractions bound to chromatin, DNA or nuclear matrix.[74] The molecules of kin17 protein move from the nucleoplasmic dispersed form to the bound form during cell proliferation.[77]

UVC Irradiation of Non-Replicating Cells Triggers the Intranuclear Accumulation of kin17 Protein

UVC-irradiation of arrested mouse or primary human fibroblasts boosts *KIN17* gene expression to its maximal values within around 16 hours.[42,78] In human cells, 50 to 75% of DNA lesions caused by UVC are eliminated within 2 hours by nucleotide excision repair (NER) in the so-called "early phase";[79] the remaining lesions are processed during a "late phase" by other mechanisms.[80,81] Since *KIN17* gene expression begins to increase 7 hours after irradiation, we conclude that this gene may be involved in transactions that helps to circumvent lesions not resolved by NER in the late phase of the response, as defined by Herrlich et al.[82] Since in these cells DNA synthesis is arrested, it remains to be determined whether kin17 protein acts directly on the lesions or if it indirectly activates the expression of repair proteins.

Identifying the Mechanism of the UVC-Induced Upregulation

The fact that UVC and ionizing radiation (IR) upregulate *KIN17* gene by two distinct pathways that are independent of the ataxia telengiectasia mutated gene (ATM) prompted us to identify other regulating factors. Two transcription factors, AP-1 and NFκβ, control several other UVC-responsive genes via the protein kinase C (PKC) pathway.[83] However, phorbol ester treatment does not affect *KIN17* RNA expression in BALB/c 3T3 or in mouse lymphoma cells.[78] Similarly, p53 protein controls the expression of nearly 100 genes implicated in the UVC response. Some of them display expression kinetics similar to those of *KIN17* gene[84-86] but normal and p53-deficient cells display a similar UV-induced upregulation of *KIN17* gene.[87,88] Finally, Weiss et al showed that the mouse *Hus1* gene, a component of the cellular machinery that responds to stalled DNA replication and DNA damage, is not involved in the UV-induced upregulation of *KIN17* RNA.[89]

The Integrity of Two DNA Repair Genes Is Required for the Upregulation of KIN17 *Gene Expression after UVC Irradiation*

We determined the expression of the human *KIN17* gene in primary fibroblasts from xeroderma pigmentosum patients, an autosomal recessive cancer-prone inherited disorder provoked by the inactivation of one of the seven XP genes (XPA to XPG). XP patients (NER-deficient) are extremely sensitive to sunlight compared to NER-proficient individuals. NER possesses two major overlapping subpathways: global genome repair (GGR) and transcription coupled repair (TCR). The earliest step of GGR is lesion recognition by a complex formed by two proteins: xeroderma pigmentosum group C (XPC) and one of the human homologues to Rad23 (HHR23B),[90] followed by the recruitment of the transcription factor II H (TFIIH) and XPG, XPA, replication protein A (RPA), excision repair complementation class-1 (ERCC-1) and XPF. Human primary fibroblasts from XPA and XPC patients are unable to trigger *KIN17* gene expression after UVC-irradiation although they present a similar upregulation of p53 protein, of *P21* and *GADD45* RNA. This observation was further confirmed using XP44RO, a cell line established from a testicular melanoma of an XPC patient that also failed to upregulate *KIN17* gene after irradiation. The introduction of a retroviral vector carrying the normal XPC cDNA in XP44RO cells fully restored their capacity to repair DNA and to trigger *KIN17* gene expression after irradiation. This indicates a direct relation between the repair capacity of the cell and the upregulation of this gene. Primary fibroblasts from an XPA patient are also unable to induce *KIN17* gene expression. These data indicate that the activities of XPA and XPC proteins are required to trigger *KIN17* gene expression after UV. This response is strictly dependent on GGR, indicating that the primary signal that leads to the upregulation of *KIN17* gene and other UVC-responsive genes is a subset of the complexes formed at the site of DNA lesions during GGR.[88] Strikingly, this is the first case of a UVC-inducible response which is strictly dependent on GGR. The fact that human XPC protein translocates very rapidly (within 5 minutes) to the sites containing UV lesions and that lesion binding seems to trigger an overall intranuclear stabilization of XPC protein further supports this hypothesis.[91]

The Knock Down of Human KIN17 *Gene Increases Radiosensitivity of Human Cells*

RKO cells present a great number of kin17 protein molecules/cell as compared with normal human fibroblasts or other tumour-derived cells. Considering this fact, we introduced an episomal vector carrying a human *KIN17* cDNA in an antisense orientation. Three stable clones presented 70-80% reduction in the level of kin17 protein were called RASK (from RKO antisense *KIN17*) and were further characterized.[67] These clones have a plating efficiency 15-fold lower than those observed for the control clones, and display a reduced proliferation rate, indicating that decreased levels of kin17 protein strongly affect cell growth. RASK cells accumulate in early and mid-S phase. Only a few cells were detected in late S phase, suggesting that low kin17 protein levels result in a better entry into S phase with some difficulties in progressing through the S phase. However, irradiation of RKO and RASK cells at 6 Gy does not affect the γ-ray-induced G_2 arrest, indicating that kin17 protein is not essential at this checkpoint.[67] As expected, RASK cells are 4- to 5-fold more radiosensitive than the parental RKO cells,[70] indicating that low levels of kin17 protein lead to important changes in the expression profile of genes relevant for cell survival.

Speculative Remarks

The interaction RecA-DNA is modulated by the C-terminal domain of the protein.[61] This domain plays its regulatory role by interacting with other domains of the RecA filament coated on DNA. Although RecA protein does not preferentially recognize curved DNA, it does bind, like kin17, to ds, ssDNA and ssRNA.[92-94] In kin17 protein, the domain homologous to the C-terminal domain of RecA participates in the preferential recognition of curved DNA. This property acquired during evolution may be helpful to detect a particular DNA (or most probably RNA) structure. During evolution, kin17 protein also acquired a KOW domain suggesting a role in transcription elongation. We hypothesize that under physiological conditions kin17 protein may be involved in elongation or splicing by recognizing particular structures on RNA. This interaction may provoke conformational changes that will increase the stability of the complex and will facilitate the recruitment of other nuclear proteins. Interestingly enough, RecA protein is able to assimilate RNA into duplex DNA leading to the formation of an R-loop.[94] Kasahara et al proposed that R-loops serve as origins of bi-directional chromosome replication.[94]

Evidence Pointing to a Role of kin17 Protein in DNA Replication

Molecular Analysis of Deletion Mutants

The deletion of the core DNA-binding domain of kin17 protein leads to the formation of large intranuclear clusters with a shape different from those formed by the wild-type protein and produces important deleterious effects.[62,68] The deletion of this region abolishes the binding to curved DNA whereas deletion of the Zn-finger domain or of the C-terminal end does not affect this interaction.[39,62] The ectopic expression of kin17 protein inhibits T-Ag-dependent DNA replication, indicating a strong interference with T-antigen or with the nuclear areas on which the synthesis of viral DNA takes place.[68] As mentioned, RKO cells presenting decreased amounts of kin17 protein display a disruption in the S phase progression, together with a significant decrease in clonogenic growth.[67] This results suggest the involvement of kin17 protein in DNA replication.

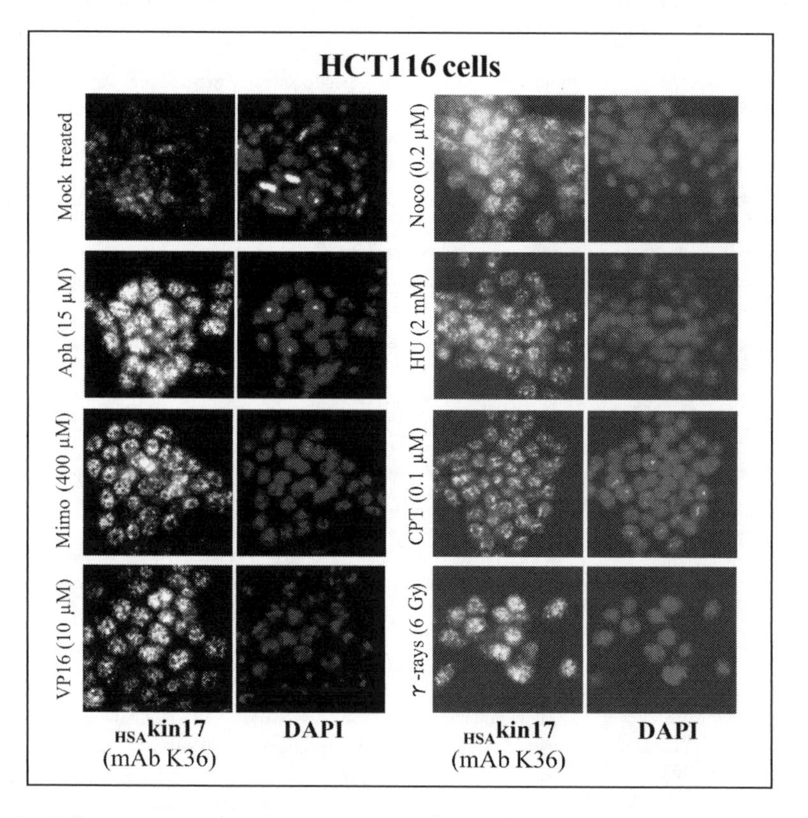

Figure 3A. Cell cycle arrest triggers the accumulation of human kin17 protein in large intranuclear clusters as shown by immunocytochemical detection of this protein and DNA. HCT116 cells were seeded for four days and thereafter treated with different drugs known to modify the cell cycle (nocodazol, aphidicolin, hydroxyurea, l-mimosine, camptothecine, etoposide) or irradiated at 6 Gy. 24 h later they were fixed with ethanol/acetone and $_{HSA}$kin17 protein was detected using purified IgG K36 coupled with a Cy3-conjugated antibody. Cells were counterstained with 4',6-diamino-2- phenylindole (DAPI; at 4 μg/ml). Immunofluorescence staining was viewed using a Zeiss Axiophot 2 epifluorescence microscope coupled to a cooled Sensys 1400 camera from Photometrics monitored by the Zeiss KS300 3.0 program. Magnification: x 315.

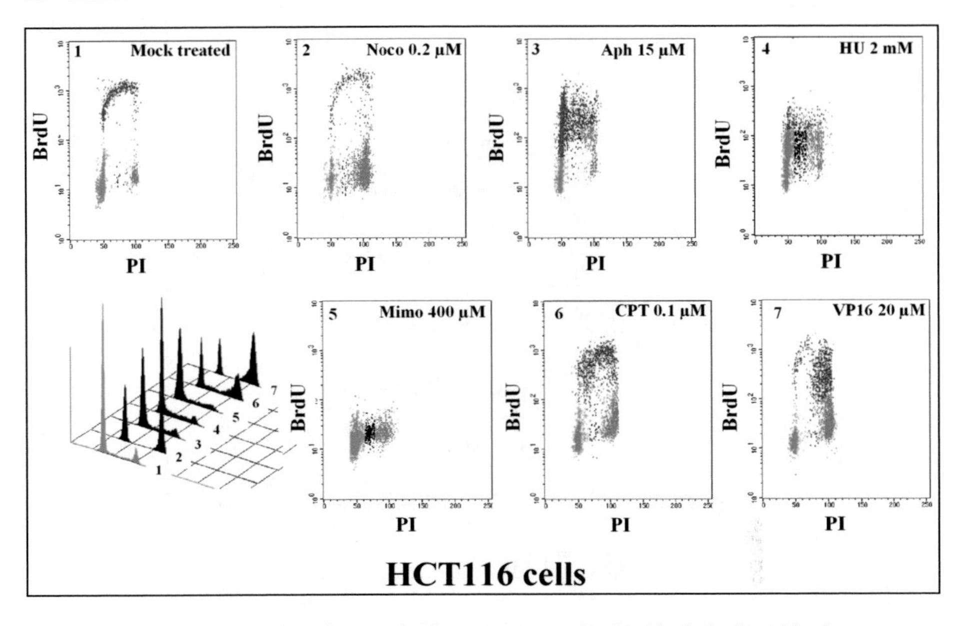

Figure 3B. Monitoring the cell cycle arrest by flow cytometry of BrdU-labelled cells. 24 h after treatment with nocodazol, aphidicolin, hydroxyurea, l-mimosine, camptothecine or etoposide, HCT116 cells were pulse-labelled with 30 mM bromodeoxyuridine (BrdU) for 15 min, washed in PBS, trypsinized and collected. The cells were resuspended in PBS and fixed with 75% ethanol. Nuclei were isolated, treated with pepsine, washed with PBS and then incubated successively with rat anti-BrdU antibody for 1 h at room tempereature. After incubation with fluorescein isothiocyanate-conjugated goat anti-rat IgG secondary antibody and staining with 25 µg/ml propidium iodide the data were collected using a FACsort flow cytometer (BD PharMingen) as previously described.[67,68]

Biochemical Detection of kin17 Protein in DNA Replication Complexes

We analyzed the properties of kin17 protein in its native state in human cells. Gel filtration of total protein extracts showed that the human kin17 protein is present in three complexes with molecular masses corresponding to Mr 400,000 (I), 600,000 (II) and 1,800,000 (III). RPA protein coelutes in complexes II and III. Treatment of human cells with HU arrests them at the G1-S border and increases the molecular weight of the kin17-containing complexes together with a relocalization of kin17 protein in large intranuclear clusters (Fig. 3). L-mimosine (MIMO) and HU trigger a similar effect in RKO cells.[77] The other drugs affecting DNA replication and producing a redistribution of kin17 protein are the following: aphidicolin (APH) which inhibits DNA polymerase α and δ; camptothecin (CPT) which interferes with the sealing activity of DNA topoisomerase I producing double strand breaks (DSBs), and etoposide (VP16), a specific inhibitor of DNA topoisomerase II. 24 hours after treatment with any of these drugs the nuclear concentration of kin17 protein increased and formed intranuclear clusters which are easily detectable by immunocytochemical detection (Fig. 3A). As expected, the drug treatment strongly modified the cell cycle as shown by the comparison of DNA synthesis in the population of mock treated and treated cells (Fig. 3B). In treated cells, the intranuclear concentration of anchored kin17 protein increased as compared with that of mock treated cells as shown by western detection of nuclear proteins like kin17, RPA 70, PCNA and p34cdc2 (Fig. 3C).

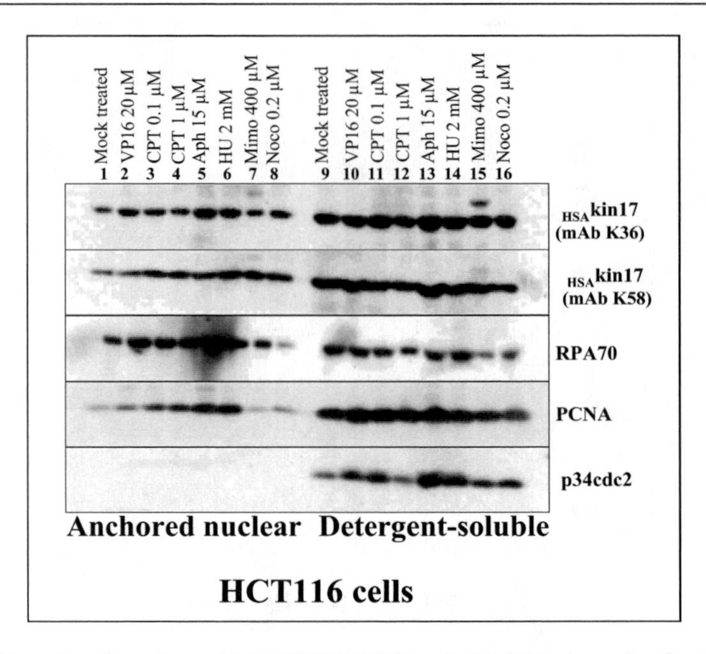

Figure 3C. Detection of proteins $_{HSA}$kin17, RPA70, PCNA and p34cdc2 in the nuclear fraction of treated cells. HCT116 cells at 50% confluence were treated with etoposide, camptothecine, aphidicolin, hydroxyurea, 1-mimosine or nocodazol at the indicated doses. 24 h later, cells were trypsinized, counted, and washed in PBS. To discriminate between chromatin-bound proteins versus detergent-soluble proteins, cells were lysed with 100 μl per 10^6 cells of buffer N (50 mM Tris-HCl [pH 7.9]/150 mM NaCl/ 1% Igepal/ 1 mM EDTA/ complete protease inhibitor cocktail from Roche). Lysates were maintained on ice for 30 min. Soluble proteins were recovered after centrifugation (20,000 g for 15 min). Remaining pellets (insoluble proteins) were directly denatured with 100 μl per 10^6 cells of 2x Laemmli buffer. Both fractions were analysed by Western blot with purified IgG K36 and K58 at the concentration of 40 ng/ml and with other antibodies as described.[67]

Intranuclear Modification of Protein Complexes during the Cell Cycle

Treatment of proliferating HeLa cells with L-mimosine for 24h followed by drug removal resulted in a synchronized population which rapidly re-entered into the S phase. A time dependent increase in kin17, cyclin B and p34cdc2 levels were observed until completion of the S phase. This should be correlated with the re-initiation of DNA replication.[74] In parallel kin17 protein is detected in DNA replication foci attached to the nuclear matrix of HeLa cells. The removal of more than 80% of total DNA did not affect the association with other nuclear proteins. The in vivo protein-protein cross-linking confirmed the association with the nuclear matrix during all the phases of the cell cycle, indicating that the "nucleoplasmic pool" of kin17 protein could serve as a "stock" that may later be associated with both chromatin and/or nuclear matrix during DNA replication.[74] It is possible that the nuclear accumulation of kin17 protein leads to an increase in the chromatin-bound fraction in order to facilitate the DNA replication process in spite of the presence of multiply damaged sites or other DNA lesions.[74] Further work will be required to precise the role of kin17 protein in replication. Although in mammals several aspects of this process remains yet to be elucidated, research on viruses indicates that replication initiation proteins belongs to the class of RNA-binding proteins involved in splicing.[95] This observation opens the possibility to use kin17 protein as a tool to precisely describe the molecular steps of replication initiation in human cells during the response to genotoxic agents.

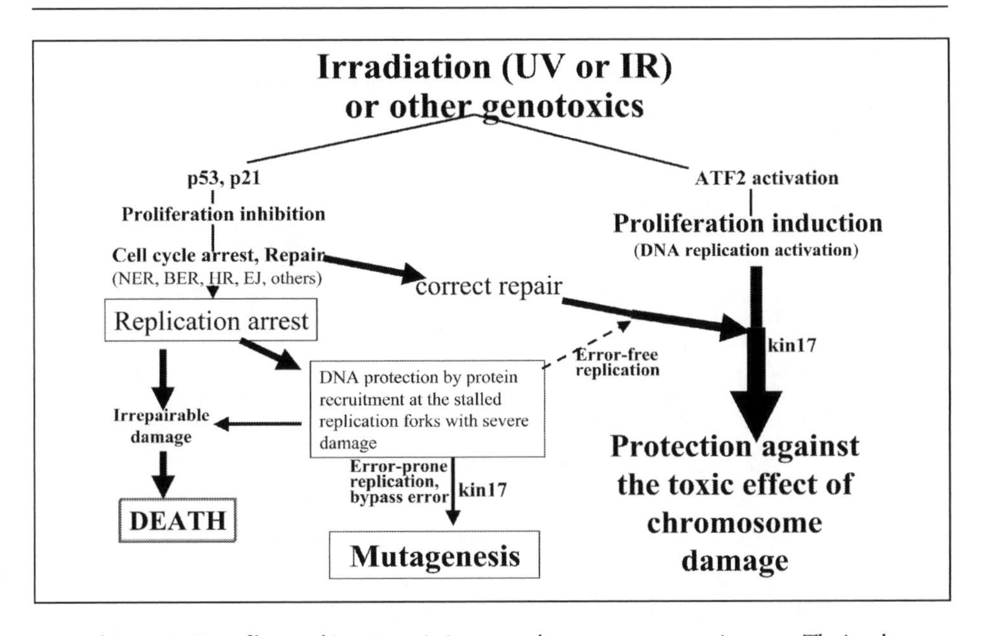

Figure 4. Participation of human kin17 protein in a general response to genotoxic agents. The involvement of kin17 protein in different responses to genotoxic agents has already been reported.[42,43,67,74,78,87,88,97] After genotoxic injury, UV or IR, there are two possibilities depending on the complexity of the lesions. At low levels of damage, ATF2 (and other transcription factors) will stimulate cell proliferation and subsequently DNA replication, eliminating the damage and leading to a normal cell division. Alternatively, when the damage level is very high, other transcription factors like p53 may arrest the cell cycle, thus allowing repair processes to occur. If the repair process performed by base excision repair (BER), homologous recombination (HR), end-joining (EJ) or other repair systems is accurately performed, then cells will proliferate following a normal cell cycle. If the cell is unable to deal with the numerous lesions, then RPA, kin17 and other proteins of DNA metabolism will be recruited and form "protection clusters" at the remaining unrepaired chromatin lesions. The "in extremis" treatment of these lesions may result in a late error-prone repair that will finally generate mutations in the descendants. A failure of this "last chance trial" to release the DNA replication arrest (due for example to the high number or complexity of the lesions) will lead to cell death.

Conclusion

We have identified a nuclear protein that participates in the response to severe lesions created by genotoxics on chromatin. Since this protein recognizes curved DNA and other particular RNA structures and forms high molecular weight complexes, we assume that it may be important for the formation of clusters around the unrepaired remaining lesions. We suggest that this process is a "last chance pathway" that gives the opportunity to restart DNA replication before activation of cell death. The recognition of topological constraints created by several types of DNA damage may perturb the accurate replication of DNA therefore decreasing fidelity and helping to generate biological diversity (see Fig. 4).[96]

Acknowledgements

The authors are indebted to B. Dutrillaux, C. Créminon, Y. Frobert, J. Grassi, F. Harper, E. Pichard and E. Puvion for advice and support, and to M. Plaisance, P. Lamourette and M.C. Nevers for their efficient help in producing monoclonal antibodies. Work in the authors' laboratory was supported by the Commissariat à l'énergie atomique (CEA) and Electricité de France Contract 8702.

References

1. Toney JH, Donahue BA, Kellett PJ et al. Isolation of cDNAs encoding a human protein that binds selectively to DNA modified by the anticancer drug *cis*-diamminedichloroplatinum(II). Proc Natl Acad Sci USA 1989; 86:8328-8332.
2. Donahue BA, Augot M, Bellon SF et al. Characterization of a DNA damage-recognition protein from mammalian cells that binds specifically to intrastrand d(GpG) and d(ApG) DNA adducts of the anticancer drug cisplatin. Biochemistry 1990; 29:5872-5880.
3. Donahue BA, Yin S, Taylor JS et al. Transcript cleavage by RNA polymerase II arrested by a cyclobutane pyrimidine dimer in the DNA template. Proc Natl Acad Sci USA 1994; 91:8502-8506.
4. Coates PJ, Save V, Ansari B et al. Demonstration of DNA damage/repair in individual cells using in situ end labelling: association of p53 with sites of DNA damage. J Pathol 1995; 176:19-26.
5. Bianchi ME, Beltrame M, Paonessa G. Specific recognition of cruciform DNA by nuclear protein HMG1. Science 1989; 243:1056-1059.
6. Pierro P, Capaccio L, Gadaleta G. The 25 kDa protein recognizing the rat curved region upstream of the origin of the L-strand replication is the rat homologue of the human mitochondrial transcription factor A. FEBS Lett 1999; 457:307-310.
7. Setlow RB, Setlow JK. Evidence that ultraviolet-induced thymine dimers in DNA cause biological damage. Proc Natl Acad Sci USA 1962; 48:1250-1257.
8. Cleaver JE. Defective repair replication of DNA in xeroderma pigmentosum. Nature 1968; 218:652-656.
9. Tanaka K, Miura N, Satokata I et al. Analysis of a human DNA excision repair gene involved in group A xeroderma pigmentosum and containing a zinc-finger domain. Nature 1990; 348:73-76.
10. Friedberg EC, Walker GC, Siede W. Human hereditary diseases with defective processing of DNA damage. In: DNA Repair and Mutagenesis. Washington DC: ASM Press, 1995:633-685.
11. Eggleston AK, West SC. Exchanging partners: recombination in E. coli. Trends Genet 1996; 12:20-26.
12. Sommer S, Bailone A, Devoret R. The appearance of the UmuD'C protein complex in Escherichia coli switches repair from homologous recombination to SOS mutagenesis. Mol Microbiol 1993; 10:963-971.
13. Cox MM. Recombinational DNA repair in bacteria and the RecA protein. Prog Nucleic Acid Res Mol Biol 1999; 63:311-366.
14. Eisen JA. The RecA protein as a model molecule for molecular systematic studies of bacteria: Comparison of trees of RecAs and 16S rRNAs from the same species. J Mol Evol 1995; 41:1105-1123.
15. Sato S, Kobayashi T, Hotta Y et al. Characterization of a mouse *recA*-like gene specifically expressed in testis. DNA Res 1995; 2:147-150.
16. Sato S, Hotta Y, Tabata S. Structural analysis of a *recA*-like gene in the genome of Arabidopsis thaliana. DNA Res 1995; 2:89-93.
17. Yoshida K, Kondoh G, Matsuda Y et al. The mouse *RecA*-like gene Dmc1 is required for homologous chromosome synapsis during meiosis. Mol Cell 1998; 1:707-718.
18. Takahashi E, Matsuda Y, Hori T et al. Chromosome mapping of the human (RECA) and mouse (*Reca*) homologs of the yeast *RAD51* and Escherichia coli *recA* genes to human (15q15.1) and mouse (2F1) chromosomes by direct R-banding fluorescence in situ hybridization. Genomics 1994; 19:376-378.
19. Brendel V, Brocchieri L, Sandler SJ et al. Evolutionary comparisons of RecA-like proteins across all major kingdoms of living organisms. J Mol Evol 1997; 44:528-541.
20. McKee BD, Ren X, Hong C. A *recA*-like gene in Drosophila melanogaster that is expressed at high levels in female but not male meiotic tissues. Chromosoma 1996; 104:479-488.
21. Pittman DL, Weinberg LR, Schimenti JC. Identification, characterization, and genetic mapping of *Rad51d*, a new mouse and human *RAD51/RecA*-related gene. Genomics 1998; 49:103-111.
22. Terasawa M, Shinohara A, Hotta Y et al. Localization of RecA-like recombination proteins on chromosomes of the lily at various meiotic stages. Genes Dev 1995; 9:925-934.
23. Yoshimura Y, Morita T, Yamamoto A et al. Cloning and sequence of the human *RecA*-like gene cDNA. Nucleic Acids Res 1993; 21:1665.
24. Morita T, Yoshimura Y, Yamamoto A et al. A mouse homolog of the Escherichia coli *recA* and *Saccharomyces cerevisiae RAD51* genes. Proc Natl Acad Sci USA 1993; 90:6577-6580.
25. Aihara H, Ito Y, Kurumizaka H et al. An interaction between a specified surface of the C-terminal domain of RecA protein and double-stranded DNA for homologous pairing. J Mol Biol 1997; 274:213-221.

26. Yang S, VanLoock MS, Yu X et al. Comparison of bacteriophage T4 UvsX and human Rad51 filaments suggests that RecA-like polymers may have evolved independently. J Mol Biol 2001; 312:999-1009.
27. Angulo JF, Moreau PL, Maunoury R et al. KIN, a mammalian nuclear protein immunologically related to E. coli RecA protein. Mutat Res 1989; 217:123-134.
28. Higashitani A, Tabata S, Ogawa T et al. ATP-independent strand transfer protein from murine spermatocytes, spermatids, and spermatozoa. Exp Cell Res 1990; 186:317-323.
29. Bashkirov VI, Loseva EF, Savchenko GV et al. Antibodies against Escherichia coli RecA protein reveal two nuclear proteins in bovine spermatocytes which interact with synaptonemal complex structures of meiotic chromosomes of various eukaryotic organisms. Genetika 1993; 29:1953-1968.
30. Cerutti H, Osman M, Grandoni P et al. A homolog of Escherichia coli RecA protein in plastids of higher plants. Proc Natl Acad Sci USA 1992; 89:8068-8072.
31. Cerutti H, Ibrahim HZ, Jagendorf AT. Treatment of pea (Pisum sativum L.) protoplasts with DNA-damaging agents induces a 39-kilodalton chloroplast protein immunologically related to Escherichia coli RecA. Plant Physiol 1993; 102:155-163.
32. Tissier A, Kannouche P, Biard DS et al. The mouse Kin-17 gene codes for a new protein involved in DNA transactions and is akin to the bacterial RecA protein. Biochimie 1995; 77:854-860.
33. Anderson LK, Offenberg HH, Verkuijlen WM et al. RecA-like proteins are components of early meiotic nodules in lily. Proc Natl Acad Sci USA 1997; 94:6868-6873.
34. Martin B, Ruellan JM, Angulo JF et al. Identification of the recA gene of Streptococcus pneumoniae. Nucleic Acids Res. 1992; 20:6412.
35. Aigle B, Holl AC, Angulo JF et al. Characterization of two Streptomyces ambofaciens recA mutants: identification of the RecA protein by immunoblotting. FEMS Microbiol Lett 1997; 149:181-187.
36. Borchiellini P, Angulo J, Bertolotti R. Genes encoding mammalian recombinases: Cloning approach with anti-RecA antibodies. Biogenic Amines 1997; 13:195-215.
37. Angulo JF, Rouer E, Benarous R et al. Identification of a mouse cDNA fragment whose expressed polypeptide reacts with anti-recA antibodies. Biochimie 1991; 73:251-256.
38. Angulo JF, Rouer E, Mazin A et al. Identification and expression of the cDNA of KIN17, a zinc-finger gene located on mouse chromosome 2, encoding a new DNA-binding protein. Nucleic Acids Res 1991; 19:5117-5123.
39. Mazin A, Timchenko T, Menissier-de-Murcia J et al. Kin17, a mouse nuclear zinc finger protein that binds preferentially to curved DNA. Nucleic Acids Res 1994; 22:4335-4341.
40. Araneda S, Angulo J, Devoret R et al. Identification of a Kin nuclear protein immunologically related to RecA protein in the rat CNS. C R Acad Sci III 1993; 316:593-597.
41. Araneda S, Angulo J, Touret M et al. Preferential expression of kin, a nuclear protein binding to curved DNA, in the neurons of the adult rat. Brain Res 1997; 762:103-113.
42. Kannouche P, Mauffrey P, Pinon-Lataillade G et al. Molecular cloning and characterization of the human KIN17 cDNA encoding a component of the UVC response that is conserved among metazoans. Carcinogenesis 2000; 21:1701-1710.
43. Biard DSF, Saintigny Y, Maratrat M et al. Enhanced expression of the kin17 protein immediately after low doses of ionizing radiation. Radiat Res 1997; 147:442-450.
44. Biard DSF, Saintigny Y, Maratrat M et al. Differential expression of the Hskin17 protein during differentiation of in vitro reconstructed human skin. Arch Dermatol Res 1997; 289:448-456.
45. Kyrpides NC, Woese CR, Ouzounis CA. KOW: a novel motif linking a bacterial transcription factor with ribosomal proteins. Trends Biochem Sci 1996; 21:425-426.
46. Ponting CP. Novel domains and orthologues of eukaryotic transcription elongation factors. Nucleic Acids Res 2002; 30:3643-3652.
47. Kazianis S, Gan L, Della Coletta L et al. Cloning and comparative sequence analysis of TP53 in Xiphophorus fish hybrid melanoma models. Gene 1998; 212:31-38.
48. de Murcia G, Menissier de Murcia J. Poly(ADP-ribose) polymerase: a molecular nick-sensor. Trends Biochem Sci 1994; 19:172-176.
49. Yu X, Egelman EH. Removal of the RecA C-terminus results in a conformational change in the RecA-DNA filament. J Struct Biol 1991; 106:243-254.
50. Eggler AL, Lusetti SL, Cox MM. The C terminus of the Escherichia coli RecA protein modulates the DNA binding competition with single-stranded DNA-binding protein. J Biol Chem 2003; 278:16389-16396.
51. Story RM, Steitz TA. Structure of the recA protein-ADP complex. Nature 1992; 355:374-376.
52. Story RM, Weber IT, Steitz TA. The structure of the E. coli recA protein monomer and polymer. Nature 1992; 355:318-325.

53. VanLoock MS, Yu X, Yang S et al. ATP-mediated conformational changes in the RecA filament. Structure (Camb) 2003; 11:187-196.
54. Yu X, Jacobs SA, West SC et al. Domain structure and dynamics in the helical filaments formed by RecA and Rad51 on DNA. Proc Natl Acad Sci USA 2001; 98:8419-8424.
55. Krivi GG, Bittner ML, Rowold E, Jr. et al. Purification of recA-based fusion proteins by immunoadsorbent chromatography. Characterization of a major antigenic determinant of Escherichia coli recA protein. J Biol Chem 1985; 260:10263-10267.
56. Benedict RC, Kowalczykowski SC. Increase of the DNA strand assimilation activity of recA protein by removal of the C terminus and structure-function studies of the resulting protein fragment. J Biol Chem 1988; 263:15513-15520.
57. Tateishi S, Horii T, Ogawa T et al. C-terminal truncated Escherichia coli RecA protein RecA5327 has enhanced binding affinities to single- and double-stranded DNAs. J Mol Biol 1992; 223:115-129.
58. Larminat F, Defais M. Modulation of the SOS response by truncated RecA proteins. Mol Gen Genet 1989; 216:106-112.
59. Sedgwick SG, Yarranton GT. Cloned truncated *recA* genes in *E. coli*. I. Effect on radiosensitivity and *recA*+ dependent processes. Mol Gen Genet 1982; 185:93-98.
60. Yarranton GT, Sedgwick SG. Cloned truncated *recA* genes in *E. coli* II. Effects of truncated gene. Mol Gen Genet 1982; 185:99-104.
61. Lusetti SL, Wood EA, Fleming CD et al. C-terminal Deletions of the *Escherichia coli* RecA Protein. Characterisation of in vivo and in vitro effects. J Biol Chem 2003; 278:16372-16380.
62. Kannouche P, Pinon-Lataillade G, Mauffrey P et al. Overexpression of kin17 protein forms intranuclear foci in mammalian cells. Biochimie 1997; 79:599-606.
63. Milot E, Belmaaza A, Wallenburg JC et al. Chromosomal illegitimate recombination in mammalian cells is associated with intrinsically bent DNA elements. EMBO J 1992; 11:5063-5070.
64. Stary A, Sarasin A. Molecular analysis of DNA junctions produced by illegitimate recombination in human cells. Nucleic Acids Res 1992; 20:4269-4274.
65. Mazin A, Milot E, Devoret R et al. KIN17, a mouse nuclear protein, binds to bent DNA fragments that are found at illegitimate recombination junctions in mammalian cells. Mol Gen Genet 1994; 244:435-438.
66. Timchenko T, Bailone A, Devoret R. Btcd, a mouse protein that binds to curved DNA, can substitute in *Escherichia coli* for H-NS, a bacterial nucleoid protein. EMBO J 1996; 15:3986-3992.
67. Biard DS, Miccoli L, Despras E et al. Ionizing radiation triggers chromatin-bound kin17 complex formation in human cells. J Biol Chem 2002; 277:19156-19165.
68. Kannouche P, Angulo JF. Overexpression of kin17 protein disrupts nuclear morphology and inhibits the growth of mammalian cells. J Cell Sci 1999; 112:3215-3224.
69. Biard DS, Kannouche P, Lannuzel-Drogou C et al. Ectopic expression of (Mm)Kin17 protein inhibits cell proliferation of human tumor-derived cells. Exp Cell Res 1999; 250:499-509.
70. Despras E, Miccoli L, Creminon C et al. Depletion of KIN17, a human DNA replication protein, increases the radiosensitivity of RKO cells. Radiat Res 2003; 159:748-758.
71. Rappsilber J, Ryder U, Lamond AI et al. Large-scale proteomic analysis of the human spliceosome. Genome Res 2002; 12:1231-1245.
72. Pinon-Lataillade G, Masson C, Bernardino-Sgherri J et al. KIN17 encodes an RNA-binding protein and is expressed during mouse spermatogenesis. J Cell Sci 2004; 117:3691-3702.
73. Araneda S, Mermet N, Verjat T et al. Expression of Kin17 and 8-OxoG DNA glycosylase in cells of rodent and quail central nervous system. Brain Res Bull 2001; 56:139-146.
74. Miccoli L, Biard DS, Frouin I et al. Selective interactions of human kin17 and RPA proteins with chromatin and the nuclear matrix in a DNA damage- and cell cycle-regulated manner. Nucleic Acids Res 2003; 31:4162-4175.
75. Miccoli L, Biard DSF, Creminon C et al. Human kin17 protein directly interacts with the SV40 large T antigen and inhibits DNA replication. Cancer Res 2002; 62:5425-5436.
76. Puvion E, Duthu A, Harper F et al. Intranuclear distribution of SV40 large T-antigen and transformation-related protein p53 in abortively infected cells. Exp Cell Res 1988; 177:73-89.
77. Biard DS, Miccoli L, Despras E et al. Participation of kin17 protein in replication factories and in other DNA transactions mediated by high molecular weight nuclear complexes. Mol Cancer Res 2003; 1:519-531.
78. Kannouche P, Pinon-Lataillade G, Tissier A et al. The nuclear concentration of kin17, a mouse protein that binds to curved DNA, increases during cell proliferation and after UV irradiation. Carcinogenesis 1998; 19:781-789.
79. Jensen KA, Smerdon MJ. DNA repair within nucleosome cores of UV-irradiated human cells. Biochemistry 1990; 29:4773-4782.

80. Dresler SL, Gowans BJ, Robinson-Hill RM et al. Involvement of DNA polymerase δ in DNA repair synthesis in human fibroblasts at late times after ultraviolet irradiation. Biochemistry 1988; 27:6379-6383.
81. Tornaletti S, Pfeifer GP. UV damage and repair mechanisms in mammalian cells. Bioessays 1996; 18:221-228.
82. Herrlich P, Blattner C, Knebel A et al. Nuclear and non-nuclear targets of genotoxic agents in the induction of gene expression. Shared principles in yeast, rodents, man and plants. Biol Chem 1997; 378:1217-1229.
83. Nishizuka Y. Intracellular signaling by hydrolysis of phospholipids and activation of protein kinase C. Science 1992; 258:607-614.
84. Maltzman W, Czyzyk L. UV irradiation stimulates levels of p53 cellular tumor antigen in nontransformed mouse cells. Mol Cell Biol 1984; 4:1689-1694.
85. Smith ML, Fornace AJ, Jr. p53-mediated protective responses to UV irradiation. Proc Natl Acad Sci USA 1997; 94:12255-12257.
86. May P, May E. Twenty years of p53 research: structural and functional aspects of the p53 protein. Oncogene 1999; 18:7621-7636.
87. Blattner C, Kannouche P, Litfin M et al. UV-induced stabilization of c-*fos* and other short-lived mRNAs. Mol Cell Biol 2000; 20:3616-3625.
88. Masson C, Menaa F, Pinon-Lataillade G et al. Global genome repair is required to activate *KIN17*, a UVC-responsive gene involved in DNA replication. Proc Natl Acad Sci USA 2003; 100:616-621.
89. Weiss RS, Enoch T, Leder P. Inactivation of mouse *Hus1* results in genomic instability and impaired responses to genotoxic stress. Genes Dev 2000; 14:1886-1898.
90. Sugasawa K, Ng JM, Masutani C et al. Xeroderma pigmentosum group C protein complex is the initiator of global genome nucleotide excision repair. Mol Cell 1998; 2:223-232.
91. Ng JM, Vermeulen W, Van Der Horst GT et al. A novel regulation mechanism of DNA repair by damage-induced and RAD23-dependent stabilization of xeroderma pigmentosum group C protein. Genes Dev 2003; 17:1630-1645.
92. Kirkpatrick DP, Rao BJ, Radding CM. RNA-DNA hybridization promoted by *E. coli* RecA protein. Nucleic Acids Res 1992; 20:4339-4346.
93. Zaitsev EN, Kowalczykowski SC. A novel pairing process promoted by *Escherichia coli* RecA protein: inverse DNA and RNA strand exchange. Genes Dev 2000; 14:740-749.
94. Kasahara M, Clikeman JA, Bates DB et al. RecA protein-dependent R-loop formation in vitro. Genes Dev 2000; 14:360-365.
95. Stenlund A. Initiation of DNA replication: Lessons from viral initiator proteins. Nat Rev Mol Cell Biol 2003; 4:777-785.
96. Aravind L, Walker DR, Koonin EV. Conserved domains in DNA repair proteins and evolution of repair systems. Nucleic Acids Res 1999; 27:1223-1242.
97. Masson C, Menaa F, Pinon-Lataillade G et al. Identification of *KIN* (*KIN17*), a human gene encoding a nuclear DNA-binding protein, as a novel component of the TP53-independent response to ionizing radiation. Radiat Res 2001; 156:535-544.

Part III
Implied Roles of Left-Handed Z-DNA, Triplex DNA, DNA Supercoiling, and Miscellaneous Alternative Conformations of DNA in Transcription

Chapter 7

Roles for Z-DNA and Double-Stranded RNA in Transcription:

Encoding Genetic Information by Shape Rather than by Sequence

Alan Herbert

Abstract

Readout of eukaryotic genomes is soft-wired, leading to many different messages from a single gene. Z-DNA and double-stranded RNA (dsRNA) are both examples where genetic information is encoded by shape rather than by sequence. The use of these two conformational motifs to produce sequence-specific changes in RNA transcripts is discussed using dsRNA editing by ADAR1 as a model. This concept is extended to other RNA-directed modifications of DNA and RNA.

Introduction

Our understanding of how genetic information is encoded within DNA has changed dramatically over the last decade. We now know that the readout of genetic information in eukaryotes is not only more complex than previously appreciated but also more dynamic. Much of the information is soft-wired, with many different messages being produced from a single gene.[1] This outcome is only possible because of extensive alternate splicing and editing of pre-mRNA.[2,3] The exact information read out from a gene depends upon cellular context.[4] It varies with the pattern of histone modification and DNA methylation according to a program transmitted epigenetically from a cell's parent.[5,6] It changes according to signals from the extracellular environment and according to the other gene products present within the cell. It involves very specific RNA-based mechanisms that direct information readout by regulating RNA stability, DNA methylation and DNA rearrangement.[7] It produces a large number of different ribotypes, but only a selection of those that could be produced from the genome.[1]

Readout of genetic information begins at the level of the gene. Again here our understanding of how this happens has undergone revision. In eukaryotes, very specific macromolecular complexes are assembled from modular components.[8] Their formation depends on the spatial proximity of appropriate DNA sequence motifs and interactions involving generic protein domains. The combination is able to target specific chromosomal regions even though, in a large genome, there may be hundreds of thousands of potential binding sites for each individual component.

Along with sequence-specific recognition of DNA, we now also appreciate that recognition of DNA shape can provide information important for the assembly of macromolecular machines. Often there is a requirement for specialized nucleotide sequences such as the TATA

DNA Conformation and Transcription, edited by Takashi Ohyama. ©2005 Eurekah.com and Springer Science+Business Media.

box, which undergoes a major distortion in shape when bound by the TATA binding protein.[9,10] This need is also true of other bent DNA conformations.[11] Specific sequences, however, are not always required, an example being Holiday junctions that form during recombination following the exchange of DNA strands between homologous duplexes.[12]

Here, we discuss biological processes that utilize Z-DNA and double-stranded RNA conformations to modify the sequence-specific read-out of genes.

Formation of Z-DNA

Z-DNA is a high energy conformer of B-DNA that is stabilized by negative supercoiling. The ease of formation varies with sequence-$(CG)_n$ is best, $(TG)_n$ is next, while a $(GGGC)_n$ repeat is better than (TA)n[13] and favored when deoxycytidine is 5-methylated.[13]

Sequences that form Z-DNA are ten times more frequent in 5' than in 3' regions of genes, reflecting the overlap of CpG rich islands with the first exon of genes.[14,15] This distribution fits with the expectation that Z-DNA formation in vivo will be associated with actively transcribed genes. As demonstrated by Liu and Wang, negative supercoils arise behind a moving RNA polymerase as it ploughs through the DNA double helix.[16] The torsional strain generated by passage of RNA polymerases thus becomes a potent source of energy to stabilize Z-DNA.

Other enzymes also can induce Z-DNA formation.[17] The SWI-SNF-like BAF complex remodels chromatin to form an open structure; in the case of the colony-stimulating factor 1 (*CSF1*) gene, this results in Z-DNA formation, either as a result of negative supercoiling generated by the processive movement of the BAF helicase or through the release of negative supercoils previously constrained by histones.[18]

Potential Roles for the Z-DNA Conformation Due to Its Physical Presence

Formation of Z-DNA in transcriptionally active regions of the genome may have many functional consequences:

Protein exclusion: Z-DNA formation could affect the placement of nucleosomes as well as the organization of chromosomal domains by providing regions from which histones or other architectural proteins are excluded.[19]

Polymerase stalling: Formation of Z-DNA behind (5') to a moving polymerase may block the following RNA polymerase from transcribing that region of a gene.[20] Thus formation of Z-DNA could ensure spatial separation between successive polymerases and perhaps minimize non-functional mis-splicing of messages.

Inhibition of DNA modification: Formation of Z-DNA may protect sequences from modification. For example, DNA methylases do not modify Z-DNA.[21,22] Also B-DNA specific restriction endonucleases do not cleave sequences in the Z-DNA conformation.[21,22]

DNA topology: Z-DNA formation could alter the phasing of recognition sites for DNA binding proteins. The helical repeat for Z-DNA is 12 base pairs, whereas that for B-DNA is 10.5. Formation of two turns of Z-DNA will cause the relative position of two binding sites at opposite ends of the Z-helix to change by about one third of a turn. Further, if the B-Z junctions are bent,[23] then the relative positions in space of the two binding sites will also change. Z-DNA formation could thus impact formation of macromolecular complexes.

DNA kinetics: Z-DNA formation could facilitate a number of processes by relieving localized topological strain. For example, sequences forming Z-DNA may favor recombination of homologous chromosomal domains at nearby sites by allowing intertwining of intact duplexes to form paranemic joints.[24] The Z-DNA forming $(CA/GT)_n$ sequence has been shown to be recombinogenic in yeast[25] but less effective than $(CG)_n$ in human cells.[26,27] Furthermore, several reports correlate chromosomal breakpoints in human tumors with potential Z-DNA forming sequences; although a causal relationship has not yet been established.[28-32]

Proof of these mechanisms would be aided by methods to demonstrate formation of Z-DNA in vivo. Sensitive techniques such as infrared Raman spectroscopy that detect a specific Z-DNA signature are limited by a requirement for a high localized concentration of the conformer.[33] Techniques that involve the use of a protein probe, such as an antibody, that is specific for Z-DNA conformation must be used with care as the protein can induce Z-DNA formation where none had been present before. A promising new technique uses chemistry that is Z-DNA specific. C2'α-hydroxylation of deoxyguanosine is favored when 5-iodouracil-containing Z-DNA is irradiated with UV light due to the spatial orientation of these two residues in the left-handed conformation.[34,35] The reaction results in formation of guanosine, which can be detected in modified DNA using ribonuclease T1. The reaction is not inhibited by protein.[35] This, or similar approaches, may be useful for tracking the formation of Z-DNA in vivo under physiological conditions.

Protein Recognition of the Z-DNA Conformation

There has been an extensive search by a number of laboratories for Z-DNA binding proteins. Early studies were unfruitful and caused widespread skepticism that Z-DNA would be associated with any biological function. Many of the positive results reported in these studies may have been due either to artefacts or misinterpretation of data. [36-41] This trend continues today and underscores the need to apply existing physicochemical methods to establish the case for Z-DNA binding.[42,43]

The first successful identification of a natural, high affinity Z-DNA binding protein depended on the development of a method that incorporated rigorous controls and was designed to show unambiguous Z-DNA binding.[44] A short linear radiolabeled probe that incorporated 5-bromodeoxycytidine was used in electrophoretic mobility shift assays (EMSA).[45] This probe flipped to the Z-DNA conformation in physiological salt supplemented with magnesium. The method allowed competition to be performed with both linear and supercoiled plasmids so that specificity of binding could be confirmed. The assay required that the protein be of high affinity for the Z-DNA conformation. Indeed the protein that was eventually purified had a nanomolar affinity for the Z-DNA.[46,47] In particular, it had a slow off-rate, around 10^{-2} s^{-1}, explaining the stability of complexes in EMSA under conditions where no magnesium was present in the gel buffer to stabilize the probe in the Z-conformation.[47] The Z-DNA binding domain of the protein, Zα (Table 1), was mapped and expressed in *Escherichia coli*, allowing its specificity for Z-DNA to be confirmed by circular dichroism[47] and Raman spectroscopy.[48] These studies demonstrated that Zα binds to Z-DNA formed by many different sequences of varying nucleotide composition[49] and induces formation of Z-RNA at 45°C in physiological salt.[50] The final demonstration that this protein was indeed specific for the Z-conformation was the co-crystallization of the Zα domain with Z-DNA at 2.1 angstrom resolution (Fig. 1) [51]

The Zα Fold

Zα belongs to the winged helix-turn-helix family of proteins that include the non-sequence-specific B-DNA binder histone H5 and the sequence-specific B-DNA binding transcription factor hepatocyte nuclear factor 3γ (HNF-3γ).[52-54] Two molecules of Zα bind to one turn of the Z-DNA helix, reflecting the stoichiometry that was first detected in EMSA[49] and confirmed in ultracentrifugation studies.[55] In contrast to B-DNA binding, the first alpha helix of Zα does not contact DNA.[51] The interaction involves conserved residues in α3 and the C-terminal β-sheet wing. The wing ensures specificity of Zα for the Z-DNA conformation in a number of ways. The wing contains two conserved prolines that directly contact the Z-DNA backbone and a conserved tryptophan that is essential to the Zα fold.[52] The tryptophan lies in a hydrophobic pocket between the wing and α3, contributing to the overall rigidity of the domain (Fig. 1).[51] The tryptophan has an indirect water mediated DNA contact. It also orientates a conserved tyrosine present in α3 through a perpendicular edge to face interaction. This tyrosine makes a direct DNA contact. Other DNA contacts are diagrammed in Figure 1. The relative importance of these contacts in Z-DNA binding have been evaluated by mutation.[52] Relate to wildtype the K_d for the K169A mutation reduced binding 37 fold, the N173A mutation 168 fold, the Y177A mutation 26 fold and the P192A mutation 13 fold.[55] In

Table 1. Sequence and structural analysis of the Zα and related domains

	α1	β1	α2		α3	β2		β3
1_hza	GEGK	ATT	AHDLSGKL	GTP	**KK**.EI**NR**VL**Y**SLAKKGK	LQKEA	GT**P**	**P**L**W**KI
2_rza	GEGK	ATT	AYALAREL	RTP	KK.DINRILYSLERKGK	LHRGV	GKP	PLWSL
3 bza	GDGK	ATT	ARDLARKL	QAP	KK.DINRVLYSLAEKGK	LHQEA	GSP	PLWRA
4_xza1	G.TK	TFT	AKALAWQF	KVE	KK.RINHFLYTFETKGL	LCRYP	GTP	PLWRV
5_xza2	GDTQ	TFT	AKALAWQF	KVK	KK.HINYFLYKFGTKGL	LCKNS	GTP	PLWKI
6_hzb	S--D	-SS	ALNLAKNI	GLT	KARDINAVLIDMERQGD	VYRQG	TTP	PIWHL
7_rzb	S--K	-SS	ALNLAKNI	GLA	KARDVNAVLIDLERQGD	VYREG	ATP	PIWYL
8_bzb	SS--	-SS	ALNLAKNI	GLT	KARDVNAVLIDLERQGD	VYRQG	TTP	PIWYL
9_xzb1	PP--	-ST	TLIIRKNV	GIS	KLPELNQILNTLEKQGE	ACKAS	TNP	VKWTL
10_xzb2	PP--	-ST	PFIIRKNV	GIS	KMPELTQILNTLEKQGE	ACKAS	TNP	VKWTL
11_e3l	GIEG	-AT	AAQLTRQL	NME	KR.EVNKALYDLQRSAM	VYSSD	DIP	PRWFM
12_var	GLEG	-VT	AVQLTRQL	NME	KR.EVNKALYDLQRSAM	VYSSD	DIP	PRWFM
13_mEST1	SDGG	PVK	IGQLVKKC	QVP	**KK**.TL**NQ**VLYRLKKEDR	VSSPE	**P**	**A**T**W**SI
14 mEST2	EANG	PHR	ALHIAKAL	GMT	TAKEVNPLLYSMRNKHL	LSYDG		QTWKI

Human Zα and Zβ (*hza, hzb,* and HSU10439A), rat Zα and Zβ (*rza, rzb,* and RNU18942), bovine Zα and Zβ (*bza, bzb,* and this paper), two Zα -related sequences (*xza1 and xza2*) and two Zβ related sequences (*xzb1 and xzb2*) present in *Xenopus* ADAR1a and ADAR1b2 (XLU88065 and XLU88066, respectively), the vaccinia E3L protein (*e3l*, S64006), and the variola equivalent (*var*, VVCGAA), as well as a mouse expressed sequence tag (*mEST1, mEST2,* AA204007) with relationship to Zα are shown. A second sequence in AA204007 related to Zα was found when the EST was sequenced. Residues that contact Z-DNA in crystal structures are bolded. The structural motifs corresponding to sequence blocks are given above the sequences. (Adapted from Herbert et al, 1997.)

solution, most of these Z-DNA contacting residues are prepositioned to bind Z-DNA.[54] The domain is extremely stable and resistant to thermal denaturation.[44]

The Zα Family of Z-DNA Binding Domains

A number of sequences related to Zα exist. The first ten sequences listed in Table 1 belong to a family of dsRNA editing enzymes, that contain a Zα domain and the related Zβ domain. The next two sequences are for E3L proteins expressed by vaccinia and the closely related variola virus early during infection and that are essential for pathogenicity. The last two sequences are from the DLM-1 protein, which was not named at the time of the original publication describing Zα.[56,57] The Zα domain from DLM-1 has been recently crystallized bound to Z-DNA.[58] Residues in Zα and DLM-1 that bind Z-DNA are bolded.

Z-DNA binding by Zβ and E3L domains has been investigated. The E3L proteins bind with lower affinity for Z-DNA than Zα (Kd, ~10^6 M^{-1}; Schade M, unpublished). The major difference is a faster off-time from Z-DNA than Zα. Experimentally, this is reflected in the requirement for both cytosine methylation and low magnesium concentration for E3L to induce a transition of linear poly$(CG)_n$ from the B- to Z-DNA conformation (unpublished). However, a fast off-rate may be of less consequence in vivo because of the tendency of E3L to multimerize through protein-protein contacts.[59] The binding of one E3L molecule to Z-DNA may initiate and stabilize the binding of a second molecule. Interestingly, deletion of the E3L Z-DNA binding domain of vaccinia virus renders the virus nonlethal in murine models of infection. Substitution of the human Zα domain for the E3L Z-DNA binding domain restores pathogenicity.[60] Mutations that diminish Z-DNA binding of the Zα-E3L fusion reduce pathogenicity, although it is possible that some of these mutations affect nuclear localization of the fusion protein, which is also essential for pathogenicity.[60]

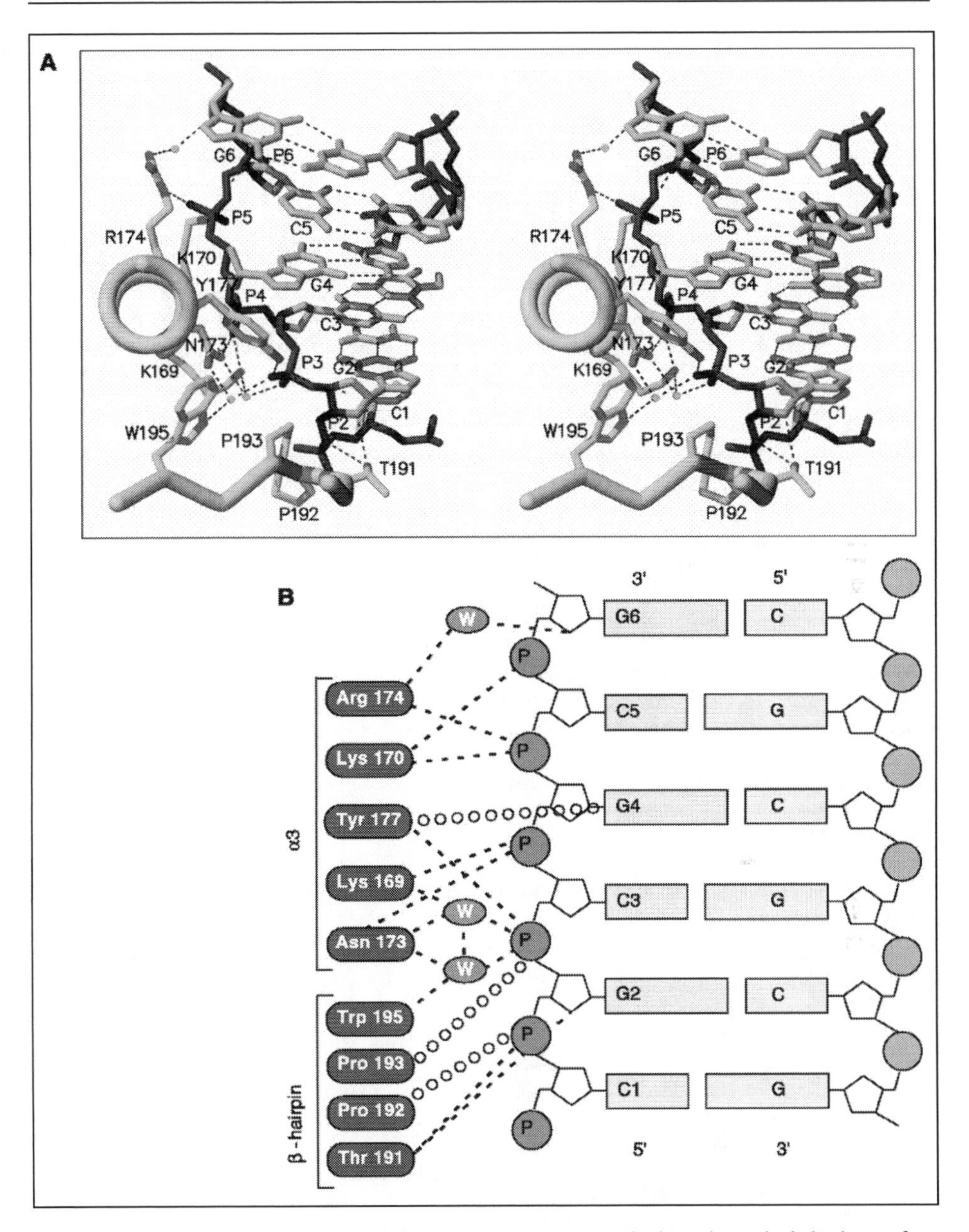

Figure 1. The interaction between Zα and Z-DNA. A) A stereoview looking down the helical axis of α3 emphasizes the residues in Zα important for binding to Z-DNA. B) A representation of the contacts between Zα domain and Z-DNA showing that these are all with one strand of the helix. Three water molecules that mediate interactions are labeled W. Hydrogen bonds are shown with dotted lines and hydrophobic contacts indicated by lines of bubbles. B) adapted from Schwartz et al, 1999.

The Zβ domain does not appear to bind Z-DNA by itself.[49,61-64] Constructs in which Zβ is fused to Zα show reduced competition by Z-DNA polymer and lower binding to alternating $(CA)_n$ sequences in supercoiled plasmids.[49,62,63] One explanation is that Zβ does not bind

Z-DNA but increases the off-rate of the fusion protein. This explanation is supported by the behavior of Zα-Zβ fusions in EMSA assays, which require a dimer with two Zα domains to bind whereas a Zα-Zα fusion binds as a monomer.[49] One key difference between Zα and Zβ is the replacement of a tyrosine in α3 by an isoleucine in Zβ (Table 1). This residue contacts Z-DNA in both the Zα and DLM-1 crystal structures.[51,58] Mutation of the isoleucine in Zβ to tyrosine results in a protein that can flip a $(CG)_6$ from the B-DNA to Z-DNA.[60] The Zβ domain of DLM-1 differs as it binds well in EMSA to a Z-DNA probe, but is not competed better by either Z- or B-form $poly(CG)_n$, suggesting that the C-terminal β-sheet may be compatible with the domain binding to both Z- and B-DNA through different modes of interaction (unpublished). In this context, it is also possible that some members of the Zα family could be optimized to bind DNA in a sequence-specific manner or to RNA. An important point is the assays designed to find Zα would not have found lower affinity Z-DNA binding proteins, or those that could bind to both B- and Z-DNA.[45]

Other Z-DNA Binding Motifs

Peptides in which every second residue is lysine will stabilize Z-DNA in vitro at micromolar concentrations.[65] This provides a simple protein motif for recognition of Z-DNA. This motif exists in a number of proteins, but it remains to be shown that such proteins interact with Z-DNA. In addition, evidence has been presented to show that topoisomerase II from *Drosophila,* humans and calf thymus recognize a number of different DNA shapes, including Z-DNA.[41,66,67] However, the domain interacting with these shapes has not yet been biochemically defined. A crystal structure of yeast topoisomerase II suggests that the two strands of DNA bound by the catalytic domain are in the B-DNA conformation.[68] The interaction of phospholipid binding domains to Z-DNA has been noted and dismissed as an artefact arising because such proteins have hydrophobic cavities lined with basic residues, conditions that favor Z-formation.[39] However, this conclusion may need re-evaluation now that the role of phospholipids in signaling within the nucleus has been firmly established. Z-DNA binding by this class of proteins could be modulated by phosphatidylinositol or its derivatives.[42,69] A binding motif with a hydrophobic cavity could also make formation of Z-RNA energetically more favorable by promoting RNA dehydration.[70]

Potential Roles for the Z-DNA Conformation Due to Protein Recognition

Z-DNA binding proteins can impact both the spatial and temporal organization of biological processes. The Zα domain and artificial Z-DNA-specific restriction nucleases constructed from it[62,71] have provided tools to demonstrate in principle how this might occur. Evidence for a number of different effects is gradually accumulating. In general, Z-DNA binding proteins may augment or modulate all those functions, listed in the section above "Potential Roles for the Z-DNA Conformation Due to Its Physical Presence", that are possible in the absence of protein-specific recognition. Z-DNA binding proteins may also target associated protein domains or macromolecular complexes to particular locations in the genome: those undergoing active transcription, those being remodeled, methylated or rearranged and those that are being silenced. These are not necessarily independent processes. The following are given as examples.

RNA editing: The Zα domain is present in the RNA editing enzyme ADAR1 along with the Zβ domain.[47] ADAR1 acts on regions of dsRNA to convert adenosine to inosine by deamination of the C6 position.[3] Inosine is recognized as guanosine by most enzymes, including the translation apparatus. One outcome is an amino acid substitution when one of the first two positions of a codon is edited.[72] ADAR1 has three dsRNA binding motifs that are not sequence specific.[73] They increase the efficiency of editing of long dsRNA substrates but are not required for it.[74] Many long dsRNA editing substrates are formed by folding introns back onto exons, requiring that editing occur before splicing.[75-77] Z-DNA, generated in regions of active transcription, could help target ADAR1 to such substrates, ensuring that they are edited before they are spliced.[13,46]

The Zα domain of ADAR1 has been shown to increase editing efficiency of short (15 bp) dsRNAs.[74] A site of Z-DNA formation immediately *cis* to short dsRNAs was not required for editing.[74] Instead, formation of Z-DNA elsewhere in the plasmid appeared sufficient to target ADAR1 to these editing substrates. The use of Z-DNA forming sites distant from the editing site has a number of advantages. It allows the use of chromatin structure to regulate editing. The only requirement is that a site of Z-DNA formation be physically close to the dsRNA target. This placement could change during development, allowing editing to be turned on or off. Indeed, editing of a particular transcript could be regulated in different tissues by localizing different Z-DNA-forming elements to the site of dsRNA transcription. Further, the efficiency of editing in this situation would not be determined by the rate at which the substrate is transcribed but rather by factors that regulate the energetics of nearby Z-DNA formation. Evolutionarily, this is a very flexible arrangement as it does not require engineering of specific sequences in *cis* to the editing site, but rather takes advantage of alternate chromatin structures stabilized by proteins with different combinations of generic binding domains.

In this example, Z-DNA acts as a flag, targeting ADAR1 to a dsRNA substrate. What is remarkable about this mechanism is the use of two shape-specific protein motifs, one for Z-DNA and the other for dsRNA, to make very specific sequence changes to an RNA transcript. Processes that depend on recognition of both Z-DNA and dsRNA shapes by ADAR1 may be impacted by other members of the Zα family, such as E3L, which also has a dsRNA binding domain in addition to a Z-DNA binding domain. By competing for interaction sites with ADAR1 and other related proteins, E3L could deregulate a number of RNA-directed processes, especially those that are part of the anti-viral response.[78] For example, during vaccinia infection, viral DNA replication occurs in the cytoplasm; in response, ADAR1 accumulates in the cytoplasm. The production of E3L may constitute a counter response by the virus.[78]

Allele-specific gene silencing: Both Z-DNA and dsRNA could play a mechanistic role in the assembly of silencing complexes. An example of this may be the dsRNA-directed methylation associated with suppression of gene expression in plants.[79] The presence of Z-DNA would help target the machinery to the appropriate site. Z-DNA would signal that the region is still transcriptionally active while the requirement for dsRNA would allow RNA generated in *trans* to target a particular transcriptional unit for inactivation. A similar mechanism would also be effective at suppressing expression of short interspersed elements (SINEs) and other retrotransposons in mammals. An interesting system in which the dependence of this process on Z-DNA formation could be investigated is the allele-specific expression of the *RT6* gene in rat, which has *RT6a* and *RT6b* alleles.[80] Both a rodent ID retroelement and the length of a Z-DNA repeat present in a promoter appear to correlate with selective methylation of the *RT6a* allele in peripheral T-cells.[80] Silencing associated with rodent B2 SINEs, which contain 15% of $(CA)_n$ repeats in the mouse genome, may also influence gene expression.[81]

Antisense gene regulation: dsRNA can be formed by overlapping genes. A recent survey of the mouse genome identified 2,431 cases where the sense-antisense overlap of mRNA transcribed from contiguous was greater than 20 bps in length.[82] The open chromatin regions associated with nascent dsRNA may also form Z-DNA, targeting enzymes like ADAR1 that modify transcripts or others that induce silencing.

Transcription: Protein-induced stabilization of Z-DNA has been shown to increase transcription from a *lac-Z* reporter construct in the yeast *Saccharomyces cerevisiae* when Z-DNA forming elements were placed in the promoter.[64] The effect was greatest when a second gene transcribing in the opposite orientation to the *lac-Z* gene was placed immediately 5′ of the Z-DNA forming segment. In this situation, the Z-DNA forming segment was embedded in an open region of chromatin within the promoter region of both genes. Interestingly, Z-DNA formation was also observed, though to a lesser extent, when the second gene were placed in the same orientation as the *lac-Z* gene i.e., the Z-DNA forming segment was 3′ to one gene and 5′ to the reporter gene. In this case, both genes were transcribed in the same direction from the same strand. With this arrangement, less Z-DNA may have been formed because the chromatin structure surrounding the Z-DNA forming insert may have been less open than when the orientation of both genes was opposite. Alternatively, the negative supercoiling generated by transcription of the *lac-Z* gene may have been partially canceled by read-through of polymerases from the upstream gene. The existence of any negative supercoiling at all implied that forward diffusion of positive supercoiling generated by upstream polymerases was quite limited, even though the two transcription units were very close together. Whether Z-DNA formation prevented the upstream polymerase from reading through the downstream gene was not evaluated in these experiments. If that were so, the primary transcript from the upstream gene may have been truncated due to stalling of polymerases in the region of Z-DNA formation.

Chromosomal remodeling: Active genes have open promoter regions in which torsional restraints imposed by histones are relaxed. The *CSF1* promoter has been shown to form Z-DNA during remodeling by the SWI-SNF-like BAF complex. Evidence for Z-DNA formation during activation of the *CSF1* gene was obtained by showing that Z-DNA forming sequences were sufficient for transcription, regardless of nucleotide composition, and that these sequences were specifically cleaved by the Zα nuclease in a transcription-dependent manner.[18] The results obtained mirror previous reports where activation of c-*myc* and a corticotrophin gene was accompanied by enhanced binding of a Z-DNA specific antibody.[83,84] Protein recognition of Z-DNA in these situations could help target remodeling and transcriptional complexes. Similar roles for Z-DNA forming sequences in modulating transcription have been proposed.[85]

DNA methylation: Methylation of CpG would favor the Z-DNA conformation and allow Z-DNA binding proteins with lower affinity than Zα to stabilize this shape. Binding of such proteins could help phase placement of histones so that these regions are accessible first to remodeling complexes during reactivation of a locus and then to the transcriptional machinery.

Nuclear architecture: Z-DNA binding domains may perform a structural role in nuclear architecture. They could tether active DNA to particular sites within the nucleus: to areas where transcription factories are already assembled; to sites where recombination enzymes are active or topoisomerases concentrated.[86] Attachment of DNA to the nuclear matrix at such locations appears to be transient and may involve type II intermediate filaments such as vimentin and glial fibrillary acidic protein.[42,87] It has been suggested that transcripts heavily modified by the Z-DNA binding enzyme ADAR1 also participate in organization of the nuclear matrix.[88]

DNA rearrangement: Another interesting example might be found in *Tetrahymena*. The ciliated protozoan has a micronucleus used for genetic reproduction and a macronucleus where transcription occurs. The transcriptionally-active macronucleus is stained exclusively with anti-Z-DNA antibodies, but not the quiescent micronucleus.[89] Interestingly, the macronucleus has extensive deletions of DNAs resolving from genomic rearrangements. Short RNAs have been isolated that appear to direct these deletions. The coordination of these rearrangement events could involve recognition of both Z-DNA and dsRNA.[90]

Z- triggers: Z-DNA binding proteins may orchestrate complex biological processes in both time and space, acting as molecular triggers. The possibility of using Z-DNA as a conformational switch has been demonstrated. For example, Z-DNA formation at one site in a closed circular plasmid diminishes Z-DNA formation at another distant site.[91] This process has been shown to inhibit transcription from a conformationally sensitive promoter by reducing negative superhelicity in the region of the promoter.[92] In a similar manner, Z-DNA binding proteins could act at a distance to switch a gene on and off.

Z-DNA binding proteins could also influence gene expression in a different fashion by altering the geometry of a chromosomal domain. By bringing spatially separated elements together or moving them apart, Z-DNA proteins may affect promoter/enhancer interactions as well as those necessary for site-specific recombination. Both of these processes require that DNA elements be correctly orientated in space and potentially could be regulated by Z-DNA binding proteins acting at a distance.[17]

A different type of Z-DNA switch has been engineered into a nano device.[93] In this system, conversion of a $(CG)_{10}$ connector from B- to Z-DNA conformation changed the relative position of two reporter fluorescent dyes by 50 to 60 angstroms. Similarly, Z-DNA binding proteins could instigate conformational rearrangements in the macromolecular machines that are involved in transcription, replication or recombination. The allosteric realignment of components in space would trigger subsequent steps in the reaction. Recognition of Z-DNA by proteins would not only amplify the signal generated by Z-DNA formation, but allow the sequence of events to be timed precisely. Related domains that bind to Z-RNA could play a similar role in RNA splicing, editing and translation.[70]

Conclusion

The Z-DNA conformation has many potential roles in transcription. We now have molecular tools to investigate these possibilities. Editing by ADAR1 shows how genetic information stored in Z-DNA and dsRNA shapes can alter the readout of information stored in sequence. In this process, generic protein domains bind these two very different nucleic acid conformations and guide the sequence-specific modification of transcripts.

References

1. Herbert A, Rich A. RNA processing and the evolution of eukaryotes. Nat Genet 1999; 21:265-269.
2. Maniatis T, Reed R. An extensive network of coupling among gene expression machines. Nature 2002; 416:499-506.
3. Bass BL. RNA editing by adenosine deaminases that act on RNA. Annu Rev Biochem 2002; 71:817-846.
4. Black DL. Mechanisms of Alternative Pre-Messenger RNA Splicing. Annu Rev Biochem 2003; 72:291-336.
5. Fischle W, Wang Y, Allis CD. Histone and chromatin cross-talk. Curr Opin Cell Biol 2003; 15:172-183.
6. Nikaido I, Saito C, Mizuno Y et al. Discovery of imprinted transcripts in the mouse transcriptome using large-scale expression profiling. Genome Res 2003; 13:1402-1409.
7. Herbert A. The 4Rs of RNA-directed evolution. Nat Genet 2004; 36:1-7.
8. Ptashne M, Gann A. Transcription initiation: imposing specificity by localization. Essays Biochem 2001; 37:1-15.
9. Kim JL, Nikolov BD, Burley SK. Co-crystal structure of TBP recognizing the minor groove of a TATA element. Nature 1993; 365:520-527.
10. Kim Y, Geiger JH, Hahn S et al. Crystal structure of a yeast TBP/TATA complex. Nature 1993; 365:512-520.
11. Ohyama T. Intrinsic DNA bends: an organizer of local chromatin structure for transcription. Bioessays 2001; 23:708-715.
12. Ho PS, Eichman BF. The crystal structures of DNA Holliday junctions. Curr Opin Struct Biol 2001; 11:302-308.
13. Herbert A, Rich A. The biology of left-handed Z-DNA. J Biol Chem 1996; 271:11595-11598.
14. Schroth GP, Chou PJ, Ho PS. Mapping Z-DNA in the human genome. Computer-aided mapping reveals a nonrandom distribution of Z-DNA-forming sequences in human genes. J Biol Chem 1992; 267:11846-11855.
15. Human Genome Sequencing Consortium. Initial sequencing and analysis of the human genome. Nature 2001; 409:860-921.
16. Liu LF, Wang JC. Supercoiling of the DNA template during transcription. Proc Natl Acad Sci USA 1987; 84:7024-7027.
17. Droge P. Protein tracking-induced supercoiling of DNA: a tool to regulate DNA transactions in vivo? Bioessays 1994; 16:91-99.
18. Liu R, Liu H, Chen X et al. Regulation of CSF1 promoter by the SWI/SNF-like BAF complex. Cell 2001; 106:309-318.
19. Garner MM, Felsenfeld G. Effect of Z-DNA on nucleosome placement. J Mol Biol 1987; 196:581-590.
20. Peck LJ, Wang JC. Transcriptional block caused by a negative supercoiling induced structural change in an alternating CG sequence. Cell 1985; 40:129-137.
21. Vardimon L, Rich A. In Z-DNA the sequence G-C-G-C is neither methylated by *Hha* I methyltransferase nor cleaved by *Hha* I restriction endonuclease. Proc Natl Acad Sci USA 1984; 81:3268-3272.
22. Zacharias W, Larson JE, Kilpatrick MW et al. *Hha*I methylase and restriction endonuclease as probes for B to Z DNA conformational changes in d(GCGC) sequences. Nucleic Acids Res 1984; 12:7677-7692.

23. Sheardy RD, Levine N, Marotta S et al. A thermodynamic investigation of the melting of B-Z junction forming DNA oligomers. Biochemistry 1994; 33:1385-1391.
24. Pohl FM. Ein Modell der DNS-struktur. Naturwissenschaften 1967; 54:616.
25. Treco D, Arnheim N. The evolutionary conserved repetitive sequence d(TG•AC)$_n$ promotes reciprocal exchange and generates unusual recombinant tetrads during yeast meiosis. Mol Cell Biol 1986; 6:3934-3947.
26. Bullock P, Miller J, Botchan M. Effects of poly[d(pGpT)•d(pApC)] and poly[d(pCpG)•d(pCpG)] repeats on homologous recombination in somatic cells. Mol Cell Biol 1986; 6:3948-3953.
27. Wahls WP, Wallace LJ, Moore PD. The Z-DNA motif d(TG)$_{30}$ promotes reception of information during gene conversion while stimulating homologous recombination in human cells in culture. Mol Cell Biol 1990; 10:785-793.
28. Aplan PD, Raimondi SC, Kirsch IR. Disruption of the SCL gene by a t(1; 3) translocation in a patient with T cell acute lymphoblastic leukemia. J Exp Med. 1992; 176:1303-1310.
29. Boehm T, Mengle-Gaw L, Kees UR et al. Alternating purine-pyrimidine tracts may promote chromosomal translocations seen in a variety of human lymphoid tumours. EMBO J 1989; 8:2621-2631.
30. Satyanarayana K, Strominger JL. DNA sequences near a meiotic recombinational breakpoint within the human HLA-DQ region. Immunogenetics 1992; 35:235-240.
31. Steinmetz M, Stephan D, Lindahl KF. Gene organization and recombinational hotspots in the murine major histocompatibility complex. Cell 1986; 44:895-904.
32. Weinreb A, Katzenberg DR, Gilmore GL et al. Site of unequal sister chromatid exchange contains a potential Z-DNA forming tract. Proc Natl Acad Sci USA 1991; 85:529-533.
33. Thamann TJ, Lord RC, Wang AH-J et al. The high salt form of poly(dG-dC)•poly(dG-dC) is left-handed Z-DNA: Raman spectra of crystals and solutions. Nucleic Acids Res 1981; 9:5443-5457.
34. Kawai K, Sugiyama H, Saito I. Photoreaction of 5-halouracil-containing Z-form DNA. Nucleic Acids Symp Ser 1997; 37:93-94.
35. Oyoshi T, Kawai K, Sugiyama H. Efficient C2′alpha-hydroxylation of deoxyribose in protein-induced Z- form DNA. J Am Chem Soc 2003; 125:1526-1531.
36. Nordheim A, Tesser P, Azorin F et al. Isolation of *Drosophila* proteins that bind selectively to left-handed Z-DNA. Proc Natl Acad Sci USA 1982; 79:7729-7733.
37. Lafer EM, Sousa RJ, Rich A. Z-DNA-binding proteins in *Escherichia coli* purification, generation of monoclonal antibodies and gene isolation. J Mol Biol 1988; 203:511-516.
38. Azorin F, Rich A. Isolation of Z-DNA binding proteins from SV40 minichromosomes: evidence for binding to the viral control region. Cell 1985; 41:365-374.
39. Krishna P, Kennedy BP, Waisman DM et al. Are many Z-DNA binding proteins actually phospholipid-binding proteins? Proc Natl Acad Sci USA 1990; 87:1292-1295.
40. Rohner KJ, Hobi R, Kuenzle CC. Z-DNA-binding proteins. Identification critically depends on the proper choice of ligands. J Biol Chem 1990; 265:19112-19115.
41. Arndt-Jovin DJ, Udvardy A, Garner MM et al. Z-DNA binding and inhibition by GTP of *Drosophila* topoisomerase II. Biochemistry 1993; 32:4862-4872.
42. Li G, Tolstonog GV, Traub P. Interaction in vitro of type III intermediate filament proteins with Z-DNA and B-Z-DNA junctions. DNA Cell Biol 2003; 22:141-169.
43. Gagna CE, Chen JH, Kuo HR et al. Binding properties of bovine ocular lens zeta-crystallin to right-handed B-DNA, left-handed Z-DNA, and single-stranded DNA. Cell Biol Int 1998; 22:217-225.
44. Herbert AG, Spitzner JR, Lowenhaupt K et al. Z-DNA binding protein from chicken blood nuclei. Proc Natl Acad Sci USA 1993; 90:3339-3342.
45. Herbert AG, Rich A. A method to identify and characterize Z-DNA binding proteins using a linear oligodeoxynucleotide. Nucleic Acids Res 1993; 21:2669-2672.
46. Herbert AG, Lowenhaupt K, Spitzner JR et al. Chicken double-stranded RNA adenosine deaminase has apparent specificity for Z-DNA. Proc Natl Acad Sci USA 1995; 92:7550-7554.
47. Herbert A, Alfken J, Kim Y-G et al. A Z-DNA binding domain present in the human editing enzyme, double-stranded RNA adenosine deaminase. Proc Natl Acad Sci USA 1997; 94:8421-8426.
48. Berger I, Winston W, Manoharan R et al. Spectroscopic characterization of a DNA-binding domain, Zα, from the editing enzyme, dsRNA adenosine deaminase: evidence for left-handed Z-DNA in the Zα-DNA complex. Biochemistry 1998; 37:13313-13321.
49. Herbert A, Schade M, Lowenhaupt K et al. The Zα domain from human ADAR1 binds to the Z-DNA conformer of many different sequences. Nucleic Acids Res 1998; 26:3486-3493.
50. Brown BA, 2nd, Lowenhaupt K, Wilbert CM et al. The Zα domain of the editing enzyme dsRNA adenosine deaminase binds left-handed Z-RNA as well as Z-DNA. Proc Natl Acad Sci USA 2000; 97:13532-13536.

51. Schwartz T, Rould MA, Lowenhaupt K et al. Specific Recognition of left-handed Z-DNA: co-crystal structure of the Zα DNA recognition motif of ADAR1 at 2.1Å resolution. Science 1999; 284:1841-1845.
52. Schade M, Turner C, Lowenhaupt K et al. Structure/function analysis of the Z-DNA binding domain Zα of ADAR1 reveals similarity to (α + β) family of helix-turn-helix proteins. EMBO J 1998; 18:470-479.
53. Schwartz T, Shafer K, Lowenhaupt K et al. Crystallizaton and preliminary studies of the DNA binding domain, Za, from ADAR1 complexed to left-handed DNA. Acta Crystallographica 1999; D55:1362-1364.
54. Schade M, Turner CJ, Kuhne R et al. The solution structure of the Zα domain of the human RNA editing enzyme ADAR1 reveals a prepositioned binding surface for Z-DNA. Proc Natl Acad Sci USA 1999; 96:12465-12470.
55. Schade M, Behlke J, Lowenhaupt K et al. A 6 bp Z-DNA hairpin binds two Zα domains from the RNA editing enzyme ADAR1. FEBS Lett 1999; 458:27-31.
56. Fu Y, Comella N, Tognazzi K et al. Cloning of DLM-1, a novel gene that is up-regulated in activated macrophages, using RNA differential display. Gene 1999; 240:157-163.
57. Rothenburg S, Schwartz T, Koch-Nolte F et al. Complex regulation of the human gene for the Z-DNA binding protein DLM- 1. Nucleic Acids Res 2002; 30:993-1000.
58. Schwartz T, Behlke J, Lowenhaupt K et al. Structure of the DLM-1-Z-DNA complex reveals a conserved family of Z-DNA-binding proteins. Nat Struct Biol 2001; 8:761-765.
59. Ho CK, Shuman S. Physical and functional characterization of the double-stranded RNA binding protein encoded by the vaccinia virus E3 gene. Virology 1996; 217:272-284.
60. Kim Y-G, Muralinath M, Brandt T et al. A role for Z-DNA binding in vaccinia virus pathogenesis. Proc Natl Acad Sci USA 2003; 100:6974-6979.
61. Schwartz T, Lowenhaupt K, Kim Y-G et al. Proteolytic dissection of Zab, the Z-DNA binding domain of human ADAR1. J Biol Chem 1999; 274:2899-2906.
62. Kim YG, Lowenhaupt K, Schwartz T et al. The interaction between Z-DNA and the Zab domain of double-stranded RNA adenosine deaminase characterized using fusion nucleases. J Biol Chem 1999; 274:19081-19086.
63. Kim YG, Lowenhaupt K, Maas S et al. The Zab domain of the human RNA editing enzyme ADAR1 recognizes Z-DNA when surrounded by B-DNA. J Biol Chem 2000; 275:26828-26833.
64. Oh DB, Kim YG, Rich A. Z-DNA-binding proteins can act as potent effectors of gene expression in vivo. Proc Natl Acad Sci USA 2002; 99:16666-16671.
65. Takeuchi H, Hanamura N, Harada I. Structural specificity of peptides in Z-DNA formation and energetics of the peptide-induced B-Z transition of poly(dG-m^{5}C). J Mol Biol 1994; 236:610-617.
66. Bechert T, Diekmann S, Arndt-Jovin DJ. Human 170 kDa and 180 kDa topoisomerases II bind preferentially to curved and left-handed linear DNA. J Biomol Struct Dyn 1994; 12:605-623.
67. Glikin CG, Jovin MT, Arndt-Jovin DJ. Interactions of *Drosophila* DNA topoisomerase II with left-handed Z-DNA in supercoiled minicircles. Nucleic Acids Res 1991; 19:7139-7144.
68. Berger JM, Gamblin SJ, Harrison SC et al. Structure and mechanism of DNA topoisomerase II. Nature 1996; 379:225-232.
69. Irvine RF. Nuclear lipid signalling. Nat Rev Mol Cell Biol 2003; 4:349-360.
70. Hall K, Cruz P, Tinoco I, Jr. et al. 'Z-RNA'-a left-handed RNA double helix. Nature 1984; 311:584-586.
71. Kim YG, Kim PS, Herbert A et al. Construction of a Z-DNA-specific restriction endonuclease. Proc Natl Acad Sci USA 1997; 94:12875-12879.
72. Sommer B, Kohler M, Sprengel R et al. RNA editing in brain controls a determinant of ion flow in glutamate-gated channels. Cell 1991; 67:11-19.
73. Kim U, Wang Y, Sanford T et al. Molecular cloning of a cDNA for double-stranded RNA adenosine deaminase, a candidate enzyme for nuclear RNA editing. Proc Natl Acad Sci USA 1994; 91:11457-11461.
74. Herbert A, Rich A. The role of binding domains for dsRNA and Z-DNA in the in vivo editing of minimal substrates by ADAR1. Proc Natl Acad Sci USA 2001; 98:12132-12137.
75. Higuchi M, Single FN, Kohler M et al. RNA editing of AMPA receptor subunit GluR-B: a base-paired intron-exon structure determines position and efficiency. Cell 1993; 75:1361-1370.
76. Kohler M, Burnashev N, Sakmann B et al. Determinants of Ca^{2+} permeability in both TM1 and TM2 of high affinity kainate receptor channels: diversity by RNA editing. Neuron 1993; 10:491-500.
77. Lomeli H, Mosbacher J, Melcher T et al. Control of kinetic properties of AMPA receptor channels by nuclear RNA editing. Science 1994; 266:1709-1713.
78. Brandt TA, Jacobs BL. Both carboxy- and amino-terminal domains of the vaccinia virus interferon resistance gene, E3L, are required for pathogenesis in a mouse model. J Virol 2001; 75:850-856.

79. Pickford A, Cogoni C. RNA-mediated gene silencing. Cell Mol Life Sci 2003; 60:871-882.
80. Rothenburg S, Koch-Nolte F, Thiele HG et al. DNA methylation contributes to tissue- and allele-specific expression of the T-cell differentiation marker RT6. Immunogenetics 2001; 52:231-241.
81. Mouse Genome Sequencing Consortium. Initial sequencing and comparative analysis of the mouse genome. Nature 2002; 420:520-562.
82. Okazaki Y, Furuno M, Kasukawa T et al. Analysis of the mouse transcriptome based on functional annotation of 60,770 full-length cDNAs. Nature 2002; 420:563-573.
83. Wolfl S, Wittig B, Rich A. Identification of transcriptionally induced Z-DNA segments in the human c-*myc* gene. Biochim Biophys Acta 1995; 1264:294-302.
84. Wolfl S, Martinez C, Rich A et al. Transcription of the human corticotropin-releasing hormone gene in NPLC cells is correlated with Z-DNA formation. Proc Natl Acad Sci USA 1996; 93:3664-3668.
85. Sharma VK, Rao CB, Sharma A et al. (TG:CA)(n) repeats in human housekeeping genes. J Biomol Struct Dyn 2003; 21:303-310.
86. Cook PR. The organization of replication and transcription. Science 1999; 284:1790-1795.
87. Cook PR. Predicting three-dimensional genome structure from transcriptional activity. Nat Genet 2002; 32:347-352.
88. Zhang Z, Carmichael GG. The fate of dsRNA in the nucleus: a $p54^{nrb}$-containing complex mediates the nuclear retention of promiscuously A-to-I edited RNAs. Cell 2001; 106:465-475.
89. Lipps HJ, Nordheim A, Lafer EM et al. Antibodies against Z-DNA react with the macronucleus but not the micronucleus of the hypotrichous ciliate *Stylonychia mytilus*. Cell 1983; 32:435-441.
90. Mochizuki K, Fine NA, Fujisawa T et al. Analysis of a *piwi*-related gene implicates small RNAs in genome rearrangement in *Tetrahymena*. Cell 2002; 110:689-699.
91. Ellison MJ, Fenton MJ, Shing PS et al. Long-range interactions of multiple DNA structural transitions within a common topological domain. EMBO J 1987; 6:1513-1522.
92. Sheridan SD, Opel ML, Hatfield GW. Activation and repression of transcription initiation by a distant DNA structural transition. Mol Microbiol 2001; 40:684-690.
93. Mao C, Sun W, Shen Z et al. A nanomechanical device based on the B-Z transition of DNA. Nature 1999; 397:144-146.

CHAPTER 8

Do DNA Triple Helices or Quadruplexes Have a Role in Transcription?

Michael W. Van Dyke

Abstract

Certain DNA sequences preferentially adopt multistranded, non-B-form structures under physiological conditions. These include three-stranded DNA triplexes and four-stranded DNA quadruplexes. Several lines of evidence suggest that multiplex structures can form in vivo, either from the addition of oligonucleotides or through the transient formation of single-stranded regions. The consequences of multiplex structures on many DNA-dependent biological processes have been described. In this chapter I will review the effects of different DNA multiplexes on the process of transcription. The influence of parameters such as multiplex type and multiplex formation conditions on different transcription mechanistic steps in organisms spanning from prokaryotes to *Xenopus* oocytes and mammalian cells will be discussed.

Introduction

Oligopurine/oligopyrimidine-rich DNA sequences have long been known to preferentially adopt multistranded structures quite different from the familiar Watson-Crick base-paired, right-handed, antiparallel-stranded, B-form double-helical structure.[1,2] Examples include triple helical DNA (triplexes) and G-quartet-containing quadruplexes (G_4), both of which can form under physiological conditions, and once formed, are extremely stable.[3-6]

Although a considerable amount of information is available about the properties of DNA multiplex structures in vitro, little is known about their existence and biological roles in vivo.[7-10] Sequences capable of forming these structures abound in all eukaryotic organisms.[11] Examples include the G-rich 3' overhangs on the ends of chromosomes and long oligopurine tracts within the promoter regions of several genes. DNA multiplexes have been invoked as necessary intermediates in many biological processes, including chromosome condensation, recombination, replication, telomere function, and transcriptional control.[12-18] Potentially deleterious multiplex structures could also form as a consequence of essential biological processes that use the DNA as a template (e.g., replication and transcription) with removal of these structures then being necessary for viability.[19-20] In this chapter, I review the literature to address these questions: (1) can DNA multiplexes such as triplexes and quadruplexes affect the process of transcription, and (2) do DNA multiplexes play a role in transcription regulation in vivo?

DNA Triplexes

As has been well known, certain nucleic acid sequences preferentially adopt a triple-helical structure under the proper conditions.[3,4] Triplex structures are characterized by a single

DNA Conformation and Transcription, edited by Takashi Ohyama. ©2005 Eurekah.com and Springer Science+Business Media.

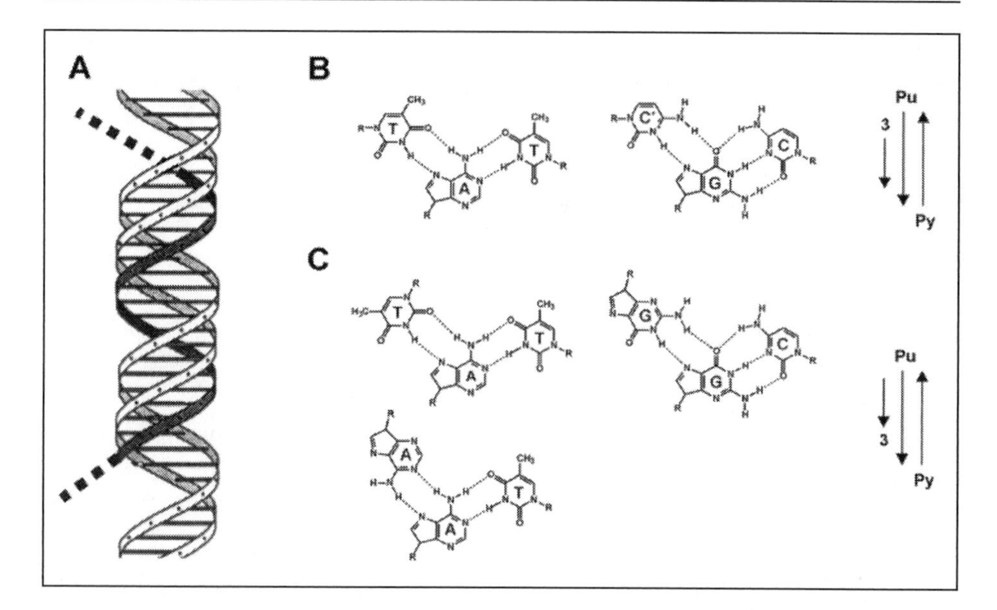

Figure 1. Triplex nucleic acids. A) Schematic representation of an intermolecular triplex. The third strand (black) is seen residing in the major groove of duplex DNA. Note that this third strand may be part of a larger molecule (dotted line extensions). B) Base triplets and strand orientations in the pyrimidine motif. C) Base triplets and strand orientations in the purine motif.

polynucleotide strand residing in the former major groove of a homopurine-homopyrimidine duplex (Fig. 1A), which are reviewed in Chapter 1 of this book. Two triplex motifs are known. The parallel- or pyrimidine-motif (Py) has a C- or T-rich third strand bound in a parallel orientation with respect to the duplex homopurine strand, while the antiparallel- or purine-motif (Pu) has the opposite orientation and a primarily A- or G-rich third strand. Both types of triplexes utilize Hoogsteen hydrogen bonding between their third strands and purines in their duplex acceptors. The primary base triplets of Py triplexes are T•A•T and C•G•C^+, while the base triplets of Pu triplexes are T•A•A, T•A•T, and C•G•G (Fig. 1B, C). Py triplexes can occur with RNA being present as any of the three strands, while Pu triplexes only occur with DNA.[21,22] Both inter- and intramolecular triplexes have been observed. The former involves a third DNA strand that originates from either a second DNA molecule or from a distal site on the same molecule, while the latter involves homopurine-homopyrimidine sequences immediately adjacent to the duplex acceptor (Fig. 2A). Four isomers of intramolecular triplexes can exist dependent on the half-element strand that serves as the third strand (Fig. 2B). Intramolecular triplexes are also known as H-DNA or H'-DNA, depending on whether they contain Py or Pu triplexes, respectively.

In theory, a homopurine-homopyrimidine duplex should be capable of forming triplexes of either motif. However, under physiological conditions, cytosine protonation is not favored, and C•G•G is the most stable base triplet in the purine motif. T-rich nucleic acids would be expected, therefore, to form Py triplexes, while G-rich DNAs would form Pu triplexes. The same is true for intramolecular triplexes, with the additional condition that the different isomers are not isoenergetic.[23,24] In both intermolecular and intramolecular triplexes, contiguous homopurine-homopyrimidine runs of at least 10 base pairs are required for the duplex acceptor, since shorter triplexes are not very stable under physiological conditions, and even single base interruptions are known to greatly destabilize triplexes.[25-27] Triplex formation is kinetically slow compared to duplex annealing.[25,28] However, once formed, triplex RNA and DNA are very stable, exhibiting half-lives on the order of days.[25,29]

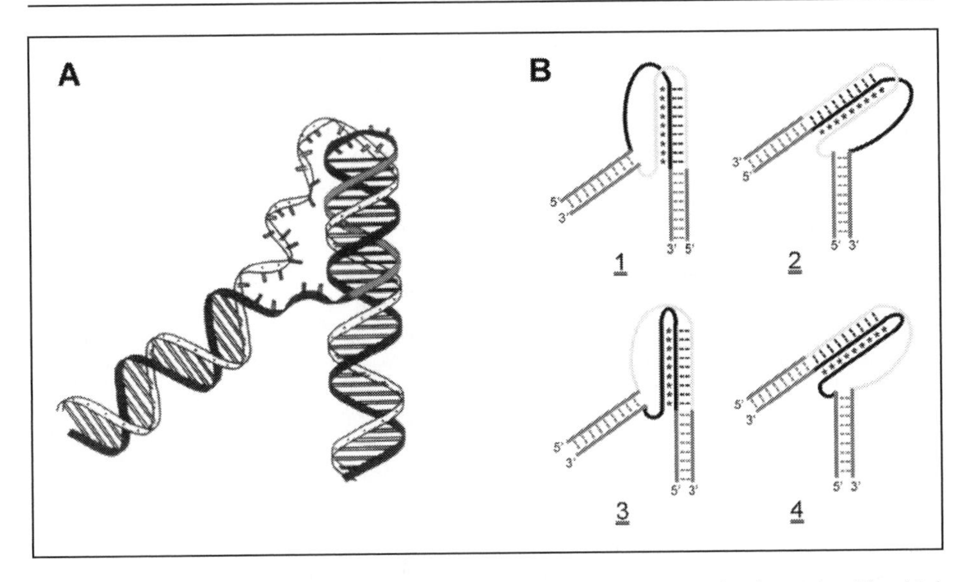

Figure 2. Intramolecular triplexes. A) Schematic representation of an intramolecular triplex. The third strand (gray) is shown residing in the major groove of duplex DNA. B) Schematic representations of H-DNA isomers H-y3 (1) and H-y5 (2), and H'-DNA isomers H-r3 (3) and H-r5 (4). In H(H') DNA nomenclature, the nature of the third strand is identified as either pyrimidine (y) or purine (r) and originating from either the 3' or 5' end of the corresponding oligopyrimidine or oligopurine strand. The purine-rich strand is shown in black, while the pyrimidine-rich strand is shown in light gray.

DNA Quadruplexes

DNAs (and RNAs) containing guanine tracts will associate in vitro to form four-stranded, right-handed helices known as quadruplexes or tetraplexes.[5,6] These G_4 nucleic acids are characterized by stacked G-quartet structures, square planar arrays of four guanines, each serving as the donor and acceptor of two Hoogsteen hydrogen bonds. Electronegative carbonyl oxygens line the center of the G-ring, where they interact with a suitably sized monovalent cation, typically Na^+ or K^+ (Fig. 3A). Several isoforms of DNA and RNA quadruplexes have been described by NMR and X-ray crystallographic studies.[30,31] The isoforms are characterized by either parallel or *cis* or *trans* antiparallel strand orientations and may be composed of either intermolecular or intramolecular or both types of hydrogen bonding (Fig. 3B). G-rich nucleic acids can be highly polymorphic, adoption of the exact G_4 structure depending on several factors including nucleotide sequence, strand concentration, and the types and concentrations of monovalent, divalent, and polyvalent cations present. Formation of G_4 nucleic acids requires one or more polynucleotide strands, each containing one or more runs of two or more contiguous guanosine nucleotides. Four parallel-stranded intermolecular G4 nucleic acids (Fig. 3B, structure 1) require only a single G-tract. However, their strand stoichiometry and very slow formation kinetics lessen the likelihood that this form of G_4 nucleic acid often occurs in vivo. More likely in vivo are G_4 multiplex species formed from polynucleotides containing multiple G-runs, which have the ability to form intramolecular Hoogsteen hydrogen bonds. These species include purely intramolecular G4' nucleic acids that require only a single DNA or RNA molecule (Fig. 3B, 2), and G'2 hairpin dimer species that can link two separate polynucleotides (Fig. 3B, 3, 4). Formation of intermolecular species may be rather slow under physiological conditions, though intermediates containing intramolecular G•G Hoogsteen base pairs (e.g., G') form quite rapidly.[32] Once formed, each of the G_4 species is quite stable, with measured enthalpies approaching -25 kcal/mole of G-quartet.[33] Thus, equilibration between different G_4 species is glacially slow under physiological conditions, and the thermodynamically favored structure is not necessarily the species that occurs in vivo.

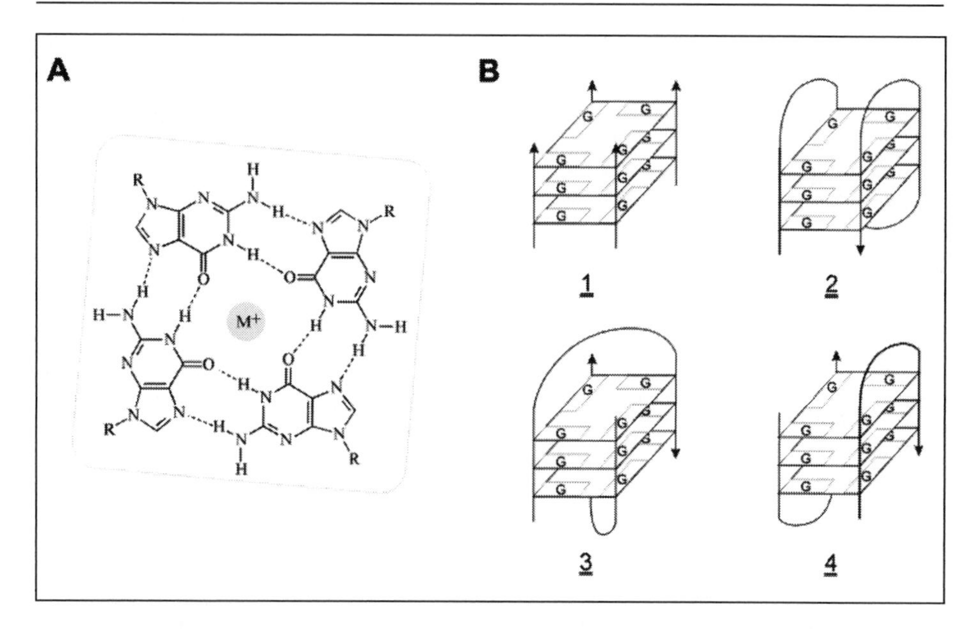

Figure 3. G_4 nucleic acids. A) Chemical structure of a Hoogsteen hydrogen-bonded G-quartet. M^+ represents a monovalent cation, typically Na^+ or K^+. B) Schematic representations of a parallel-strand intermolecular tetraplex (G4, 1), a monomeric intramolecular quadruplex (G4', 2), and two forms of dimeric hairpin quadruplexes (G'2, 3,4). Both RNA and DNA and mixtures of these nucleic acids can form G_4 structures. A G' structure contains only Hoogsteen hydrogen-bonded G•G base pairs and would correspond to a single hairpin (indicated in bold) in representation 4.

Multiplexes and Transcription

A considerable body of evidence indicates that multiplex nucleic acids may affect transcription. Briefly, transcription requires a start site (+1), usually indicated by an arrow in most schematic representation, to define where transcription begins and in which direction it proceeds. In addition, transcription requires an RNA polymerase, which is the enzyme that catalyzes the template-directed sequential condensation of ribonucleotides to generate a product RNA. In many cases, the RNA polymerase itself does not directly recognize the +1 site but relies on auxiliary proteins for this purpose. In addition, these proteins and/or RNA polymerase do not typically interact directly with the +1 site but rather recognize nearby sequences known as the promoter. Once these proteins and RNA polymerase have assembled on a promoter, addition of ribonucleotides will allow transcription to begin. The process of transcription initiation is shown schematically in Figure 4A. Afterward, the transcribing RNA polymerase can proceed downstream of the +1 site and generate an RNA transcript (wavy line) in a process known as elongation (Fig. 4B). Note that transcription is a multistep process: promoter recognition, initiation, elongation, and that the overall rates of transcription depend on the efficiencies of these different steps. These steps are often affected by a class of nucleic acid binding proteins (specific transcription factors), which can greatly modulate transcription.

Multiplex structures are believed to interfere with transcription primarily through two different mechanisms: promoter occlusion and elongation arrest.[34] In promoter occlusion (Fig. 4C), a DNA multiplex interferes with the binding of a transcription factor to a gene promoter. Note that, for occlusion to occur, the sites of transcription factor binding and multiplex formation need to overlap, and the extent of overlap necessary depending on the transcription factor and multiplex structure used. As shown in this example (Fig. 4C), the typical occluded protein is a specific transcription factor that normally stimulates transcription initiation or

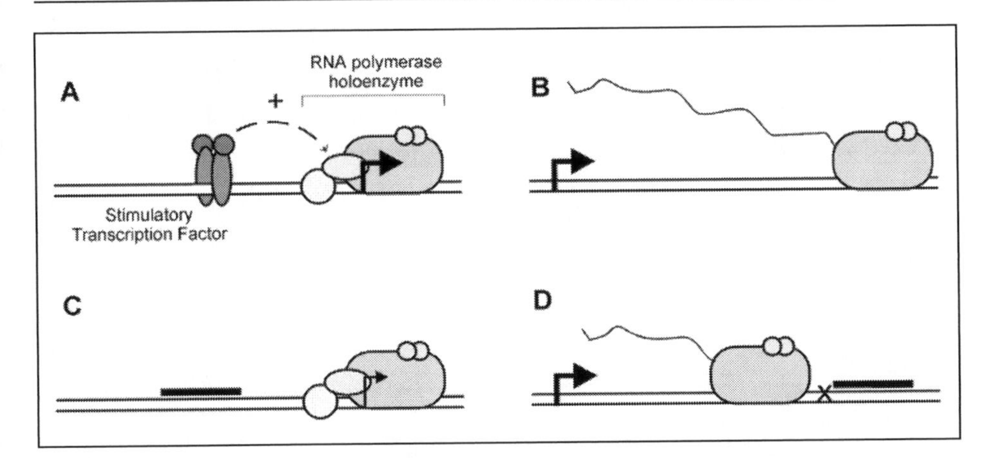

Figure 4. Transcription regulation by multiplexes. Schematic representations of transcription and its inhibition by multiplexes. A) A distal homodimeric specific transcription factor (left, dark gray) positively affects the function of basic transcription factors (center, light gray) and RNA polymerase II (right, medium gray) bound proximally to the start site of transcription, thereby promoting initiation (rightward bold arrow). B) An elongating RNA polymerase II proceeds downstream from the start site of transcription, synthesizing an RNA transcript (tail) complementary to the DNA template strand. C) A promoter-bound multiplex (bold line) occludes the binding of a specific transcription factor, thereby diminishing transcription initiation. D) A multiplex located downstream of the transcription start site positions a covalent DNA crosslink (X) that impedes elongation by RNA polymerase II.

elongation. However, it is also possible to inhibit transcription of a targeted promoter by occluding a DNA-binding basic transcription factor (e.g., TFIID). Likewise, it is possible to stimulate transcription through protein occlusion, if the occluded protein is a transcriptional repressor. In elongation arrest (Fig. 4D), a post-initiation RNA polymerase II has its progress impeded by a downstream-situated multiplex. Note that a multiplex alone usually cannot effectively impede elongation by an RNA polymerase, especially eukaryotic RNA polymerases that normally function in a chromatin environment. Thus, unless the multiplex is located immediately downstream of a transcription pause or termination site, it is usually necessary for the multiplex to direct a subsequent covalent modification of the template (e.g., a cross-link or strand break) that renders it unsuitable for elongation.

Conceivably, there are several other mechanisms by which a multiplex structure might affect transcription. Some are shown schematically in Figure 5. For example, multiplex-forming oligonucleotides could themselves adopt structures that bind proteins involved in transcription (Fig. 5A). Note that these could include proteins directly involved in RNA synthesis (e.g., transcription factors) as well as proteins that ultimately modulate their activity (e.g., signal transduction proteins). Given the appropriate sequence homology, multiplex-forming oligonucleotides could bind to RNA transcripts through conventional Watson-Crick base pairing, thereby leading to transcript degradation (and apparent loss) through endogenous RNase H activity (Fig. 5B). Alternatively, multiplexes could inhibit transcription through the delivery of nonspecific and specific inhibitors of transcription, instead of through the direct occlusion of stimulatory transcription factors (Fig. 5C). Discerning these possible mechanisms relies on the use of adequate and sufficient controls, including mutagenesis of multiplex-forming sequences, order-of-addition experiments, and physical verification of multiplex structures.

Intermolecular Triplexes and Transcription

Most studies on the modulation of transcription by multiplexes have been done with oligonucleotides and intermolecular triplexes, because of the ease of forming such structures, the variety of controls that can be performed, and the flexibility possible through use of chemically

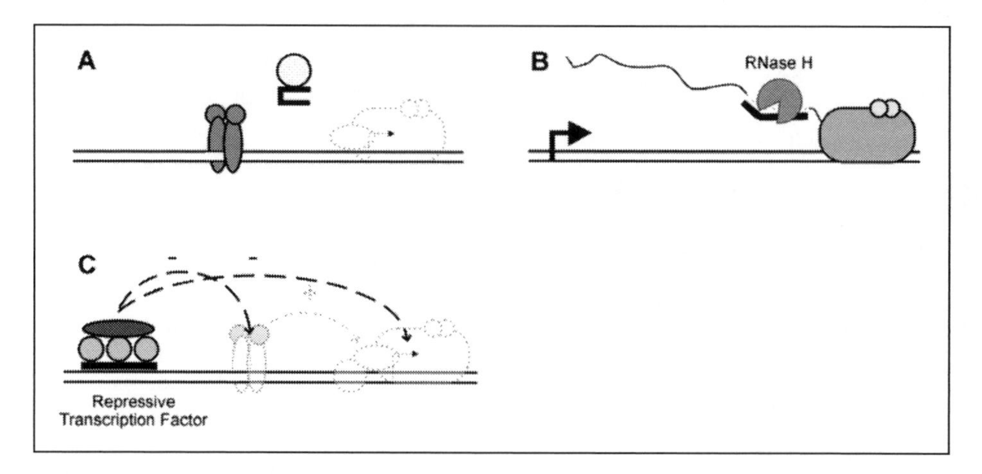

Figure 5. Alternative mechanisms for transcription regulation by multiplexes. Schematic representations of transcription regulation by multiplexes and multiplex-forming oligonucleotides. A) A multiplex-forming oligonucleotide folds into a structure that serves as an aptamer, which competes for promoter binding by a general transcription factor (center, light gray). B) A multiplex-forming oligonucleotide binds to an RNA transcript and directs its degradation by RNase H. C) An upstream promoter multiplex binds proteins that actively interfere with the function of specific and general transcription factors.

modified triplex-forming oligonucleotides (TFOs). Studies on intermolecular triplexes have been performed both in vitro and in vivo, "in vivo" referring to any living organism, including cultured cells.

Intermolecular triplex effects on transcription have been investigated in vitro for a number of model systems, including prokaryotic, eukaryotic, and various hybrid systems. A list of representative studies is presented in Table 1. Both triplex motifs, purine and pyrimidine, have been explored, as have binding modes that are less well defined. Occlusion of specific transcription factors or general transcription factor/RNA polymerase binding has been proposed and/or reported in many studies.[35,39,42-45,47,51,54,56,71] Typical observed results have been in the range of 50% to 105% transcription inhibition when 0.2 to 50 μM TFO was present. Control reactions usually involved oligonucleotides (ODN) that were not capable of triplex formation or templates that lacked TFO binding sites. Some unusual findings include the demonstration that a TFO targeting an upstream stimulatory transcription factor could apparently inhibit transcript appearance through partial hybridization to these transcripts and RNase H-mediated RNA degradation, and that transcription was inhibited when triplexes were located distal to transcription factor binding sites.[39,88] Promotion of transcription has also been described in vitro, through the direct delivery of transcription activators by hybrid TFOs.[85] Inhibition of transcription elongation has been observed in vitro as well.[38,40,49,52,58,60,61,68,72] Typical effects range from 60% to 95% transcription inhibition, depending on several factors including the location of the intermolecular triplex relative to the start site of transcription, the type of RNA polymerase investigated, and whether the TFO was noncovalently bound or whether it directed a covalent modification of the DNA template. Taken together, these data demonstrated that many types of intermolecular triple helices can specifically and effectively inhibit several types of transcription through multiple mechanisms in vitro.

Given the observed successes with intermolecular triplexes in vitro, several research groups have investigated the effects of intermolecular triplexes on transcription in vivo (see Table 1). Both transcription factor occlusion and polymerase elongation mechanisms of triplex action have been investigated in vivo, with reports of efficiencies in excess of -90% reported in some circumstances, depending on oligonucleotide type, delivery method, and target site. Substantial transcription stimulation in vivo mediated by an activation domain peptide/triplex-forming oligonucleotide hybrid has also been reported.[85] These findings suggest that intermolecular triplexes appear to be an efficient means of inhibiting specific gene transcription in vivo.

In experiments performed on in vivo targets, researchers encounter complications not found with in vitro experiments, including maintaining oligonucleotide stability in the presence of serum and cellular nucleases, delivering adequate concentrations of TFO to the proper cellular compartment (nucleus), and ensuring that triplex-formation actually occurs. Each of these difficulties has been addressed by a variety of means. Stability questions have been addressed by chemical modifications of the TFO termini and/or its phosphodiester backbone.[36,41,42,44,46,47,53,55,62,66,69,70,75-82,84] Delivery difficulties have been surmounted by transfection with cationic lipids, electroporation, microinjection, or synthesis in situ.[48,53,55-57,65,67,70,76,80,81,83,85,86] Even triplex formation, which can be highly problematic in an intracellular milieu with its high protein concentrations and surfeit of nonspecific nucleic acid targets, has been overcome by first preforming triplexes on their plasmid targets in vitro and then introducing the entire complex into cells.[41,44,46,50,63,64,66,69,71,73-75,78,79,82,87] Note that these ex vivo experiments, although successful at addressing particular aspects of triplex-mediated transcription modulation, do not completely address the overall feasibility of triplexes in vivo. In addition, since very few investigators have actually demonstrated triplex formation in vivo, and oligonucleotides can affect cells through multiple specific and nonspecific mechanisms, most studies supporting triplex effects in vivo are not as compelling as they could be.

Intramolecular Triplexes and Transcription

Oligopurine•oligopyrimidine sequences have long been understood to play an instrumental role in the regulation of transcription for many genes.[1] It has also been well known that certain oligopurine•oligopyrimidine sequences, especially those possessing mirror repeats, can form intramolecular triplexes in vitro under conditions of low pH or increased negative superhelicity.[3] Thus it has been tempting to speculate that intramolecular triplexes are responsible for the transcriptional regulation observed at these sites. There is some evidence that intramolecular triplexes can form in vivo, albeit under less than physiological conditions in prokaryotic systems.[89,90] Additionally, in triplex-specific antibody studies, cross-reactive structures have been identified near the centromeres of chromosomes.[12,91] However, most reports in the literature regarding the involvement of H-DNA (or its purine-motif counterpart, H'-DNA) on transcription are only suppositions; few researchers have tested whether these sequences actually form intramolecular triplexes, and most physical studies have been performed in vitro. Nonetheless, a few exemplary studies have been done to investigate the possible role of intramolecular triplex structures on transcriptional regulation. Some are presented in Table 2.

Intramolecular triplexes may affect transcription from two locales: either proximally upstream the transcription start site or at any distance downstream. In the former case, intramolecular triplexes located within gene promoters are believed to arise in response to increased negative superhelical tension, which can result from nearby transcription. Such triplexes could then inhibit subsequent transcription events by displacing necessary transactivating proteins (Fig. 4C) or by recruiting repressive proteins (Fig. 5C). An alternative view is that the single-stranded region resulting from intramolecular triplex formation could serve as an entry point for RNA polymerase and thus serve as an activator of transcription.[105] In the latter case, downstream intramolecular triplexes could arise as a result of processes that locally denature the DNA template (e.g., replication, transcription). These downstream triplexes would then either impede subsequent transcription elongation (Fig. 4D) or inhibit transcription elongation by sequestering essential proteins (Fig. 5A) or by delivering repressive proteins (Fig. 5C).

Table 1. Intermolecular triplexes and transcription

Target[a]	Multiplex[b]	Triplex Proof[c]	Transcription Assay[d]	Maximum Effect[e]	Controls[f]	Ref.
human *c-myc* P1 promoter, linear plasmid	27mer GAT PO TFO, "parallel"	EMSA, DNase I FP	in vitro, HeLa NE, c-myc P2 runoff	-95% @ 160 nM TFO	complement ODN	35
human IL2Rα promoter, SRE, chromosome	28mer PO TFO, 3'-amine, parallel	REPA	in vivo, PBMC cells, northern, media TFO	-55% @ 15 μM TFO	31mer GT ODN / c-myc or β-actin mRNA	36
human *c-myc* P1 promoter, chromosome	27mer GAT PO TFO, Pu3	EMSA / chromatin DNase I hyper -sensitivity	in vivo, HeLa cells, northern, media TFO	-90% @ 125 μM TFO	complement ODN / β-actin mRNA	37
syn., +180 G-free cassette, Ad2 MLP, linear plasmid	15mer MT PO TFO, 3'-C*, Py3	DMS FP	in vitro, Jurkat NE, G-free, pre. 3plx	-90% @ 10 μM TFO, covalent	MT TFO noncovalent, scrambled ODN covalent	38
syn., 1-5 copies, Sp1 site, Ad E4 core promoter, supercoiled plasmid	21mer $T^{5me}C$ PO TFO, Py3	n.s. (prior work)	in vitro, *Drosophila* K_c cell NE, primer ext., pre. 3plx	-90% @ 2.5 μM TFO, pre. 3plx, basal transcription (-Sp1)	scrambled ODN / *ftz* promoter (co-transcribed)	39
E. coli bla gene, +22, DNA fragment	13mer $T^{5me}C$ PO TFO, 5'-psoralen, Py3	DNase I FP / DNase I FP	in vitro, *E. coli* RNA pol holoenzyme, runoff, pre. 3plx	-80% @ 50 μM TFO, covalent	unmod. TFO, nonspecific pso-16-mer (-20% each) / *tetR* gene	40
human IL-2Rα, enhancer, NF-κB site, plasmid	15mer $T^{5me}C$ PO TFO, 3'-acridine, Py3	EMSA, REPA, DNase I FP	in vivo, HSB2 cells, CAT assay, pre. 3plx	-90% @ 10 μM TFO	HIV LTR CAT	41
HIV-1 promoter, Sp1 and +1 sites, plasmid	31 & 38mer GT PO TFO, 3'-amine, parallel	EMSA, CD	in vitro, HeLa NE, smear; in vivo, U937/ HIV-1 cells, media TFO	-80% (in vitro) & -90% (in vivo)@ 10 μM TFO	scrambled ODN	42
syn., ϕ10 promoter, +1, linear plasmid	21mer $T^{5me}C$ PO TFO, Py3; 22mer GT GU PO TFOs, Pu3	DNase I & MPE FP	in vitro, T7 pol., runoff	-92% @ 1 μM $T^{5me}C$; -99% @ 1 μM GT TFO	scrambled ODN / wt ϕ10 promoter (co-transcribed)	43

Table continued on next page

Table 1. Continued

Target[a]	Multiplex[b]	Triplex Proof[c]	Transcription Assay[d]	Maximum Effect[e]	Controls[f]	Ref.
human IL-2Rα, enhancer, NF-κB site, plasmid	15mer $T^{5me}C$ PO TFO, 3'-acridine, Py3	DNase I FP	in vitro, C8166 NE, primer extension, pre. 3plx; in vivo, HSB2 cells, CAT assay, pre. 3plx	-80% @ 1 μM TFO covalent (in vitro) -80% @ 0.2 μM TFO, covalent (in vivo)	mutant 3plx site / HIV LTR	44
syn., ϕ10 promoter, +1, linear plasmid	21mer $T^{5me}C$ PO TFO, Py3; 22mer GT GU PO TFOs, Pu3	n.s. (prior work)	in vitro, T7 pol., runoff, pre. 3plx	-90% @ 1 μM $T^{5me}C$, GT, GU TFO	wt ϕ10 promoter (co-transcribed)	45
human IL-2Rα, enhancer, NF-κB site, plasmid	15mer $T^{5me}C$ PO TFO, 3'-acridine, Py3	DNase I FP	in vivo, HSB2 cells, CAT assay, pre. 3plx?	-90% @ 5 μM TFO	unmod. TFO, mutant mod. ODN / mutant 3plx site	46
syn., PRE site, Lov core promoter, plasmid	38mer GT PO TFO,3'-cholesterol, Pu3	EMSA, DNase I FP	in vitro, HeLa NE, G-free, pre. 3plx; in vivo, CV-1 cells, CAT assay, media TFO	-105% @ 200 nM TFO (in vitro); -50% @ 20 μM TFO, (in vivo)	(in vitro) mut. ODN (-50% @ 200 nM) / mut. promoter, Ad2 MLP (co-transcribed); (in vivo) mut. ODN (-15% @ 20 μM) / SV2CAT	47
human 6-16 promoter, IRE, plasmid	21mer AGT PO TFO, Pu3	EMSA, DNase I & Cu-phen. FP	in vivo, HeLa cells, CAT, co-transfect	-99% @ 1.8 μM TFO, (-50% for controls)	GA ODN / SV40 or minimal TK promoter	48
syn. +63 or +103, linear plasmid	53mer GT PO TFO, 3' -amino, Pu3	EMSA	in vitro, T3 or T7 RNA pol., truncated transcript, pre. 3plx	-80% @ 0.2 μM TFO, (-30% @ 4 μM reverse ODN)	reverse ODN	49
syn., +204, plasmid	15mer $T^{5me}C$ PO TFO, 5'-psoralen, Py3	PCR / PCR	in vivo, HeLa or XP2Y0(SV) cells, β-gal assay, pre. 3plx	-75% @ 1.8 μM TFO (24h HeLa), -85% (72h XP)	mut. ODN	50
human H-*ras* promoter, -8, linear plasmid	21mer GC PO TFO, Pu3	EMSA, DNase I FP	in vitro, HeLa NE, runoff, pre. 3plx	-50% @ 15 μM TFO	reverse ODN / CMV promoter	51

Table continued on next page

Table 1. Continued

Target[a]	Multiplex[b]	Triplex Proof[c]	Transcription Assay[d]	Maximum Effect[e]	Controls[f]	Ref.
syn., +46, linear plasmid	11mer CT PO PS TFO, Py3	T_m, CD	in vitro, T7 RNA pol., runoff, pre. 3plx	-95% @ 5 μM PO TFO	11mer CT ODN, -TFO	52
IgH 3' α enhancer, Pax5 site, plasmid & chromosomal	41mer GT PO TFO, 3'-amine, Pu3	EMSA / in vivo DMS FP	in vivo, CH12.LX.A2 cells, luciferase, northern, (co)transfect TFO	+160% @ 25 μg TFO	mut. ODN/ mut. enhancer	53
human c-*myc*, P2 promoter, MAZ & E2F sites, linear plasmid	23mer GAT PO, Pu3	EMSA, DNase I FP	in vitro, HeLa NE, runoff, pre. 3plx	-80% @ 20 μM TFO	reverse ODN / CMV promoter	54
human *ALDH₂* downstream promoter, chromosome	21mer GT PO, PS, & PO/PS TFO, Pu3	EMSA	in vivo, HepG2 cells, RT-PCR, transfect TFO	-90% @ 600 nM PS TFO	albumin mRNA	55
rat α1(I) collagen promoter, -138, fragment or plasmid	30mer GA PO TFO, parallel	EMSA	in vitro, RCF NE, runoff, pre. 3plx; in vivo, RCF cells, CAT, post transfect TFO	-95% @ 250 μM TFO, (in vitro); -60% @ 1 μM TFO (in vivo)	20mer AGCT ODN / mut. & CMV IE promoter	56
human *HER2* core promoter, chromosome	28mer GAC PO TFO, Pu3	EMSA, FTIR	in vivo, MCF7 cells, ELISA, northern, transfect TFO	-28% protein, -49% mRNA @ 0.22 μM TFO, 6 h post transfect.	mut., complement, scrambled / EGFR protein, GAPDH mRNA	57
HIV1 *nef* gene, linear plasmid	11mer TGC PN TFO, Py3	REPA, T_m	in vitro, HeLa NE, truncated transcript, pre. 3plx	-60% @ 2 μM TFO	PO & other modif. TFOs	58
human *GM-CSF* promoter, NF-κB site, plasmid & chromosome	15mer GT PO TFO, Pu3	EMSA, DNase I FP	in vivo, Jurkat cells, luciferase, RNase protect, ELISA, media TFO	-70% @ 2.4 μM TFO	random GT ODN / IL3 promoter, mRNA	59
HIV1 *nef* gene, linear plasmid	15mer TG^{5me}C PO TFO, 5'-acridine, Py3, stabilized with BePI	T_m	in vitro, HeLa NE or SP6 or T7 RNA pol., truncated transcript, pre. 3plx	-77% @ 10 μM TFO (pol II), -60% @ 10 μM TFO + 25 μM BEPI (SP6)	unsub. TFO & random ODN	60

Table continued on next page

Table 1. Continued

Target[a]	Multiplex[b]	Triplex Proof[c]	Transcription Assay[d]	Maximum Effect[e]	Controls[f]	Ref.
HIV1 *nef* gene, linear plasmid	16mer TC or $T^{5me}C$ PN TFO, Py3	n.s.	in vitro, HeLa NE or SP6 RNA pol., truncated transcript, pre. 3plx	-77% @ 1 μM TC TFO (pol II), -60% @ 10 μM $T^{5me}C$ TFO (SP6)	titration	61
human *TNF* gene, 3rd intron, chromosome	27mer GT PO TFO, 3'-cholesterol, Pu3	EMSA	in vivo, THP-1 cells, TNF bioassay, RT -PCR, media TFO	-75% @ 0.6 μM TFO	other GT ODNs / IL-β mRNA	62
syn., +400, plasmid	19mer GT PO TFO, 5'-psoralen, Pu3	REPA, PCR/ REPA, PCR	in vivo, HeLa cells, luciferase, pre. 3plx	-79% @ 0.6 μM TFO, covalent	-pre. 3plx & -crosslink / β-gal	63
murine c-Ki-*ras* promoter, -290, plasmid	20mer GA PO TFO, Pu3	EMSA, T_m, CD	in vivo, NIH 3T3 cells, CAT, pre. 3plx	-90% @ 1 μg TFO (-40% @ 1 μg non-specific ODN)	non-specific ODN	64
human *IGF-I* gene, 1st exon, chromosome	RNA cont. 23 nt. GA sequence, Pu3?	n.s. for RNA	in vivo, C6 cells, stable clones, northern, in situ TFO	-95%	complement RNA	65
rat α1(I) collagen promoter, -141, plasmid	18mer GA PS TFO, Pu3	EMSA, DNase I FP	in vivo, 2TK cells, CAT, pre. 3plx	-80% @ 50 μg TFO	18mer ACGT ODN	66
rat *IGF-IR* gene, 3' (+4504), chromosome	RNA cont. 24 nt. GA sequence, Pu3?	n.s.	in vivo, C6(t1) cells, stable clones, northern, in situ TFO	-75%?, (similar inhibition IGF-I mRNA)	complement RNA / β-actin	67
syn., +185, linear plasmid	15mer $TG^{5me}C$ PO TFO, Py3	n.s.	in vitro, HeLa NE, runoff	-90% @ 10 μM TFO, crosslink	-UV	68
human, c-*myc*, P1 promoter PuF site & P2 promoter MAZ site, plasmid	26 & 23mer GAT PS TFOs, Pu3	EMSA	in vivo, HeLa cells, luciferase, pre. 3plx	90% @ 100 μM, both TFOs	reverse ODNs	69
human *cyclin D1* promoter, Sp1, chromosome	18mer GT PS TFO, Pu3	EMSA, DNase I FP	in vivo, HeLa cells, luciferase, transfect TFO	-60% @ 10 μM TFO	reverse ODN/ CMV luc	70
human α1(I) collagen promoter, -140, fragment or plasmid	30mer GAT PO TFO, Pu3	EMSA	in vitro, HeLa NE, runoff, pre. 3plx; in vivo, CK-Y cells, GFP, pre. 3plx	-75% @ 200 ng TFO, (in vitro); -55% @ 30 nM TFO (in vivo)	mut. & control? ODN / CMV IE promoter	71

Table continued on next page

Table 1. Continued

Target[a]	Multiplex[b]	Triplex Proof[c]	Transcription Assay[d]	Maximum Effect[e]	Controls[f]	Ref.
human *HER2*, +205, linear plasmid	23mer ^{PP}GAT PO TFO, 5'-pam, Pu3	EMSA	in vitro, HeLa NE, truncated transcript, pre. 3plx	-80% @ 1 µM TFO	titration, diff. TFOs	72
syn., -40, minimal c-*fos* promoter, plasmid	20mer GA PO TFO, 5' -or 3' -peptide, Pu3	EMSA, co-migrate	in vivo, NIH 3T3 cells, luciferase, pre. 3plx	+250% @ 1:1 TFO:plasmid (+10% @ 2:1)	titration, unmod. TFO	73
human c-*src* promoter, Sp1 & SPY sites, plasmid	11mer GA PO TFO, Pu3	n.s.	in vivo, 10T1/2 cells, CAT, pre. 3plx	-70% @ 500:1 TFO:plasmid	GA & ACGT ODNs	74
CAT gene, +578, plasmid	19mer GAT PD TFO, Pu3	n.s.	in vivo, *Xenopus* oocyte, CAT, northern, pre. 3plx	-90% @ 180 µM TFO	+/- TFO / Cyclin B1 mRNA	75
syn., +113, plasmid & chromosome	15mer TG^{5me}C PN TFO, 5'-acridine or psoralen, Py3	EMSA / REPA	in vivo, P4 cells or nuclei, luciferase, northern, co-transfect TFO or steptolysin permeabilize	-70% @ 0.5 µM TFO (5' acridine, cell, plasmid); -50% @ 0.5 µM TFO (5' psoralen, nuclei, chromosome	mut. & reverse ODN / mut. site	76
human c-*myc*, P2 promoter, MAZ site, chromosome	23mer GT PO TFO, 3'-amine, Pu3	EMSA, DNase I FP	in vivo, CEM cells, northern, RT-PCR, media TFO	-63% @ 20 µM TFO	GA & scrambled ODNs / GAPDH mRNA	77
human rhodopsin gene, 1st or 2nd intron, plasmid	18 or 28mer GA(T or C) PO TFO, 5'-psoralen & 3'-amine, Pu3	n.s. (prior)	in vivo, HT1080 cells, IFM, pre. 3plx	-90% @ 1 µM, both TFOs	-TFO, -UV	78
human *bcl-2* 3' UTR, + 1946, plasmid	18mer GAT PO(PS ends) TFO, 3'-amine, Pu3	EMSA, co-migrate	in vivo, Tet-On HeLa, western, pre. 3plx	-90% @ 12.5:1 TFO:plasmid	random, scrambled ODN / actin protein	79
murine Ki-*ras* promoter, plasmid	20mer, PO/PS GA TFO, Pu3	EMSA, DMS & DNase I FP	in vivo, 293 cells, CAT, co-transfect TFO	-75% @ 5 µM TFO	-TFO	80

Table continued on next page

Table 1. Continued

Target[a]	Multiplex[b]	Triplex Proof[c]	Transcription Assay[d]	Maximum Effect[e]	Controls[f]	Ref.
syn., HIV-1 PPT, +113, plasmid	15mer TG^{5me}C PN TFO, Py3	n.s. / REPA	in vivo, P4 cells, single-cell luciferase, microinject TFO	-85% @ 32 μM TFO	reverse ODN, *Renilla* luciferase	81
human *IgE* promoter, STAT6 & Pu.1/NF-κB sites, plasmid	21mer TC AE TFO, Py3	EMSA	in vivo, DG75 cells, luciferase, pre. 3plx	-80% @ 200:1 TFO: plasmid, (-20% mut. AE ODN)	random PO ODN, mut. AE ODN	82
human *ICAM-1* gene, 3rd intron, chromosome	16mer GT PO TFO, 3'- or 5'-psoralen, Py3	EMSA / capture-PCR	in vivo, A431 cells, FACS, northern, transfect TFO	-60% @ 3 μM TFO (northern), +/- UV?	scrambled ODN, HLA-DR expression/ GAPDH mRNA	83
human *bcr* promoter, +1, chromosome	13mer GA PO TFO, 3'-PEG, Pu3	EMSA, DMS FP	in vivo, K562 cells, RT-PCR, media TFO	-35% @ 15 μM TFO, (random ODN, -15%)	G-rich PEG'd random ODN / abl mRNA	84
syn., 5 sites upstream core Ad E4 promoter, linear plasmid	22mer, GT PO TFO, 3' -peptide, Pu3	co-migrate	in vitro, HeLa NE, runoff, pre. 3plx; in vivo, HeLa cells, CAT, co-transfect TFO	+1000% @ 2.5 nM TFO in vitro (+100% @ 50 nM?); +3000% @ 50 nM TFO in vivo	GAL4 E4 promoter	85
human *Ets2* promoter, Sp1, plasmid & chromosome	25mer GT PO TFO, Pu3	EMSA	in vivo, DU145 cells, luciferase, RT-PCR, northern, co-transfect TFO	-80% @ 250 nM TFO (plasmid), -62% @ 400 nM TFO (chromosome)	scrambled ODN, c-src or c-myc reporters	86
human *HER2*, core promoter, plasmid	25mer ^{PP}GAT PO TFO, 5'- / 3'-pam, Pu3	EMSA, alkylation/ Southern	in vivo, HeLa cells, luciferase, pre. 3plx	-70% @ 2 μM TFO (5' & 3' pam)	different modified TFOs	87

[a] Target information is indicated as gene name, site, and whether located on an extrachromosomal DNA (e.g., plasmid) or on a chromosome. Syn., synthetic target. Ad, adenovirus. MLP, major late promoter. [b] Triplex information is indicated as length, base composition (in order of abundance), backbone composition, modifications (indicated by 5'- or 3'-), auxiliary compounds ("stabilized with..."), and triplex motif. Bases include N^6-methyl-8-oxo-adenine (M), 5-methylcytidine (^{5me}C), and pyrazolopyrimidine guanine (^{PP}G). Backbone compositions include phosphodiester (PO), phosphorothioate (PS), N3'-P5' phosphoramidate (PN), 2'-aminoethoxyribose (AE), and *N,N*-diethylethylenediamine phosphoramidate (PD). Modifications include N^4,N^4-ethano-2'-deoxycytidine (C*), phenylacetate mustard (pam), polyethylene glycol (PEG). Motifs include purine (Pu3), pyrimidine (Py3), and parallel. Additional abbreviations include triplex-forming oligonucleotide (TFO), containing (cont.), nucleotide (nt.). [c] Assays used to demonstrate triplex formation include electrophoretic mobility shift assays (EMSA), footprinting (FP) with enzymes like DNase I or with chemicals including dimethyl sulfate (DMS), copper-1,10 phenanthroline (Cu-phen.), and methidiumpropyl-EDTA-iron (MPE),

Table continued on next page

Table 1. Continued

circular dichroism (CD), melting temperature (T_m), Fourier transform infrared spectroscopy (FTIR), and restriction endonuclease protection assays (REPA). Assays shown after a slash (e.g., PCR / PCR) indicate those assays performed after transcription has occurred. [d] Transcription assay information is presented as extract (e.g., nuclear, NE) or polymerase (pol.), template, and assay type for in vitro studies and cell type, assay type, and TFO administration method for in vivo studies. In vitro assays include runoff, G-free cassette, primer extension, and truncated transcript. In vivo assays include RNA-based (e.g., northern, reverse transcriptase PCR [RT-PCT]), immunological (e.g., enzyme-linked immunosorption [ELISA], immunofluorescence microscopy [IFM], fluorescence-activated cell sorting [FACS]), and enzymatic activity assays (e.g., chloramphenicol acetyltransferase [CAT], β-galactosidase [β-gal.], luciferase [luc.]). Preformed triplexes (pre. 3plx). [e] Maximum effect of triplex on transcription. Minus and plus signs indicate triplex-dependent transcription inhibition and stimulation, respectively. Unexpected or unusual findings are indicated in parentheses. Mut., mutant. [f] Controls used to demonstrate specificity of triplex effects on transcription. Controls involving oligonucleotides are shown before the slash, while controls involving targets are shown after (e.g., scrambled ODN / mutant promoter). Additional abbreviations include: unmodified (unmod.), cytomegalovirus immediate-early promoter (CMV IE), and unsubstituted (unsub.).

Promoter-based intramolecular triplex effects on transcription have been reported to be quite variable, with magnitudes ranging from highly stimulatory to no effect to moderately inhibitory.[15,92,94,96,97,102-104] More telling have been the results of the corresponding control experiments, which in the majority of studies showed no correlation between intermolecular triplex formation and transcriptional strength.[92,94,97,102-104] For downstream intramolecular triplexes, significant inhibitory effects have been consistently reported both in vitro and in vivo, though their exact correlation with a specific triplex structure has been somewhat weak.[98-101] All in all, these studies suggested that the transcriptional effects ascribed to relatively short polypurine•polypyrimidine sequences located upstream of many genes is most likely not the result of intramolecular triplex formation, whereas the transcriptional effects observed with very long downstream polypurine•polypyrimidine sequences may well involve some form of intramolecular triplex, especially of the H' variety.

Quadruplexes and Transcription

While quadruplexes, especially of the G-quartet variety, have primarily been invoked as playing a role in the biogenesis of chromosome telomeres, recent studies have suggested that they may also have a role in the transcriptional regulation of certain genes.[9,10] G-rich sequences capable of forming quadruplex structures in vitro have been identified in the immunoglobulin switch region, the c-*myc* promoter, and upstream of the insulin gene.[32,106,107] Use of a single-chain antibody fragment probe specific for guanine quadruplexes has led to identification of cross-reactive species in the macronucleus but not the micronucleus of *Stylonychia lemnae*, suggesting that quadruplexes do exist in vivo.[17] Less clear is how such quadruplex structures arise, although arguments concerning the formation of intramolecular triplexes, including local negative superhelical tension in the promoter region, chromatin remodeling, and the consequence of transcription and/or replication events may also apply here.[9,19] At present only a few studies directly describe quadruplex effects on transcription (see Table 3). Effects are believed to occur at the level of transcription factor occlusion and/or transcription factor recruitment, and significant effects, both stimulatory and repressive, have been observed.[108-111] One major weakness of all these studies is the lack of in vivo characterization of G_4 structures, which makes ascribing transcriptional effects to bona fide quadruplexes a bit tenuous.

Table 2. Intramolecular triplexes and transcription

Target[a]	Multiplex[b]	Triplex Proof[c]	Transcription Assay[d]	Maximum Effect[e]	Controls[f]	Ref.
Drosophila hsp26 promoter -89, chromosome	25 bp mirror repeat, H	oligo hybrid., DEPC FP, S1 nuclease/ DEPC FP	in vivo, *Drosophila*, *hsp26-lacZ* transgene, β-gal	-67% del. H, +2% mut. H⁻, -68% H'⁺	del. H, mut. H⁻, H'	92
syn., promoter, plasmid	$(G \bullet C)_n$, H'	CAA FP	in vivo, LTK⁻ cells, TK CAT, direct & competitions	+970% $(G)_{30}$, +90% $(G)_{18}$, +50% $(G)_{35}$	number of repeats	15
syn., +120, plasmid	38 bp mirror repeat, H or H'?	chloroquine 2D gel electro-phoresis	in vivo, *E. coli*, *lacZ*, β-gal	-80%	mut. H⁻	93
murine c-Ki-*ras* promoter, plasmid	27 bp mirror repeat, H	CAA FP	in vivo, HepG2 cells, CAT	-50% mut. H⁻, -50% mut. H⁺	mut. H⁻, multi mut. H⁺	94
syn., +17, plasmid	69 bp mirror repeat, stabilized with BePI, Hy-5	CAA FP, mung-bean nuclease	in vivo, *E. coli*, viability	-53% (-74% @ 5 μM BePI)	H⁻ / *tet*[R]	95
chicken malic enzyme promoter, plasmid	49 bp Py/ Pu, H'	S1 & P1 nuclease	in vivo, chicken hepatocytes, CAT	+900%, (+0% reversed orientation)	H⁻, H⁺ reversed	96
murine *metallothionein-1* promoter, plasmid	128 bp, H-y3	CAA, DEPC, DMS FP	in vivo, NIH 3T3 cells, luciferase	no effect, basal or Cd^{2+}-induced	H⁻, reverse orientation	97
syn., human *frataxin* 1st intron GAA repeat, +100? plasmid	$(GAA \bullet TTC)_n$ repeats, H'?	n.s.	in vivo, COS-7 cells, RNase protection, β-gal	-91% @ n = 230 repeats	number of repeats, reverse orientation	98
syn., human *frataxin* 1st intron GAA repeat, +100? plasmid	$(GAA \bullet TTC)_n$ repeats, H'?	DEPC (ss DNA tested!)	in vitro, T7 RNA pol., runoff	-90% @ n = 88 repeats	number of repeats	99
syn., human *frataxin* 1st intron GAA repeat, +100? plasmid	$(GAA)_{88}$, H'?	n.s.	in vitro, T7 RNA pol., runoff	-90% (-25% @ 2.5 μM complement $[TTC]_7$ ODN)	$(GGA)_7$ ODN / $(CUG)_{88}$ template	100

Table continued on next page

Table 2. Continued

Target[a]	Multiplex[b]	Triplex Proof[c]	Transcription Assay[d]	Maximum Effect[e]	Controls[f]	Ref.
syn., human *frataxin* 1st intron GAA repeat, +100? plasmid	$(GAA)_{150}$, (multi-H' (sticky DNA)	gel mobility / gel mobility	in vitro, T7 or SP6 RNA pol. or HeLa NE, runoff	-99% $r[UUC]_{150}$, -95% co-transcribed control	number of repeats, orientation / co-transcribed control template	101
human *HMGA2* promoter, -25, plasmid	60 bp Py/Pu, H?	S1 nuclease, EMSA	in vivo, NIH 3T3, luciferase	+400% H^+ supercoiled, (+200% mut. H^-)	mut. H^-, linear or supercoiled/ CMV β-gal	102
syn., *E. coli* PIT sequences, promoter & downstream, plasmid	33 nt multimers, H'	n.s.	in vivo, *E. coli*, *lacZ*, β-gal	-0% (promoter PIT); +300% (downstream 2 x PIT), mRNA translation effect	PIT^-, mut. PIT	103
Drosophila hsp26 promoter, chromosome	25 bp mirror repeat, H	DMS, kethoxal, CMCT, $KMnO_4$ & UV FP / REPA	in vivo, *Drosophila*, *hsp26-lacZ* transgene, β-gal	-0% H^-, -40% mut. H^+	mut. H^- & multi-mut. H^+	104

[a] Target information is indicated as gene name, site, and whether located on an extrachromosomal DNA (e.g., plasmid) or on a chromosome. Syn., synthetic target. [b] Triplex information is indicated as length of the triplex-forming region, sequence of repeating element, auxiliary compounds ("stabilized with..."), and triplex motif (H or H'). BePI, (3-methoxy- 7H-8-methyl-11-[(3'-amino)propylamino] benzo[e]pyrido [4,3-b]indole). [c] Assays used to demonstrate triplex formation include electrophoretic mobility shift assays (EMSA), oligonucleotide hybridization, footprinting (FP) with enzymes like S1 nuclease or with chemicals including chloroacetaldehyde (CAA), 1-cyclohexyl-3-(2-morpholinoethyl)-carbodiimide metho-*p*-toluene sulfonate (CMCT), diethylpyrocarbonate (DEPC), and dimethyl sulfate (DMS), or with ultraviolet radiation (UV) and restriction endonuclease protection assays (REPA). Assays indicated after a slash (e.g., UV FP / REPA) indicate those assays performed after transcription has occurred or in vivo. [d] Transcription assay information is presented as extract (e.g., nuclear, NE) or polymerase (e.g., T7 RNA pol.), template, assay type (e.g., runoff) for in vitro studies, and cell type, template, and assay type for in vivo studies. In vivo assays include RNA-based (e.g., RNase protection) and enzymatic activity assays (e.g., chloramphenicol acetyltransferase [CAT], β-galactosidase [β-gal.], luciferase [luc]). [e] Maximum effect of triplex on transcription. Minus and plus sign preceding percentage effect indicate triplex-dependent transcription inhibition and stimulation, respectively. H^- and H^+ refer to mutations that either disrupt or promote H(H')-DNA formation. Del., deletion. Mut., mutation. Unexpected or unusual findings are indicated in parentheses. [f] Controls used to demonstrate specificity of triplex effects on transcription. Controls involving template DNA are shown before the slash, while controls involving alternative targets are shown after (e.g., H^- / tet^R).

Table 3. Quadruplexes and transcription

Target[a]	Multiplex[b]	Quad. Proof[c]	Transcription Assay[d]	Maximum Effect[e]	Controls[f]	Ref.
human insulin gene, -363, plasmid	$(G_9A_2T_2C)_n$, intra- or inter-molecular G_4	EMSA, DMS FP	in vivo, HeLa cells, luciferase	-80% with Pur-1	mut. G_4^-, multi-mut. G_4^+, Pur-1 expression	109
syn.	G_4 RNA?	n.s.	in vitro, T7 RNA pol. + GTP only, PAGE	+100% termination @ 13-14 nt.	7-deaza-GTP	110
human c-*myc* promoter, chromosome	22mer $G_{18}A_2T_2$, G4'	n.s.	in vivo, Ramos cells, proliferation, G4' ODN, media	-70% @ 100 nM G4 ODN, (-55% @ 100 nM ss G-ODN)	random ODN	111
human c-*myc* promoter, -115 P1, plasmid or chromosome	27mer $G_{20}A_4T_3$, stimulated by TMPyP4, "chair" G4'	DMS FP, TMPyP4 UV FP, *Taq* DNA pol. stop assay	in vivo, HeLa cells, luciferase or Ramos cells, RT-PCR	+200% chair mut.	TmPyP2, alternative G_4^- mut., CA46 cells	112

[a] Target information is indicated as gene name, site, and whether located on an extrachromosomal DNA (e.g., plasmid) or on a chromosome. Syn., synthetic target. [b] Quadruplex information is indicated as length of the quadruplex-forming region, composition or sequence, auxiliary compounds ("stabilized with..."), and quadruplex form. TMPyP4, cationic porphyrin. [c] Assays used to demonstrate quadruplex formation include electrophoretic mobility shift assays (EMSA), footprinting (FP) with dimethyl sulfate (DMS) or ultraviolet radiation (UV), and polymerase (pol.) stop assays. Data not shown (n.s.). [d] Transcription assay information is presented as assay type for in vitro studies and as cell type, template, and assay type for in vivo studies. PAGE, polyacrylamide gel electrophoresis. RT-PCR, reverse transcriptase PCR. [e] Maximum effect of quadruplex on transcription. Minus and plus sign preceding percentage effect indicate quadruplex-dependent transcription inhibition and stimulation, respectively. Abbreviations include nucleotide (nt.), single-stranded (ss), and mutation (mut.). Unexpected or unusual findings are indicated in parentheses. [f] Controls used to demonstrate specificity of triplex effects on transcription. G_4^- and G_4^+ refer to mutations that either disrupt or promote G_4 formation.

Conclusions

Do DNA multiplexes affect transcription? From the aforementioned studies, the following conclusions can be made: (1) Some intermolecular triplexes can significantly repress transcription in vitro. However, their effectiveness in vivo often requires triplex preassembly in vitro. (2) Intramolecular triplexes may be responsible for impeding transcription on long, repeated sequences. (3) G-quadruplexes may affect transcription.

Acknowledgements

This work was supported by a grant from the Robert A. Welch Foundation (G-1199), and is dedicated to the memory of Claude Hélène (1938-2003).

References

1. Wells RD, Collier DA, Hanvey JC et al. The chemistry and biology of unusual DNA structures adopted by oligopurine•oligopyrimidine sequences. FASEB J 1988; 2:2939-2949.
2. Sinden, RR, ed. DNA Structure and Function. San Diego: Academic Press, 1994.
3. Mirkin SM, Frank-Kamenetskii MD. H-DNA and related structures. Annu Rev Biophys Biomol Struct 1994; 23:541-576.
4. Frank-Kamenetskii MD, Mirkin SM. Triplex DNA structures. Annu Rev Biochem 1995; 64:65-95.
5. Sen D, Gilbert W. The structure of telomeric DNA:DNA quadruplex formation. Curr Opin Struct Biol 1991; 1:435-438.
6. Williamson JR. G-quartet structures in telomeric DNA. Annu Rev Biophys Biomol Struct 1994; 23:541-576.
7. Guntaka RV, Varma BR, Weber KT. Triplex-forming oligonucleotides as modulators of gene expression. Int J Biochem Cell Biol 2003; 35:22-31.
8. Zain R, Sun JS. Do natural DNA triple-helical structures occur and function in vivo? Cell Mol Life Sci 2003; 60:862-870.
9. Arthanari H, Bolton PH. Functional and dysfunctional roles of quadruplex DNA in cells. Chem Biol 2001; 8:221-230.
10. Schafer RH, Smirnov I. Biological aspects of DNA/RNA quadruplexes. Biopolymers 2001; 56:209-227.
11. Behe MJ. An overabundance of long oligopurine tracts occurs in the genomes of simple and complex eukaryotes. Nucleic Acids Res 1995; 23:689-695.
12. Agazie YM, Burkholder GD, Lee JS. Triplex DNA in the nucleus: direct binding of triplex-specific antibodies and their effects on transcription, replication and cell growth. Biochem J 1996; 316:461-466.
13. Rooney SM, Moore PD. Antiparallel, intramolecular triplex DNA stimulates homologous recombination in human cells. Proc Natl Acad Sci USA 1995; 92:2141-2144.
14. Bianchi A, Wells RD, Heintz NH et al. Sequences near the origin of replication of the DHFR locus in Chinese hamster ovary cells adopt left-handed Z-DNA and triplex structures. J Biol Chem 1990; 21789-21796.
15. Kohwi Y, Kohwi-Shigamatsu T. Altered gene expression correlates with DNA structure. Genes Dev 1991; 5:2547-2554.
16. Dempsey LA, Sun H, Hanakahi LA et al. G4 DNA binding by LR1 and its subunits, nucleolin and hnRNP D: a role for G-G pairing in immunoglobulin switch recombination. J Biol Chem 1999; 274:1066-1071.
17. Schaffitzel C, Berger I, Postberg J et al. In vitro generated antibodies specific for telomeric guanine-quadruplex DNA react with *Stylonychia lamnae* macronuclei. Proc Natl Acad Sci USA 2001; 98:8572-8577.
18. Catasti P, Chen X, Moyzis RK et al. Structure-function correlations of the insulin-linked polymorphic region. J Mol Biol 1996; 264:534-545.
19. Sun H, Bennett RJ, Maizels N. The *Saccharomyces cerevisiae* Sgs1 helicase efficiently unwinds G-G paired DNAs. Nucleic Acids Res 1999; 27:1978-1984.
20. Fry M, Loeb LA. Human Werner's syndrome DNA helicase unwinds tetrahelical structures of the fragile X syndrome repeat sequence $d(CGG)_n$. J Biol Chem 1999; 274:12797-12802.
21. Roberts RW, Crothers DM. Stabilities and properties of double and triple helices: dramatic effects of RNA and DNA backbone composition. Science 1992; 258:1463-1466.
22. Semerad CL, Maher LJ. Exclusion of RNA strands from a purine motif triple helix. Nucleic Acids Res 1994; 22:5321-5325.
23. Htun H, Dahlberg JE. Topology and formation of triple-stranded H-DNA. Science 1989; 243:1571-1576.
24. Kohwi Y, Kohwi-Shigematsu T. Magnesium ion-dependent triple-helix structure formed by homopurine-homopyrimidine sequences in supercoiled plasmid DNA. Proc Natl Acad Sci USA 1988; 85:3781-3785.
25. Cheng AJ, Van Dyke, MW. Monovalent cation effects on intermolecular purine-purine-pyrimidine triple-helix formation. Nucleic Acids Res 1993; 21:5630-5635
26. Cheng AJ, Van Dyke, MW. Oligodeoxyribonucleotide length and sequence effects on intermolecular purine-purine-pyrimidine triple-helix formation. Nucleic Acids Res 1994; 22:4742-4747.
27. Orson FM, Klysik J, Bergstrom DE et al. Triple helix formation: binding avidity of acridine-conjugated AG motif third strands containing natural, modified and surrogate bases opposed to pyrimidine interruptions in a polypurine target. Nucleic Acids Res 1999; 27:810-816.

28. Paes HM, Fox KR. Kinetic studies on the formation of intermolecular triple helices. Nucleic Acids Res 1997; 25:3269-3274.
29. Hoyne PR, Gacy AM, McMurray CT et al. Stabilities of intrastrand pyrimidine motif DNA and RNA triple helices. Nucleic Acids Res 2000; 28:770-775.
30. Simonsson T. G-quadruplex DNA structures — variations on a theme. Biol Chem 2001; 382:621-628.
31. Deng J, Xiong Y, Sundaralingam M. X-ray analysis of an RNA tetraplex (UGGGGU)(4) with divalent Sr^{2+} ions at subatomic resolution (0.61 Å). Proc Natl Acad Sci USA 2001; 98:13665-13670.
32. Sen D, Gilbert W. A sodium-potassium switch in the formation of four-stranded G4-DNA. Nature 1990; 344:410-414.
33. Lu M, Guo Q, Kallenbach NR. Thermodynamics of G-tetraplex formation by telomeric DNAs. Biochemistry 1993; 32:598-601.
34. Praseuth D, Guieysse AL, Helene C. Triple helix formation and the antigene strategy for sequence-specific control of gene expression. Biochim Biophys Acta 1999; 1489:181-206.
35. Cooney M, Czernuszewicz G, Postel EH et al. Site-specific oligonucleotide binding represses transcription of the human c-*myc* gene in vitro. Science 1988; 241:456-459.
36. Orson FM, Thomas DW, McShan WM et al. Oligonucleotide inhibition of IL2R α mRNA transcription by promoter region collinear triplex formation in lymphocytes. Nucleic Acids Res 1991; 19:3435-3441.
37. Postel EH, Flint SJ, Kessler DJ et al. Evidence that a triplex-forming oligodeoxyribonucleotide binds to the c-*myc* promoter in HeLa cells, thereby reducing c-*myc* mRNA levels. Proc Natl Acad Sci USA 1991; 88:8227-8231.
38. Young SL, Krawczyk SH, Matteucci MD et al. Triple helix formation inhibits transcription elongation in vitro. Proc Natl Acad Sci USA 1991; 88:10023-10026.
39. Maher LJ, Dervan PB, Wold B. Analysis of promoter-specific repression by triple-helical DNA complexes in a eukaryotic cell-free transcription system. Biochemistry 1992; 31:70-81.
40. Duval-Valentin G, Thuong NT, Helene C. Specific inhibition of transcription by triple helix-forming oligonucleotides. Proc Natl Acad Sci USA 1992; 89:504-508.
41. Grigoriev M, Praseuth D, Robin P et al. A triple helix-forming oligonucleotide-intercalator conjugate acts as a transcriptional repressor via inhibition of NFκB binding to interleukin-2 receptor α-regulatory sequence. J Biol Chem 1992; 267:3389-3395.
42. McShan WM, Rossen RD, Laughter AH et al. Inhibition of transcription of HIV-1 in infected human cells by oligodeoxynucleotides designed to form DNA triple helices. J Biol Chem 1992; 267:5712-5721.
43. Maher LJ. Inhibition of T7 RNA polymerase initiation by triple-helical DNA complexes: a model for artificial gene repression. Biochemistry 1992; 31:7587-7594.
44. Grigoriev M, Praseuth D, Guieysse AL et al. Inhibition of gene expression by triple helix-directed DNA cross-linking at specific sites. Proc Natl Acad Sci USA 1993; 90:3501-3505.
45. Skoog JU, Maher LJ. Repression of bacteriophage promoters by DNA and RNA oligonucleotides. Nucleic Acids Res 1993; 21:2131-2138.
46. Grigoriev M, Praseuth D, Guieysse AL et al. Inhibition of interleukin-2 receptor α-subunit gene expression by oligonucleotide-directed triple helix formation. CR Acad Sci III 1993; 316:492-495.
47. Ing NH, Beekman JM, Kessler DJ et al. In vivo transcription of a progesterone-responsive gene is specifically inhibited by a triplex-forming oligonucleotide. Nucleic Acids Res 1993; 21:2789-2796.
48. Roy C. Inhibition of gene transcription by purine rich triplex forming oligodeoxyribonucleotides. Nucleic Acids Res 1993; 21:2845-2852.
49. Rando RF, DePaolis L, Durland RH et al. Inhibition of T7 and T3 RNA polymerase directed transcription elongation in vitro. Nucleic Acids Res 1994; 22:678-685.
50. Degols G, Clarenc JP, Lebleu B et al. Reversible inhibition of gene expression by a psoralen functionalized triple helix forming oligonucleotide in intact cells. J Biol Chem 1994; 269:16933-16937.
51. Mayfield C, Ebbinghaus S, Gee J et al. Triplex formation by the human Ha-*ras* promoter inhibits Sp1 binding and in vitro transcription. J Biol Chem 1994; 269:18232-18238.
52. Xodo L, Alunni-Fabbroni M, Manzini G et al. Pyrimidine phosphorothioate oligonucleotides form triple-stranded helices and promote transcription inhibition. Nucleic Acids Res 1994; 22:3322-3330.
53. Neurath MF, Max EE, Strober W. Pax5 (BSAP) regulates the murine immunoglobulin 3' α enhancer by suppressing binding of NF-α P, a protein that controls heavy chain transcription. Proc Natl Acad Sci USA 1995; 92:5336-5340.
54. Kim HG, Miller DM. Inhibition of in vitro transcription by a triplex-forming oligonucleotide targeted to human c-*myc* P2 promoter. Biochemistry 1995; 34:8165-8171.

55. Tu GC, Cao QN, Israel Y. Inhibition of gene expression by triple helix formation in hepatoma cells. J Biol Chem 1995; 270:28402-28407.
56. Kovacs A, Kandala JC, Weber KT et al. Triple helix-forming oligonucleotide corresponding to the polypyrimidine sequence in the rat α1(I) collagen promoter specifically inhibits factor binding and transcription. J Biol Chem 1996; 271:1805-1812.
57. Porumb H, Gousset H, Letellier R et al. Temporary ex vivo inhibition of the expression of the human oncogene HER2 (NEU) by a triple helix-forming oligonucleotide. Cancer Res 1996; 56:515-522.
58. Escude C, Giovannangeli C, Sun JS et al. Stable triple helices formed by oligonucleotide N3'->P5' phosphoramidates inhibit transcription elongation. Proc Natl Acad Sci USA 1996; 93:4365-4369.
59. Kochetkova M, Shannon MF. DNA triplex formation selectively inhibits granulocyte-macrophage colony-stimulating factor gene expression in human T cells. J Biol Chem 1996; 271:14438-14444.
60. Giovannangeli C, Perrouault L, Escude C et al. Specific inhibition of in vitro transcription elongation by triplex-forming oligonucleotide-intercalator conjugates targeted to HIV proviral DNA. Biochemistry 1996; 35:10539-10548.
61. Giovannangeli C, Perrouault L, Escude C et al. Efficient inhibition of transcription elongation in vitro by oligonucleotide phosphoramidates targeted to proviral HIV DNA. J Mol Biol 1996; 261:386-398.
62. Aggarwal BB, Schwarz L, Hogan ME et al. Triple helix-forming oligodeoxyribonucleotides targeted to the human tumor necrosis factor (TNF) gene inhibit TNF production and block the TNF-dependent growth of human glioblastoma tumor cells. Cancer Res 1996; 56:5156-5164.
63. Musso M, Wang JC, Van Dyke MW. In vivo persistence of DNA triple helices containing psoralen-conjugated oligodeoxyribonucleotides. Nucleic Acids Res 1996; 24:4924-4932.
64. Alunni-Fabbroni M, Pirulli D, Manzini G et al. (A,G)-oligonucleotides form extraordinary stable triple helices with a critical R·Y sequence of the murine c-Ki-*ras* promoter and inhibit transcription in transfected NIH 3T3 cells. Biochemistry 1996; 35:16361-16369.
65. Shevelev A, Burfeind P, Schulze E et al. Potential triple helix-mediated inhibition of IGF-I gene expression significantly reduces tumorigenicity of glioblastoma in an animal model. Cancer Gene Ther 1997; 4:105-112.
66. Joseph J, Kandala JC, Veerapanane D et al. Antiparallel polypurine phosphorothioate oligonucleotides form stable triplexes with the rat α1(I) collagen gene promoter and inhibit transcription in cultured rat fibroblasts. Nucleic Acids Res 1997; 25:2182-2188.
67. Rininsland F, Johnson TR, Chernicky CL et al. Suppression of insulin-like growth factor type I receptor by a triple-helix strategy inhibits IGF-I transcription and tumorigenic potential of rat C6 glioblastoma cells. Proc Natl Acad Sci USA 1997; 94:5854-5859.
68. Wang Z, Rana TM. DNA damage-dependent transcriptional arrest and termination of RNA polymerase II elongation complexes in DNA template containing HIV-1 promoter. Proc Natl Acad Sci USA 1997; 94:6688-6693.
69. Kim HG, Reddoch JF, Mayfield C et al. Inhibition of transcription of the human c-*myc* protooncogene by intermolecular triplex. Biochemistry 1998; 37:2299-2304.
70. Kim HG, Miller DM. A novel triplex-forming oligonucleotide targeted to human cyclin D1 (*bcl*-1, proto-oncogene) promoter inhibits transcription in HeLa cells. Biochemistry 1998; 37:2666-2672.
71. Nakanishi M, Weber KT, Guntaka RV. Triple helix formation with the promoter of human α1(I) procollagen gene by an antiparallel triplex-forming oligodeoxyribonucleotide. Nucleic Acids Res 1998; 26:5218-5222.
72. Ebbinghaus SW, Fortinberry H, Gamper HB. Inhibition of transcription elongation in the HER-2/neu coding sequence by triplex-directed covalent modification of the template strand. Biochemistry 1999; 38:619-628.
73. Kuznetsova S, Ait-Si-Ali S, Nagibneva I et al. Gene activation by triplex-forming oligonucleotide coupled to the activating domain of protein VP16. Nucleic Acids Res 1999; 27:3995-4000.
74. Ritchie S, Boyd FM, Wong J et al. Transcription of the human c-*Src* promoter is dependent on Sp1, a novel pyrimidine binding factor SPy, and can be inhibited by triplex-forming oligonucleotides. J Biol Chem 2000; 275:847-854.
75. Bailey C, Weeks DL. Understanding oligonucleotide-mediated inhibition of gene expression in *Xenopus laevis* oocytes. Nucleic Acids Res 2000; 28:1154-1161.
76. Faria M, Wood CD, Perrouault L et al. Targeted inhibition of transcription elongation in cells mediated by triplex-forming oligonucleotides. Proc Natl Acad Sci USA 2000; 97:3862-3867.
77. Catapano CV, McGuffie EM, Pacheco D et al. Inhibition of gene expression and cell proliferation by triple helix-forming oligonucleotides directed to the c-*myc* gene. Biochemistry 2000; 39:5126-5138.

78. Intody Z, Perkins BD, Wilson JH et al. Blocking transcription of the human rhodopsin gene by triplex-mediated DNA photocrosslinking. Nucleic Acids Res 2000; 28:4283-4290.
79. Shen C, Buck A, Mehrke G et al. Triplex forming oligonucleotide targeted to 3'UTR downregulates the expression of the *bcl-2* proto-oncogene in HeLa cells. Nucleic Acids Res 2001; 29:622-628.
80. Cogoi S, Rapozzi V, Quadrifoglio F et al. Anti-gene effect in live cells of AG motif triplex-forming oligonucleotides containing an increasing number of phosphorothioate linkages. Biochemistry 2001; 40:1135-1143.
81. Faria M, Wood CD, White MR et al. Transcription inhibition induced by modified triple helix-forming oligonucleotides: a quantitative assay for evaluation in cells. J Mol Biol 2001; 306:15-24.
82. Stutz AM, Hoeck J, Natt F et al. Inhibition of interleukin-4- and CD40-induced IgE germline gene promoter activity by 2'-aminoethoxy-modified triplex-forming oligonucleotides. J Biol Chem 2001; 276:11759-11765.
83. Besch R, Giovannangeli C, Kammerbauer C et al. Specific inhibition of ICAM-1 expression mediated by gene targeting with triplex-forming oligonucleotides. J Biol Chem 2002; 277:32473-32479.
84. Rapozzi V, Cogoi S, Spessotto P et al. Antigene effect in K562 cells of a PEG-conjugated triplex-forming oligonucleotide targeted to the *bcr/abl* oncogene. Biochemistry 2002; 41:502-510.
85. Stanojevic D, Young RA. A highly potent artificial transcription factor. Biochemistry 2002; 41:7209-7216.
86. Carbone GM, McGuffie EM, Collier A et al. Selective inhibition of transcription of the Ets2 gene in prostate cancer cells by a triplex-forming oligonucleotide. Nucleic Acids Res 2003; 31:833-843.
87. Ziemba AJ, Reed MW, Raney KD et al. A bis-alkylating triplex forming oligonucleotide inhibits intracellular reporter gene expression and prevents triplex unwinding due to helicase activity. Biochemistry 2003; 42:5013-5024.
88. Praseuth D, Guieysse AL, Itkes AV et al. Unexpected effect of an anti-human immunodeficiency virus intermolecular triplex-forming oligonucleotide in an in vitro transcription system due to RNase H-induced cleavage of the RNA transcript. Antisense Res Dev 1993; 3:33-44.
89. Kohwi Y, Malkhosyan SR, Kohwi-Shigematsu T. Intramolecular dG•dG•dC triplex detected in *Escherichia coli* cells. J Mol Biol 1992; 223:817-822.
90. Ussery DW, Sinden RR. Environmental influences on the in vivo level of intramolecular triplex DNA in *Escherichia coli.* Biochemistry 1993; 32:6206-6213.
91. Ohno M, Fukagawa T, Lee JS et al. Triplex-forming DNAs in the human interphase nucleus visualized in situ by polypurine/polypyrimidine DNA probes and antitriplex antibodies. Chromosoma 2002; 111:201-213.
92. Glaser RL, Thomas GH, Siegfried E et al. Optimal heat-induced expression of the *Drosophila hsp26* gene requires a promoter sequence containing $(CT)_n$•$(GA)_n$ repeats. J Mol Biol 1990; 211:751-761.
93. Sarkar PS, Brahmachari SK. Intramolecular triplex potential sequence within a gene down regulates its expression in vivo. Nucleic Acids Res 1992; 20:5713-5718.
94. Raghu G, Tevosian S, Anant S et al. Transcriptional activity of the homopurine-homopyrimidine repeat of the c-Ki-*ras* promoter is independent of its H-forming potential. Nucleic Acids Res 1994; 22:3271-3279.
95. Duval-Valentin G, de Bizemont T, Takasugi M et al. Triple-helix specific ligands stabilize H-DNA conformation. J Mol Biol 1995; 247:847-858.
96. Xu G, Goodridge AG. Characterization of a polypyrimidine/polypurine tract in the promoter of the gene for the chicken malic enzyme. J Biol Chem 1996; 271:16008-16019.
97. Becker NA, Maher LJ. Characterization of a polypurine/polypyrimidine sequence upstream of the mouse metallothionein-I gene. Nucleic Acids Res 1998; 26:1951-1958.
98. Ohshima K, Montermini L, Wells RD et al. Inhibitory effects of expanded GAA•TTC triplet repeats from intron I of the Friedreich ataxia gene on transcription and replication in vivo. J Biol Chem 1998; 273:14588-14595.
99. Grabczyk E, Usdin K. The GAA•TTC triplet repeat expanded in Friedreich's ataxia impedes transcription elongation by T7 RNA polymerase in a length and supercoil dependent manner. Nucleic Acid Res 2000; 28:2815-2822.
100. Grabczyk E, Usdin K. Alleviating transcript insufficiency caused by Friedreich's ataxia triplet repeats. Nucleic Acid Res 2000; 28:4930-4937.
101. Sakamoto N, Ohshima K, Montermini L et al. Sticky DNA, a self-associated complex formed at long GAA•TTC repeats in intron 1 of the frataxin gene, inhibits transcription. J Biol Chem 2001; 276:27171-27177.

102. Rustighi A, Tessari MA, Vascotto F et al. A polypyrimidine/polypurine tract within the *Hmga2* minimal promoter: a common feature of many growth-related genes. Biochemistry 2002; 41:1229-1240.
103. Hoyne PR, Maher LJ. Functional studies of potential intrastrand triplex elements in the *Escherichia coli* genome. J Mol Biol 2002; 318:373-386.
104. Lu Q, Teare JM, Granok H et al. The capacity to form H-DNA cannot substitute for GAGA factor binding to a $(CT)_n$•$(GA)_n$ regulatory site. Nucleic Acids Res 2003; 31:2483-2494.
105. Weintraub H. High-resolution mapping of S1- and DNase I-hypersensitive sites in chromatin. Mol Cell Biol 1985; 5:1538-1539.
106. Hammond-Kosack MC, Kilpatrick MW, Docherty K. Analysis of DNA structure in the human insulin gene-linked polymorphic region in vivo. J Mol Endocrinol 1992; 9:221-225.
107. Katahira M, Fukuda H, Kawasumi H et al. Intramolecular quadruplex formation of the G-rich strand of the mouse hypervariable minisatellite Pc-1. Biochem Biophys Res Commun 1999; 264:327-333.
108. Simonsson T, Pecinka P, Kubista M. DNA tetraplex formation in the control region of c-*myc*. Nucleic Acids Res 1998; 26:1167-1172.
109. Lew A, Rutter WJ, Kennedy GC. Unusual DNA structure of the diabetes susceptibility locus *IDDM2* and its effect on transcription by the insulin promoter factor Pur-1/MAZ. Proc Natl Acad Sci USA 2000; 12508-12512.
110. Kuzmine I, Gottlieb PA, Martin CT. Structure in nascent RNA leads to termination of slippage transcription by T7 RNA polymerase. Nucleic Acids Res 2001; 29:2601-2606.
111. Simonsson T, Henriksson M. C-*myc* suppression in Burkitt's lymphoma cells. Biochem Biophys Res Comm 2002; 290:11-15.
112. Siddiqui-Jain A, Grand CL, Bearss DJ, Hurley LH. Direct evidence for a G-quadruplex in a promoter region and its targeting with a small molecule to repress c-*MYC* transcription. Proc Natl Acad Sci USA 2002; 99:11593-11598.

Chapter 9

Nucleic Acid Structures and the Transcription Defects in Fragile X Syndrome and Friedreich's Ataxia

Karen Usdin

Abstract

Fragile X mental retardation syndrome (FXS) and Friedreich ataxia (FRDA) belong to a group of genetic disorders known as the Repeat Expansion Diseases. These diseases all result from expansion of a specific tandem repeat. These repeats form a variety of secondary structures that have been suggested to play a role in this expansion. In addition, the properties of these structures suggest ways in which the expanded repeat could contribute to disease pathology. The FXS and FRDA repeats are transcribed but not translated, and expansion leads to aberrant transcription of the affected genes. This chapter discusses the types of nucleic acid structures formed by these repeats and their potential consequences for disease pathology.

Introduction

The Repeat Expansion Diseases arise from expansion of a specific tandem array. Expansion occurs on intergenerational transfer of so-called premutation alleles that have repeat numbers above the normal threshold and results in alleles with even higher numbers of repeats (full mutations). The consequences of expansion depend on the affected gene and its pattern of expression, the location of the repeat within this gene, and the sequence of the repeat unit. Diseases are known where the repeat is not transcribed, where it is transcribed but not translated and where it forms part of the final protein (Fig. 1). The disease-causing repeats identified to date include the triplets CGG•CCG, CTG•CAG, and GAA•TTC, as well as the tetramer CCTG•CAGG, the pentamer ATTCT•AGAAT, and the dodecamer C_4GC_4GCG•$CGCG_4CG_4$. Where the repeat is part of the coding sequence the connection between expansion and disease is relatively straightforward: the repeat negatively affects some property of the protein, either by disrupting its normal function, by conferring toxic properties on the protein or by some combination of the two. Where the repeat is not transcribed, it presumably affects promoter initiation in some way.[1] Where the repeat is transcribed but not translated the molecular mechanisms responsible for disease symptoms are somewhat less obvious.

This chapter will focus on two disorders in this category, Fragile X syndrome and Friedreich ataxia. FXS, the most common heritable cause of mental retardation is caused by expansion of a CGG•CCG-repeat. The repeat is located in the 5′ untranslated region (5′ UTR) of the fragile X mental retardation 1 gene (*FMR1*) which encodes an RNA binding protein, FMRP, thought to be involved in translational control.[2] Cognitive difficulties are often accompanied by behavior problems, anxiety, insomnia, depression, attention deficit hyperactivity, connective tissue

DNA Conformation and Transcription, edited by Takashi Ohyama. ©2005 Eurekah.com and Springer Science+Business Media.

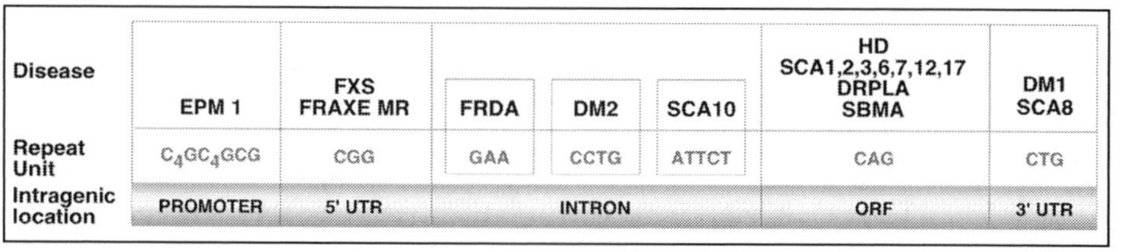

Disease	EPM 1	FXS FRAXE MR	FRDA	DM2	SCA10	HD SCA1,2,3,6,7,12,17 DRPLA SBMA	DM1 SCA8
Repeat Unit	C_4GC_4GCG	CGG	GAA	CCTG	ATTCT	CAG	CTG
Intragenic location	PROMOTER	5' UTR	INTRON			ORF	3' UTR

Figure 1. The repeat expansion diseases. The currently known repeat expansion diseases are listed with the sequence of the Watson strand of the responsible repeat shown below. The intragenic location of these repeats is illustrated on a diagrammatic representation of a generic gene. EPM1: Progressive myoclonus epilepsy Type 1, FXS: fragile X syndrome, FRAXE MR: FRAXE mental retardation, FRDA: Friedreich ataxia, DM: myotonic dystrophy, SCA: spinocerebellar ataxia, HD: Huntington disease, DRPLA: dentatorubropallidoluysian atrophy, SBMA: spinal and bulbar muscular atrophy.

abnormalities, and enlarged testicles in males.[3] Female carriers of premutation alleles show a much higher incidence of premature ovarian failure than carriers of full mutations.[4-9] Older premutation carriers of both sexes have an increased incidence of cerebellar degeneration manifesting in ataxic gait, intention tremor, and both bowel and urinary incontinence.[10-13] This is accompanied by the presence of ubiquitinated intranuclear neuronal inclusions,[14] and characteristic MRI findings including increased T2 signal intensities in the middle cerebellar peduncle.[12] Expansion affects both transcription and translation. Expansion into the full mutation range causes methylation[15] and heterochromatinization[16,17] of the promoter. Expansion also results in the stalling of the 40S ribosomal subunit on any residual *FMR1* mRNA.[18] The net result being an FMRP deficiency. Paradoxically carriers of premutation alleles often have higher than normal amounts of *FMR1* mRNA despite the fact that the stability of this RNA is unchanged.[19,20] Since FMRP levels in these individuals are close to normal it is thought that their symptoms are due to some toxic effect of RNA with expanded repeats, rather than FMRP insufficiency.

FRDA results from expansion of the triplet GAA•TTC. The repeat is located in the first intron of the frataxin gene which encodes frataxin, a mitochondrial protein thought to be involved in iron homeostasis.[21] FRDA is a relentlessly progressive disorder involving the loss of large sensory neurons, and hypertrophic cardiomyopathy.[21] Diabetes or glucose intolerance is common.[21] Affected individuals often become wheelchair bound in adolescence, and early mortality is common primarily due to cardiac failure. In these individuals there is a severe deficit of frataxin mRNA.[22] No aberrantly spliced transcripts have been detected and the stability of the frataxin RNA seems unaffected.[22]

Unusual Nucleic Acid Structures Are Formed by the FXS and FRDA Repeats

The repeats responsible for the Repeat Expansion Diseases are all structurally polymorphic. In addition to fully Watson-Crick (WC) base-paired duplexes with unusual helical parameters, the FXS and FRDA repeats form structures containing a number of non-Watson-Crick interactions. These non-canonical interactions enable the individual strands of the repeats to form a variety of inter-strand and intra-strand structures. The different classes of structures that can be formed by these repeats is illustrated in Figure 2.

The FXS-Repeat DNA Structures

While the CGG•CCG-tract is a fully base-paired, right-handed double helix, it differs from a double helix comprised of a mixed sequence in having an increased flexibility,[23,24] causing the repeat to be more writhed. As a result, the average superhelical density of a DNA domain containing the FXS-repeat would be higher than adjacent domains.[23,24] CGG•CCG-duplexes also have an unusual stable radius of curvature.[25]

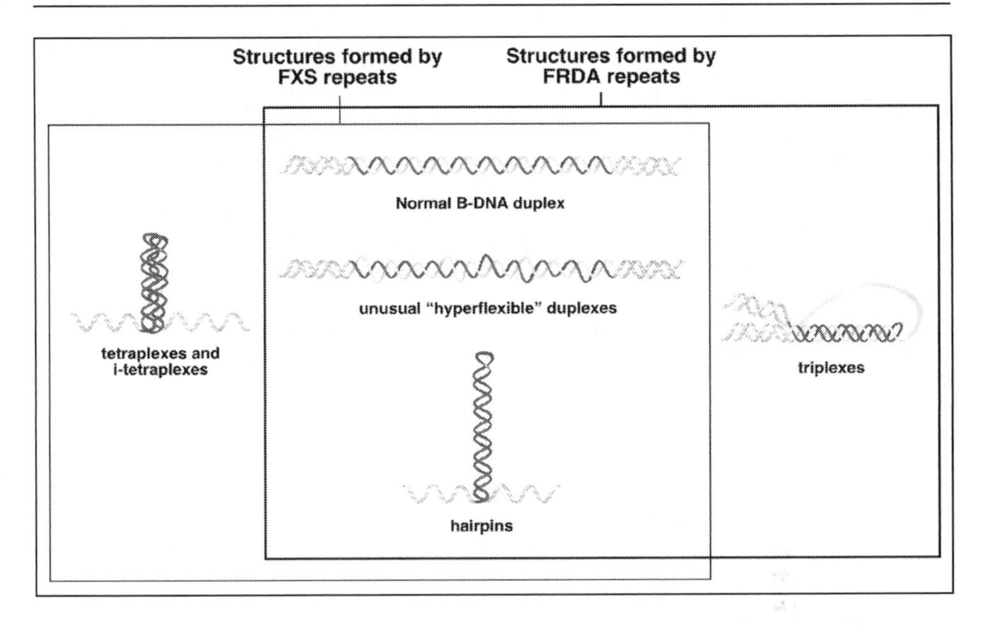

Figure 2. Structures formed by FXS and FRDA repeats. A diagrammatic representation of the different types of nucleic acid structures formed by the FXS repeats is shown in the light box shown on the left. The structures formed by the FRDA repeats are shown in the box on the right. The structures shown in the region of overlap of the 2 boxes are formed by both repeats. For reasons of space, only a single example of a triplex is provided, in this case a Pu•Pu•Py triplex, and slipped DNA structures which form when the 2 strands of the repeat slip relative to each other, resulting in multiple looped out regions which can then form hairpins and tetraplexes are not shown.

The CGG-strand of the repeat forms anti-parallel homoduplexes,[26] and hairpins containing a mixture of WC G•C and Hoogsteen G•G base pairs.[27,28] These sequences can also form inter- or intramolecular quadruplexes (tetraplexes) under physiologically reasonable conditions.[29-32] Quadruplexes are structures held together primarily by Hoogsteen hydrogen bonding between 4 G residues from different strands or different parts of the same strand (see Chaper 8).

Like its complementary strand, CCG-repeats can form antiparallel homoduplexes.[26] Short CGG-tracts form *e*-motif structures containing WC G•C base pairs and extrahelical cytosines.[26] Longer CCG-tracts form hairpins that fold in such a way as to maximize WC G•C pairs.[28,33] The Cs not involved in G•C pairs are intrahelical and well-stacked in the stem and probably form single hydrogen-bonded C•C mispairs.[34] However, these mispairs are more susceptible to open-closure than the G•C base pairs, and thus their Cs can be "flipped out" of the helix more easily.[34] At neutral and slightly acid pH, GCC-strands form tetraplexes held together by $C•C^+$ pairs.[35] Below pH 5 intercalated cytosine tetraplexes (*i*-tetraplexes) are formed. These structures consist of two pairs of parallel DNA strands held together in an antiparallel configuration via the intercalation of hemiprotonated $C•C^+$ pairs.[35]

The simple repetitive nature of these repeats also allows the slippage of the two complementary strands relative to one another. The looped out strands then have the potential to form the folded structures outlined above resulting in slipped DNA or S-DNA (see Chapter 1).

FRDA Repeat Structures

Molecular dynamic simulations suggest that like the FXS B-DNA duplex, the FRDA B-DNA duplex is also unusually flexible.[36] In addition the complementary DNA strands of the FRDA repeat can also form parallel duplexes,[37] hairpins,[38] and both pyrimidine:purine:pyrimidine

(Py•Pu•Py), [37,39-41] and purine:purine:pyrimidine (Pu•Pu•Py) triplexes[42,43] (see Chapter 1). Pu•Pu•Py triplex formation between separate but directly oriented GAA•TTC-tracts generates a structure known as "sticky DNA".[44]

The Potential Biological Significance of These Structures

Some of these structures have properties that suggest a role in expansion. For example the FXS and FRDA triplet repeats impede flap endonuclease-1 (FEN-1) processing of the 5′ flap of Okazaki fragments generated during lagging strand DNA synthesis.[45,46] FEN-1 is inhibited by secondary structures in the flap DNA and the effect of the FXS and FRDA is thought to involve the formation of such structures. A role for FEN-1 inhibition in expansion in humans is suggested by the fact that mutations in rad27p, the yeast homolog of FEN-1, increase the CGG•CCG-repeat expansion frequency.[47] The structures formed by these repeats also block the progress of DNA polymerase in vitro,[31] and cause stalling of the replication fork in bacteria[48] and yeast.[49] Thus it is also possible that subsequent repeat-induced repeated strand slippage or attempts by the cell to repair the fork contributes to expansion.[49] Alternatively, since structures with regions of single-strandedness are prone to strand breakage or strand invasion,[50] it may be that expansion results from repeat-mediated recombination instead.[51]

In addition to the role they play in FXS, long CGG•CCG-repeat tracts form folate-sensitive fragile sites,[52-54] which appear microscopically like prematurely packaged chromatin. These sites may result from the propensity of the FXS repeats to exclude nucleosomes,[55] an effect that may be related to the unusual properties of duplexes containing these repeats. However, since these sites are induced by agents that deplete intracellular nucleotide pools,[56] they may result from a problem with DNA replication instead. The FXS repeats form structures that block DNA synthesis much more effectively than other triplet repeats which do not form fragile sites.[48,49] Depleted nucleotide pools may slow replication thus allowing these structures to form more readily. This in turn may lead to the failure to complete replication before packaging is initiated.

In addition to potential effects on DNA stability and chromosome fragility, the structures formed by the repeats may also be responsible for the transcription defect in FRDA, and the symptoms seen in carriers of both FXS premutation and full mutation alleles. This issue will be the subject of the remainder of this chapter.

DNA Structures and Elevated FMR1 *mRNA Levels in FXS Premutation Carriers*

A point mutation in FMRP resulting in the substitution of asparagine for isoleucine at position 340, leads to a very severe FXS phenotype that is not accompanied by an increase in *FMR1* mRNA.[20] This suggests that the elevated levels of RNA seen in carriers of premutation alleles is not due to feedback regulation of the promoter in response to inadequate levels of functional FMRP. The failure to see increased transcription on naked DNA templates,[57,58] suggests that chromatin structure may be involved. Nucleosome exclusion together with the propensity of these repeats to act as a sink for superhelical energy,[23] may lead to a more open chromatin and DNA conformation that facilitates initiation of transcription in premutation carriers as illustrated in Figure 3. This could be accomplished not only by promoting melting of the transcription start site by RNA polymerase II but also by increasing the access of various transcription factors to the promoter.

DNA Hairpins and Aberrant Promoter Methylation in Individuals with FXS

The importance of DNA methylation for gene silencing in FXS is demonstrated by the fact that 5-azadeoxycytidine (5-azadC), an inhibitor of DNA methyltransferase (DNMT), can reactivate the *FMR1* gene in cells from fragile X patients.[59] DNA methylation also directly blocks binding of the transcription factor nuclear respiratory factor 1 (NRF-1) leading to a large drop in promoter activity.[60]

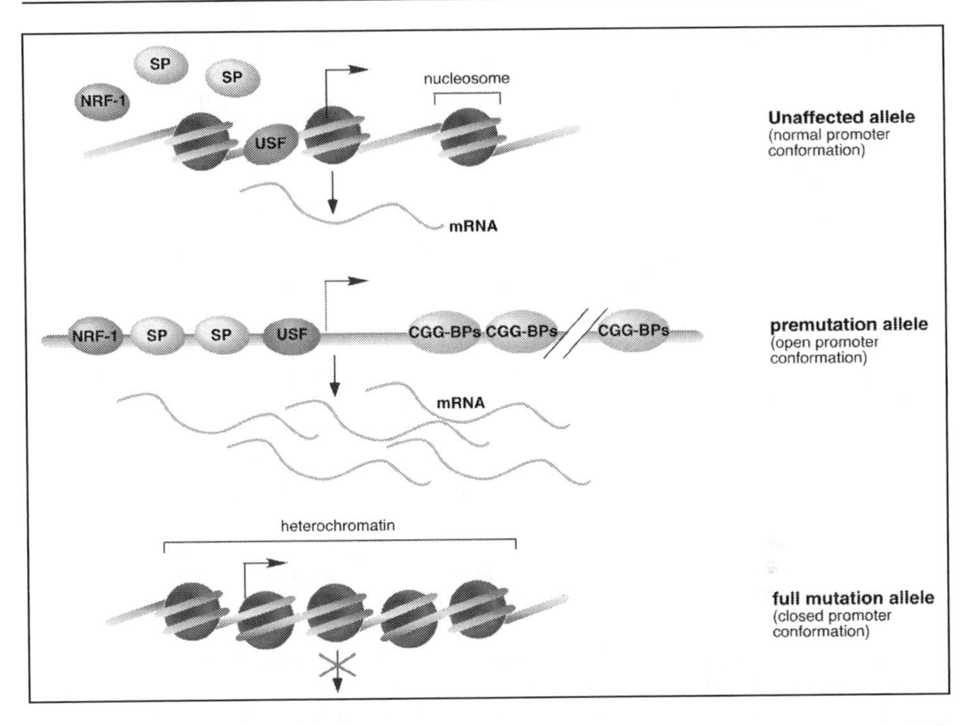

Figure 3. Diagrammatic representation of the *FMR1* promoter in normal and in FXS premutation and full mutation alleles. In alleles with fewer than 60 repeats, the promoter and the repeat bind histones normally, and these histones have modifications characteristic of actively transcribing chromatin including acetylation of histone H3 at lysine 9. The promoter is thus accessible to the transcription factors necessary for optimal promoter activity. In "premutation" alleles which contain 60-200 CGG•CCG-repeats, the flexibility of the repeat region is exacerbated, as is the peculiar radius of curvature. The repeat tracts exclude nucleosomes effectively and this together with the enhanced writhe of these repeats may lead to a more open promoter conformation that is more freely accessible to the transcription factors that have been shown to be important for the full activity of this gene. This includes NRF-1, members of the SP family of proteins, and heterodimers of upstream stimulatory factors 1 and 2 (USF). Furthermore the additional repeats may lead to a high level of binding of proteins that bind to CGG•CCG-repeats (CGG-binding proteins [CGG-BPs]) and can act as transcriptional activators. This leads to an increase in promoter activity. The net result of this is an increase in the amount of an RNA containing a long CGG-repeat tract. Evidence suggests this RNA responsible for the cerebellar and ovarian dysfunction seen in carriers of these alleles, perhaps by sequestering important proteins or by otherwise affecting normal gene expression. In full mutation alleles, a threshold for heterochromatinization is exceeded with the result that both the promoter and H3 histones associating with the promoter are methylated. Other proteins important for gene silencing are recruited to the promoter either by binding to the methylated cytosines in the DNA or the modified histones. The promoter is compacted and buried within the heterochromatin complex thereby preventing access of transcription factors to their binding sites. In addition, the DNA methylation blocks binding of NRF-1 an important transcription factor. The net result is a significant drop in the activity of the *FMR1* promoter and the *FMR1* transcript deficit that is responsible in large part for the FMRP insufficiency that leads to the symptoms of FXS.

Slipped structures containing CCG-hairpins are 10-15 times more efficient substrates for human DNMTs than either the corresponding Watson-Crick duplexes or CCG-hairpins.[61] This substrate efficiency is thought to be due to a combination of the CCG-hairpin in which cytosines are more readily flipped out of the helix into the active site of the enzyme than WC G•C pairs, and the ability of the CCG-hairpin in these three-way junctions to move along the WC arms thereby converting WC CpG sites into sites that are better methylase substrates.[61]

According to this view, structure formation by the FXS repeat is the trigger for DNA methylation which then spreads into the adjacent promoter. DNA methylation leads to the recruitment of a variety of proteins including histone methyltransferases (HMTs) that methylate histone H3 at lysine 9 (K9) generating one of the hallmarks of transcriptionally silent chromatin.

Triplexes and the Transcription Defect in FRDA

Sequences with triplex-forming potential within a transcription unit block transcription elongation both in vitro,[62,63] and in *Escherichia coli.*[64] There is also a wealth of data to support the idea that intermolecular triplex formation is able to down-regulate transcription elongation in mammalian cells.[65] This has led to the suggestion that one of the triplexes formed by the FRDA repeats may account for the transcription deficit in FRDA patients.[22,40,42,66-68] Triplexes that have been invoked include Pu•Pu•Py triplexes in which the RNA transcript acts as the third strand,[40] and the "sticky DNA" which forms when two direct repeats of GAA•TTC-repeats are present in a single supercoiled plasmid.[44]

In vitro transcription studies show that these repeats have the intrinsic ability to decrease the yield of full length transcript, an effect exacerbated by template supercoiling.[42,68] Transcripts are truncated at the distal end of the repeat and the polymerase remains attached to the template. This suggests a model in which a triplex is formed during transcription as illustrated in Figure 4. In this model, a Pu•Pu•Py triplex forms between the repeats that have already reannealed behind the polymerase and the free purine strand resulting from the occlusion of the template strand by the transcribing polymerase. The polymerase is trapped at the triplex:duplex junction, and the pyrimidine strand that would normally form a duplex with what is now the third strand in the triplex, is unpaired allowing an RNA: DNA hybrid to form. Support for this model comes from the observation that the transcription deficit can be alleviated by the addition of a single stranded oligonucleotide containing TTC-repeats. The oligonucleotide anneals to the GAA-rich strand, thus blocking triplex formation.[68]

A Unified Theory for Gene Silencing in FXS and FRDA?

However, recent work raises the intriguing possibility that the transcription defects in FXS and FRDA may share a common mechanism, at least in part. In *Drosophila* gene silencing occurs despite the absence of DNA methylation,[69] and in *Neurospora crassa* and *Arabidopsis thaliana* loss of the HMT responsible for H3K9 methylation leads to a loss of DNA methylation.[70-72] That chromatin modification is also a crucial early event in gene silencing in mammals is suggested by the fact that in mice mutation of LSH, a member of the SNF2 family of chromatin remodeling proteins, leads to a loss of DNA methylation.[73] Moreover, during X-chromosome inactivation, H3K9 methylation and gene silencing occur well before DNA methylation.[74] In FXS therefore, it may be that the repeats lead to gene silencing by directly recruiting the enzymes necessary for the histone modifications that generate transcriptionally silent chromatin.

A role for HMTs in FRDA as well is suggested by the recent finding that GAA•TTC-repeats cause heterochromatinization of a linked transgene in mice.[75] Since these repeats are devoid of CpG residues, this process presumably occurs independently of DNA methylation. Since the FRDA repeat is only ~1.7 kb away from the start of transcription in the frataxin gene,[76] heterochromatinization of the FRDA repeat may affect frataxin promoter activity. It may thus be that both FXS and FRDA result, at least in part, from gene silencing in which the primary epigenetic signal is repeat-induced histone modification.

By analogy with AT hook gene silencing proteins which are thought to recognize some feature of the minor groove of their target sequences rather than the primary DNA sequence,[77] the unusual duplexes formed by these repeats may allow binding by a protein that facilitates heterochromatinization. Alternatively, comparison with the dodeca-centromeric satellite of *Drosophila* may be more apt: one strand of this satellite forms a hairpin leaving the other strand available to bind *Drosophila* dodeca-satellite-binding protein 1 (DDP1), a single-stranded binding protein thought to be involved in silencing.[78]

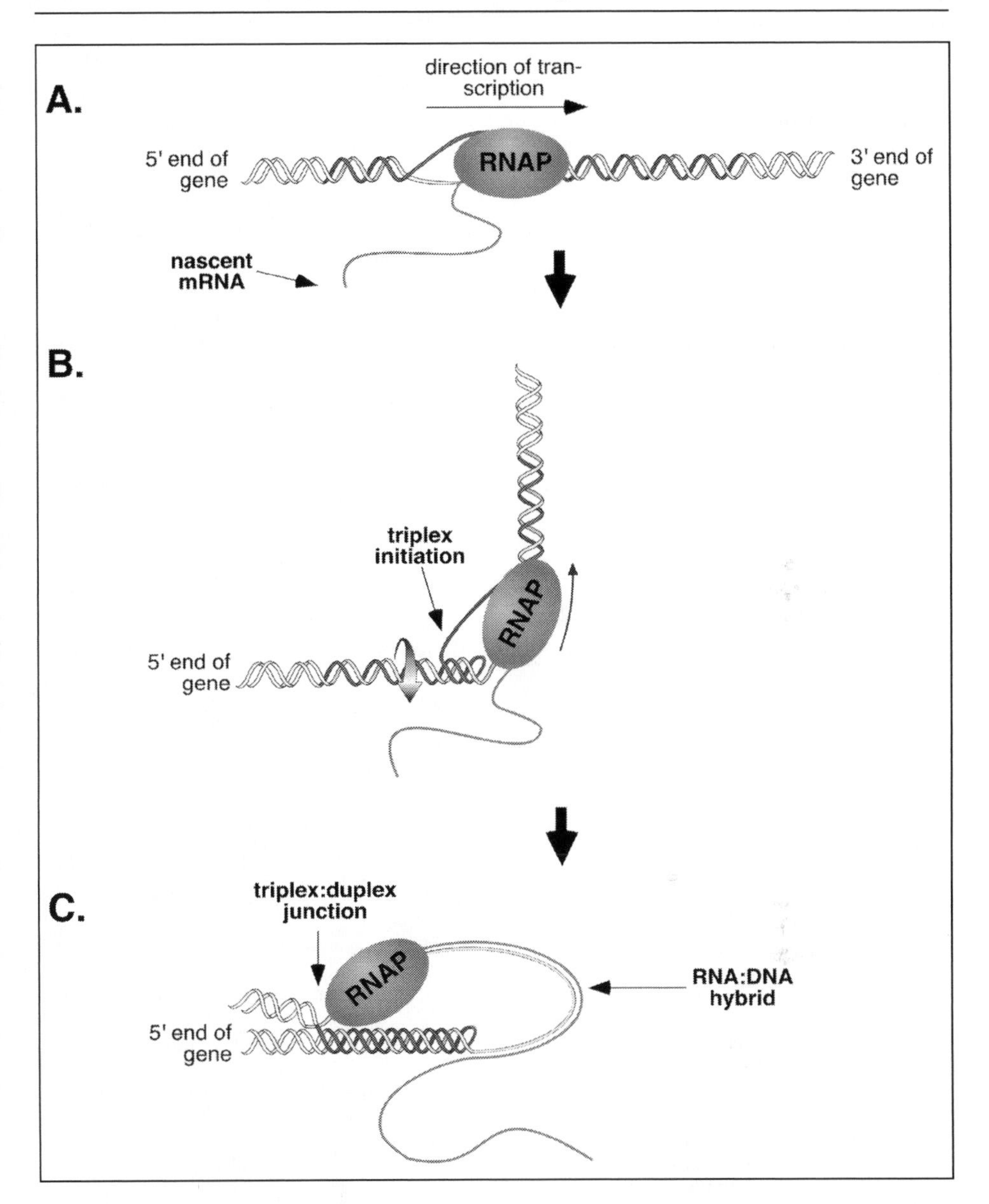

Figure 4. Triplex model for the transcription deficit in FRDA. The elongating transcription complex proceeds into the repeat region separating the template and non-template strand as it goes, and drawing the template strand into the channel of the RNA polymerase (RNAP) (A). As the polymerase moves further into the repeat, the template which has already been transcribed emerges from the interior of the polymerase and is now able to reanneal with the cognate region of the non-template strand. At the same time, the region complementary to the region of the template currently being transcribed is free to hydrogen bond with this duplex initiating the formation of an Pu•Pu•Py triplex (B). Transcription continues in this fashion forming a longer and longer triplex structure behind the polymerase, until the end of the repeat is reached (C). At this point most of the non-template strand is involved in a triplex, leaving the 5′ end of the template strand unpaired. This allows the nascent RNA strand to anneal to form an RNA:DNA hybrid. The polymerase is unable to proceed beyond this point resulting in the failure to produce a full-length transcript and reducing the amount of polymerase available to initiate further rounds of RNA synthesis. Adapted from Grabczyk and Usdin.[68]

However a wide variety of different repeats are prone to heterochromatinization and they often evolve rapidly without significant sequence conservation.[79] Moreover some of these repeats can be interrupted by different transposable elements without affecting silencing.[80] While it is possible that there is some aspect of DNA conformation that is shared by all of these sequences, recent findings suggest another explanation. In yeast, mutations in genes in the RNAi pathway lead to loss of gene silencing.[81] In plants transgene silencing by double-stranded RNA (dsRNA) is associated with methylation of the region homologous to the dsRNA,[82,83] and a mutation in argonaute 4, a protein involved in RNA-mediated gene silencing in *Drosophila* and fission yeast, reactivates silenced genes and decreases both DNA methylation and H3K9 methylation.[84] Given the many parallels between yeast, plants and mammals in various aspects of RNAi, dsRNA may also be a signal for H3K9 methylation in mammals.[85] dsRNA can be generated in vivo in 2 ways: by the generation of overlapping transcripts generated from both DNA strands and by transcription of palindromic and quasi-palindromic sequences that are able to form fold-back structures. We have shown that CGG-RNA can form hairpins analogous to its DNA counterpart,[86] and it may well be that GAA-RNA behaves the same way. Thus the unusual structures formed by the repeat containing RNA rather than the DNA may make the FXS and FRDA repeats particularly prone to heterochromatinization.

Conclusion

The FXS and FRDA repeats form a variety of different nucleic acid structures in vitro. These structures have properties that may account for both expansion and disease pathology. Whether this is in fact the case, and if so, which particular structure or structures are involved remains to be seen.

References

1. Greene E, Handa V, Kumari D et al. Transcription defects induced by repeat expansion: fragile X syndrome, FRAXE mental retardation, progressive myoclonus epilepsy type 1, and Friedreich ataxia. Cytogenet Genome Res 2003; 100:65-76.
2. Jin P, Warren ST. New insights into fragile X syndrome: from molecules to neurobehaviors. Trends Biochem Sci 2003; 28:152-158.
3. Warren ST, Sherman SL. The fragile X syndrome. In: Scriver CR, Beaudet AL, Sly WS, Valle D, eds. The Metabolic and Molecular Basis of Inherited Disease. Vol 8. New York: McGraw-Hill, 2001:257-289.
4. Conway GS, Hettiarachchi S, Murray A et al. Fragile X premutations in familial premature ovarian failure. Lancet 1995; 346:309-310.
5. Partington MW, Moore DY, Turner GM. Confirmation of early menopause in fragile X carriers. Am J Med Genet 1996; 64:370-372.
6. Macpherson J, Murray A, Webb J et al. Fragile X syndrome: of POF and premutations. J Med Genet 1999; 36:171-172.
7. Allingham-Hawkins DJ, Babul-Hirji R, Chitayat D et al. Fragile X premutation is a significant risk factor for premature ovarian failure: the international collaborative POF in Fragile X study-preliminary data. Am J Med Genet 1999; 83:322-325.
8. Vianna-Morgante AM. Twinning and premature ovarian failure in premutation fragile X carriers. Am J Med Genet 1999; 83:326.
9. Uzielli ML, Guarducci S, Lapi E et al. Premature ovarian failure (POF) and fragile X premutation females: from POF to fragile X carrier identification, from fragile X carrier diagnosis to POF association data. Am J Med Genet 1999; 84:300-303.
10. Jacquemont S, Hagerman RJ, Leehey M et al. Fragile X premutation tremor/ataxia syndrome: molecular, clinical, and neuroimaging correlates. Am J Hum Genet 2003; 72:869-878.
11. Leehey MA, Munhoz RP, Lang AE et al. The fragile X premutation presenting as essential tremor. Arch Neurol 2003; 60:117-121.
12. Brunberg JA, Jacquemont S, Hagerman RJ et al. Fragile X premutation carriers: characteristic MR imaging findings of adult male patients with progressive cerebellar and cognitive dysfunction. Am J Neuroradiol 2002; 23:1757-1766.
13. Hagerman RJ, Leehey M, Heinrichs W et al. Intention tremor, parkinsonism, and generalized brain atrophy in male carriers of fragile X. Neurology 2001; 57:127-130.

14. Greco CM, Hagerman RJ, Tassone F et al. Neuronal intranuclear inclusions in a new cerebellar tremor/ataxia syndrome among fragile X carriers. Brain 2002; 125:1760-1771.
15. Oberle I, Rousseau F, Heitz D et al. Instability of a 550-base pair DNA segment and abnormal methylation in fragile X syndrome. Science 1991; 252:1097-1102.
16. Coffee B, Zhang F, Warren ST et al. Acetylated histones are associated with *FMR1* in normal but not fragile X-syndrome cells. Nat Genet 1999; 22:98-101.
17. Coffee B, Zhang F, Ceman S et al. Histone modifications depict an aberrantly heterochromatinized *FMR1* gene in fragile X syndrome. Am J Hum Genet 2002; 71:923-932.
18. Feng Y, Zhang F, Lokey LK et al. Translational suppression by trinucleotide repeat expansion at *FMR1*. Science 1995; 268:731-734.
19. Tassone F, Hagerman RJ, Taylor AK et al. Elevated levels of *FMR1* mRNA in carrier males: a new mechanism of involvement in the fragile-X syndrome. Am J Hum Genet 2000; 66:6-15.
20. Kenneson A, Zhang F, Hagedorn CH et al. Reduced FMRP and increased *FMR1* transcription is proportionally associated with CGG repeat number in intermediate-length and premutation carriers. Hum Mol Genet 2001; 10:1449-1454.
21. Pandolfo M. The molecular basis of Friedreich ataxia. Adv Exp Med Biol 2002; 516:99-118.
22. Bidichandani SI, Ashizawa T, Patel PI. The GAA triplet-repeat expansion in Friedreich ataxia interferes with transcription and may be associated with an unusual DNA structure. Am J Hum Genet 1998; 62:111-121.
23. Bacolla A, Gellibolian R, Shimizu M et al. Flexible DNA: genetically unstable CTG•CAG and CGG•CCG from human hereditary neuromuscular disease genes. J Biol Chem 1997; 272:16783-16792.
24. Chastain PD, 2nd, Eichler EE, Kang S et al. Anomalous rapid electrophoretic mobility of DNA containing triplet repeats associated with human disease genes. Biochemistry 1995; 34:16125-16131.
25. Chastain PD, Sinden RR. CTG repeats associated with human genetic disease are inherently flexible. J Mol Biol 1998; 275:405-411.
26. Zheng M, Huang X, Smith GK et al. Genetically unstable CXG repeats are structurally dynamic and have a high propensity for folding. An NMR and UV spectroscopic study. J Mol Biol 1996; 264:323-336.
27. Mitas M, Yu A, Dill J et al. The trinucleotide repeat sequence $d(CGG)_{15}$ forms a heat-stable hairpin containing Gsyn•Ganti base pairs. Biochemistry 1995; 34:12803-12811.
28. Mariappan SV, Catasti P, Chen X et al. Solution structures of the individual single strands of the fragile X DNA triplets (GCC)n•(GGC)n. Nucleic Acids Res 1996; 24:784-792.
29. Fry M, Loeb LA. The fragile X syndrome d(CGG)n nucleotide repeats form a stable tetrahelical structure. Proc Natl Acad Sci USA 1994; 91:4950-4954.
30. Kettani A, Kumar RA, Patel DJ. Solution structure of a DNA quadruplex containing the fragile X syndrome triplet repeat. J Mol Biol 1995; 254:638-656.
31. Usdin K, Woodford KJ. CGG repeats associated with DNA instability and chromosome fragility form structures that block DNA synthesis in vitro. Nucleic Acids Res 1995; 23:4202-4209.
32. Patel PK, Bhavesh NS, Hosur RV. Cation-dependent conformational switches in d-TGGCGGC containing two triplet repeats of Fragile X Syndrome: NMR observations. Biochem Biophys Res Commun 2000; 278:833-838.
33. Yu A, Barron MD, Romero RM et al. At physiological pH, $d(CCG)_{15}$ forms a hairpin containing protonated cytosines and a distorted helix. Biochemistry 1997; 36:3687-3699.
34. Mariappan SV, Silks LA, 3rd, Chen X et al. Solution structures of the Huntington's disease DNA triplets, (CAG)n. J Biomol Struct Dyn 1998; 15:723-744.
35. Fojtik P, Vorlickova M. The fragile X chromosome (GCC) repeat folds into a DNA tetraplex at neutral pH. Nucleic Acids Res 2001; 29:4684-4690.
36. Jithesh PV, Singh P, Joshi R. Molecular dynamics studies of trinucleotide repeat DNA involved in neurodegenerative disorders. J Biomol Struct Dyn 2001; 19:479-495.
37. LeProust EM, Pearson CE, Sinden RR et al. Unexpected formation of parallel duplex in GAA and TTC trinucleotide repeats of Friedreich's ataxia. J Mol Biol 2000; 302:1063-1080.
38. Heidenfelder BL, Makhov AM, Topal MD. Hairpin formation in Friedreich's ataxia triplet repeat expansion. J Biol Chem 2003; 278:2425-2431.
39. Hanvey JC, Shimizu M, Wells RD. Intramolecular DNA triplexes in supercoiled plasmids. Proc Natl Acad Sci USA 1988; 85:6292-6296.
40. Mariappan SV, Catasti P, Silks LA, 3rd et al. The high-resolution structure of the triplex formed by the GAA/TTC triplet repeat associated with Friedreich's ataxia. J Mol Biol 1999; 285:2035-2052.
41. Gacy AM, Goellner GM, Spiro C et al. GAA instability in Friedreich's Ataxia shares a common, DNA-directed and intraallelic mechanism with other trinucleotide diseases. Mol Cell 1998; 1:583-593.

42. Grabczyk E, Usdin K. The GAA·TTC triplet repeat expanded in Friedreich's ataxia impedes transcription elongation by T7 RNA polymerase in a length and supercoil dependent manner. Nucleic Acids Res 2000; 28:2815-2822.
43. Jain A, Rajeswari MR, Ahmed F. Formation and thermodynamic stability of intermolecular (R*R·Y) DNA triplex in GAA/TTC repeats associated with Freidreich's ataxia. J Biomol Struct Dyn 2002; 19:691-699.
44. Vetcher AA, Napierala M, Iyer RR et al. Sticky DNA, a long GAA•GAA•TTC triplex that is formed intramolecularly, in the sequence of intron 1 of the frataxin gene. J Biol Chem 2002; 277:39217-39227.
45. Spiro C, Pelletier R, Rolfsmeier ML et al. Inhibition of FEN-1 processing by DNA secondary structure at trinucleotide repeats. Mol Cell 1999; 4:1079-1085.
46. Henricksen LA, Tom S, Liu Y et al. Inhibition of flap endonuclease 1 by flap secondary structure and relevance to repeat sequence expansion. J Biol Chem 2000; 275:16420-16427.
47. White PJ, Borts RH, Hirst MC. Stability of the human fragile X (CGG)(n) triplet repeat array in *Saccharomyces cerevisiae* deficient in aspects of DNA metabolism. Mol Cell Biol 1999; 19:5675-5684.
48. Samadashwily GM, Raca G, Mirkin SM. Trinucleotide repeats affect DNA replication in vivo. Nat Genet 1997; 17:298-304.
49. Pelletier R, Krasilnikova MM, Samadashwily GM et al. Replication and expansion of trinucleotide repeats in yeast. Mol Cell Biol 2003; 23:1349-1357.
50. Balakumaran BS, Freudenreich CH, Zakian VA. CGG/CCG repeats exhibit orientation-dependent instability and orientation-independent fragility in *Saccharomyces cerevisiae*. Hum Mol Genet 2000; 9:93-100.
51. Jakupciak JP, Wells RD. Genetic instabilities of triplet repeat sequences by recombination. IUBMB Life 2000; 50:355-359.
52. Parrish JE, Oostra BA, Verkerk AJ et al. Isolation of a GCC repeat showing expansion in FRAXF, a fragile site distal to FRAXA and FRAXE. Nat Genet 1994; 8:229-235.
53. Knight SJ, Flannery AV, Hirst MC et al. Trinucleotide repeat amplification and hypermethylation of a CpG island in FRAXE mental retardation. Cell 1993; 74:127-134.
54. Jones C, Mullenbach R, Grossfeld P et al. Co-localisation of CCG repeats and chromosome deletion breakpoints in Jacobsen syndrome: evidence for a common mechanism of chromosome breakage. Hum Mol Genet 2000; 9:1201-1208.
55. Wang YH, Gellibolian R, Shimizu M et al. Long CCG triplet repeat blocks exclude nucleosomes: a possible mechanism for the nature of fragile sites in chromosomes. J Mol Biol 1996; 263:511-516.
56. Sutherland GR. The role of nucleotides in human fragile site expression. Mutat Res 1988; 200:207-213.
57. Parsons MA, Sinden RR, Izban MG. Transcriptional properties of RNA polymerase II within triplet repeat-containing DNA from the human myotonic dystrophy and fragile X loci. J Biol Chem 1998; 273:26998-27008.
58. Chandler SP, Kansagra P, Hirst MC. Fragile X (CGG)n repeats induce a transcriptional repression in cis upon a linked promoter: evidence for a chromatin mediated effect. BMC Mol Biol 2003; 4:3.
59. Chiurazzi P, Pomponi MG, Willemsen R et al. In vitro reactivation of the *FMR1* gene involved in fragile X syndrome. Hum Mol Genet 1998; 7:109-113.
60. Kumari D, Usdin K. Interaction of the transcription factors USF1, USF2, and α-Pal/Nrf-1 with the *FMR1* promoter. Implications for Fragile X mental retardation syndrome. J Biol Chem 2001; 276:4357-4364.
61. Chen X, Mariappan SV, Moyzis RK et al. Hairpin induced slippage and hyper-methylation of the fragile X DNA triplets. J Biomol Struct Dyn 1998; 15:745-756.
62. Ashley C, Lee JS. A triplex-mediated knot between separated polypurine-polypyrimidine tracts in circular DNA blocks transcription by *Escherichia coli* RNA polymerase. DNA Cell Biol 2000; 19:235-241.
63. Grabczyk E, Fishman MC. A long purine-pyrimidine homopolymer acts as a transcriptional diode. J Biol Chem 1995; 270:1791-1797.
64. Sarkar PS, Brahmachari SK. Intramolecular triplex potential sequence within a gene down regulates its expression in vivo. Nucleic Acids Res 1992; 20:5713-5718.
65. Giovannangeli C, Helene C. Triplex technology takes off. Nat Biotechnol 2000; 18:1245-1246.
66. Ohshima K, Montermini L, Wells RD et al. Inhibitory effects of expanded GAA•TTC triplet repeats from intron I of the Friedreich ataxia gene on transcription and replication in vivo. J Biol Chem 1998; 273:14588-14595.
67. Sakamoto N, Larson JE, Iyer RR et al. GGA•TCC-interrupted triplets in long GAA•TTC repeats inhibit the formation of triplex and sticky DNA structures, alleviate transcription inhibition, and reduce genetic instabilities. J Biol Chem 2001; 276:27178-27187.

68. Grabczyk E, Usdin K. Alleviating transcript insufficiency caused by Friedreich's ataxia triplet repeats. Nucleic Acids Res 2000; 28:4930-4937.
69. Urieli-Shoval S, Gruenbaum Y, Sedat J et al. The absence of detectable methylated bases in *Drosophila melanogaster* DNA. FEBS Lett 1982; 146:148-152.
70. Tamaru H, Zhang X, McMillen D et al. Trimethylated lysine 9 of histone H3 is a mark for DNA methylation in *Neurospora crassa*. Nat Genet 2003; 34:75-79.
71. Tamaru H, Selker EU. A histone H3 methyltransferase controls DNA methylation in *Neurospora crassa*. Nature 2001; 414:277-283.
72. Jackson JP, Lindroth AM, Cao X et al. Control of CpNpG DNA methylation by the KRYPTONITE histone H3 methyltransferase. Nature 2002; 416:556-560.
73. Dennis K, Fan T, Geiman T et al. Lsh, a member of the SNF2 family, is required for genome-wide methylation. Genes Dev 2001; 15:2940-2944.
74. Loebel DA, Johnston PG. Methylation analysis of a marsupial X-linked CpG island by bisulfite genomic sequencing. Genome Res 1996; 6:114-123.
75. Saveliev A, Everett C, Sharpe T et al. DNA triplet repeats mediate heterochromatin-protein-1-sensitive variegated gene silencing. Nature 2003; 422:909-913.
76. Campuzano V, Montermini L, Molto MD et al. Friedreich's ataxia: autosomal recessive disease caused by an intronic GAA triplet repeat expansion. Science 1996; 271:1423-1427.
77. Girard F, Bello B, Laemmli UK et al. In vivo analysis of scaffold-associated regions in *Drosophila*: A synthetic high-affinity SAR binding protein suppresses position effect variegation. EMBO J 1998; 17:2079-2085.
78. Cortes A, Huertas D, Fanti L et al. DDP1, a single-stranded nucleic acid-binding protein of *Drosophila*, associates with pericentric heterochromatin and is functionally homologous to the yeast Scp160p, which is involved in the control of cell ploidy. EMBO J 1999; 18:3820-3833.
79. Henikoff S, Ahmad K, Malik HS. The centromere paradox: stable inheritance with rapidly evolving DNA. Science 2001; 293:1098-1102.
80. Lee C, Wevrick R, Fisher RB et al. Human centromeric DNAs. Hum Genet 1997; 100:291-304.
81. Volpe TA, Kidner C, Hall IM et al. Regulation of heterochromatic silencing and histone H3 lysine-9 methylation by RNAi. Science 2002; 297:1833-1837.
82. Mette MF, Aufsatz W, van der Winden J et al. Transcriptional silencing and promoter methylation triggered by double-stranded RNA. EMBO J 2000; 19:5194-5201.
83. Jones L, Ratcliff F, Baulcombe DC. RNA-directed transcriptional gene silencing in plants can be inherited independently of the RNA trigger and requires Met1 for maintenance. Curr Biol 2001; 11:747-757.
84. Zilberman D, Cao X, Jacobsen SE. ARGONAUTE4 control of locus-specific siRNA accumulation and DNA and histone methylation. Science 2003; 299:716-719.
85. Allshire R. Molecular biology. RNAi and heterochromatin-a hushed-up affair. Science 2002; 297:1818-1819.
86. Handa V, Saha T, Usdin K. The Fragile X syndrome repeats form RNA hairpins that do not activate the interferon-inducible protein kinase, PKR, but are cut by Dicer. Nucleic Acids Res 2003; 31:6243-6248.

Chapter 10

Possible Roles of DNA Supercoiling in Transcription

Susumu Hirose and Kuniharu Matsumoto

Abstract

Transcription and supercoiling of the template DNA are closely related each other. DNA supercoiling affects transcription and transcription affects supercoiling of the template DNA. Furthermore, packaging of genomic DNA into chromatin in eukaryotes raises another type of relation. DNA supercoiling can affect transcription through modulation of the chromatin structure.

Introduction

Typical double-stranded DNA consists of a helix with a pitch of one turn per 10.5 base pairs. Both underwinding and overwinding of the DNA double helix induce twisting and coiling of the helix unless the DNA strand can rotate freely. The coils thus formed are termed negative and positive supercoils, respectively (see Chapter 1 for details). Current views of the mechanisms underlying transcriptional regulation rely on a concept originally proposed by Jacob and Monod: regulation through *cis*-elements on DNA and *trans*-acting factors that bind to the elements.[1] However, DNA supercoiling can modulate accessibility of *trans*-acting factors to *cis*-elements. Transcription is an asymmetric process: only one strand of the DNA double helix is copied into RNA. To achieve this, the double helix must be locally unwound. Therefore, negative supercoiling of DNA is thought to facilitate the step. Moreover, DNA supercoiling can affect transcription in chromatin context in eukaryotes. This chapter discusses roles of DNA supercoiling in transcriptional regulation.

Prokaryotic Transcription

In prokaryotes, there are two major topoisomerases that act toward opposite direction.[2] DNA gyrase can generate negative supercoils into relaxed DNA and relax positively supercoiled DNA. In contrast, topoisomerase I can relax negatively supercoiled DNA but not positively supercoiled DNA. The superhelical state of cellular DNA in prokaryotes appears to be under equilibrium between actions of these topoisomerases. Measurements of psoralen binding averaged globally across the *Escherichia coli* genome have detected unconstrained negative supercoils with a superhelical density of -0.05.[3]

Consistent with the finding of unconstrained negative supercoils in genomic DNA, it has been established that DNA supercoiling functions as a regulator of prokaryotic transcription.[4] First, supercoiling affects transcription in vitro. Some promoters have an optimum level of supercoiling for their transcription and the level is different for different promoters. Second, supercoiling plays a regulatory role in vivo. Genetic studies have shown that mutations in topoisomerase I[5] or DNA gyrase[6] affect transcription in vivo.

DNA Conformation and Transcription, edited by Takashi Ohyama. ©2005 Eurekah.com and Springer Science+Business Media.

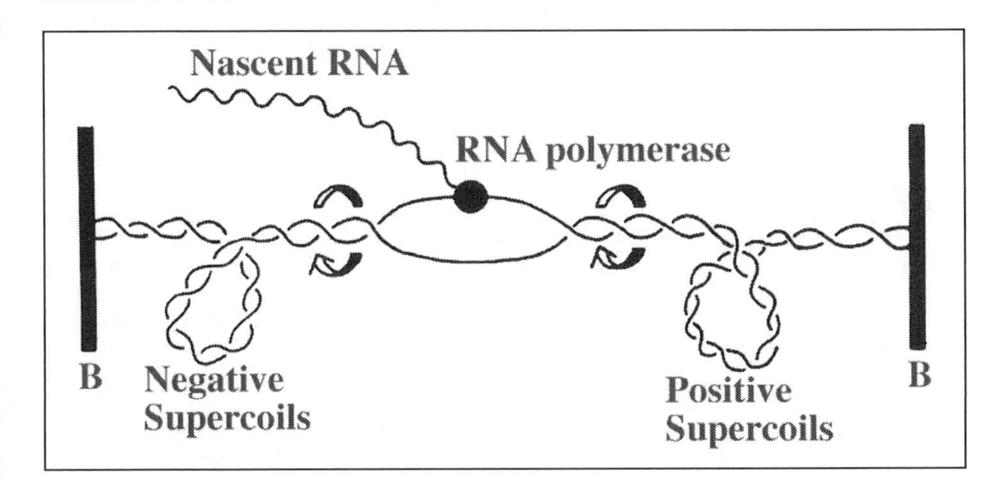

Figure 1. Transcription-driven supercoils. Tracking RNA polymerase generates positive supercoils in the template DNA ahead of it and negative supercoils behind it. "B" represents a topological barrier. It does not necessarily require attachment of DNA to cellular structures. Certain sequence-specific binding proteins can form the topological barrier upon binding to DNA.[35] Modified from Hirose S, Ohta T. Cell Struct Funct 1990;15:133-135.

Two rate limiting steps are known for prokaryotic transcription. One is formation of an RNA polymerase-DNA open complex and the other is promoter clearance. Because negative supercoiling favors the unwinding of the DNA double helix that is required for formation of the open complex, it is expected to increase the rate of transcription for promoters in which open complex formation is rate limiting. Indeed most genes are activated by increased negative supercoiling. However, transcription of *gyr A* and *gyr B* encoding the subunits of DNA gyrase is induced by relaxation of DNA. It has been proposed that promoter clearance but not open complex formation is the rate-limiting step for these promoters.[7]

The superhelical state of DNA is known to change depending on the growth conditions of cells.[8] For example, nutrient downshift and stationary growth phase cause a decrease in the extent of negative supercoiling, while high osmolarity leads to an increase in the negative supercoiling of DNA. In agreement with these findings, transcription by RNA polymerase harboring the stationary phase-specific σ^S appears to be enhanced on templates with decreased superhelicity.[9] DNA supercoiling also changes transiently during heat shock.[10] The heat stress induces rapid relaxation of negative supercoils and then DNA topology returns back to the original state with negative supercoiling. In response to the relaxation, most genes are repressed but some specific or stress genes are induced.[11]

Transcription of a double helical DNA requires a rotation of an RNA polymerase and its nascent RNA chain relative to DNA. The velocity of the rotation is calculated to be a few hundred rounds per minute since a pitch of the helix is about 10 base pairs and the rate of RNA chain elongation is a few thousand nucleotides per minute. It seems difficult for the RNA polymerase and its nascent RNA to rotate around the template DNA at such a high velocity. Instead the DNA must turn around its axis during transcription. Then the tracking RNA polymerase generates positive supercoils in the template DNA ahead of it and negative supercoils behind it (Fig. 1). These supercoils will be relaxed by DNA topoisomerases. Liu and Wang summarized the concept as the twin-supercoiled-domain model.[12] The model predicts the followings. First, negative supercoils should accumulate in the absence of a negative supercoil-relaxing enzyme. Second, positive supercoils should accumulate in the absence of a positive supercoil-relaxing enzyme. These predictions are fulfilled by two earlier observations. First, pBR322 DNA harboring high degrees of negative supercoiling has been isolated from *topA* mutants of *E. coli* and

Salmonella typhimurium, and the presence of the highly negatively supercoiled DNA was dependent on the transcription of the *tetA* gene.[13] Second, highly positively supercoiled pBR322 DNA has been isolated from *E. coli* treated with gyrase inhibitors.[14] According to the model, negative and positive supercoils would accumulate in the intergenic regions of two divergent and convergent transcription units, respectively. Such supercoils can in turn affect transcription.

Eukaryotic Transcription

In eukaryotes, psoralen-binding assays on human HeLa and *Drosophila* Schneider cell lines have shown that the bulk of each genomic DNA is torsionally relaxed within nuclei.[3] However, it did not necessarily exclude a possibility that there were negatively supercoiled micro domains within these genomes. Indeed unconstrained negative supercoils have been demonstrated in the *hsp70* and 18S-ribosomal RNA genes of the Schneider cell line,[15] and in the dihydrofolate reductase gene[16] and the hygromycin resistance transgenes[17] of cultured human cells.

To test whether the twin-supercoiled-domain model is applicable to eukaryotic transcription, Giaever and Wang constructed a yeast plasmid carrying the coding sequence of the *E. coli topA* gene placed downstream of an inducible yeast promoter.[18] The plasmid DNA became positively supercoiled in yeast *Δtop1 top2ts* cells at the restrictive temperature when the *E. coli* DNA topoisomerase I was expressed. The generation of positive supercoils was observed only during transcription. Because neither one of the yeast DNA topoisomerases I and II can be functional under these conditions and because the *E. coli* enzyme can relax only negative supercoils, these results verify the model. Recently Matsumoto and Hirose[19] have visualized transcription-coupled, unconstrained negative supercoils of DNA in approximately 150 loci on polytene chromosomes of *Drosophila melanogaster*. The results demonstrate that transcription-coupled negative supercoils of DNA exist within a cell even in the presence of active topoisomerases. These negative supercoils can affect transcription.

As described for prokaryotic transcription, supercoiling modulates in vitro transcription of eukaryotic genes.[20] Hirose and Suzuki,[21] and Mizutani et al[22] have shown that transcription of the *Bombyx mori* fibroin gene increases and plateaus from templates of increasing negative supercoiling, and transcription from the adenovirus type 2 major late promoter (Ad2MLP) rises and then falls, while transcription of the *Drosophila hsp70* gene remains unchanged (Fig. 2). Dissection of transcription revealed that formation of a preinitiation complex on the

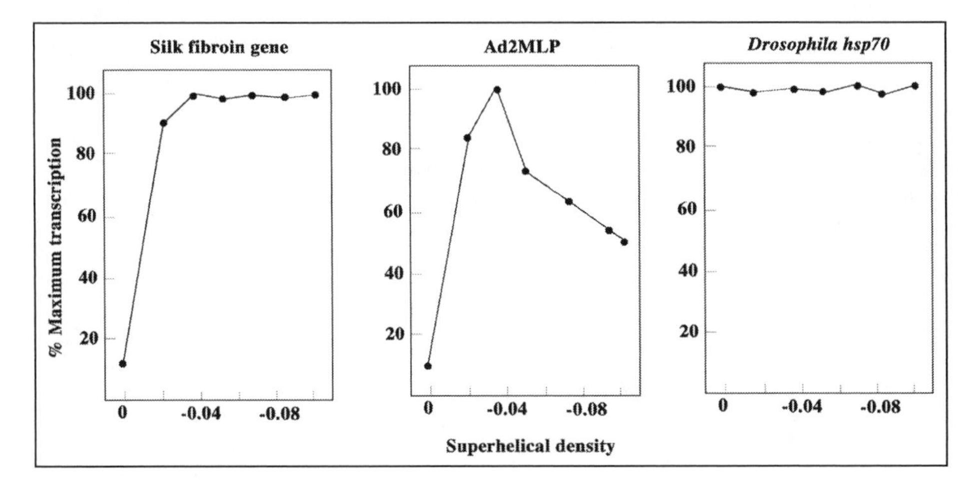

Figure 2. DNA superhelicity affects transcription of eukaryotic genes differently. In vitro transcription activities of indicated promoters were measured on plasmid DNAs with various superhelical densities. Reprinted from Mizutani M, Ura K, Hirose S. Nucl Acids Res 1991;19:2907-2911 with permission from Oxford University Press.

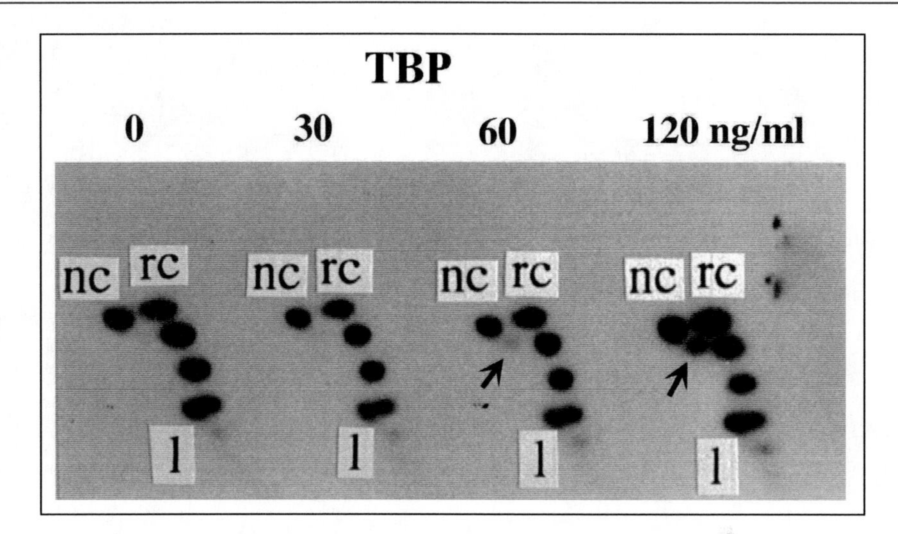

Figure 3. Underwinding of DNA upon binding of TBP to the TATA element. Negatively supercoiled plasmid DNA carrying Ad2MLP was incubated with indicated concentration of yeast TBP at 30°C and then treated with DNA topoisomerase I. DNA was purified and analyzed by two-dimensional electrophoresis. The mark "rc", relaxed closed circular DNA; "nc", nicked circular DNA; "l", linear DNA. Reprinted from Tabuch H, Handa H, Hirose S. Biochem Biophys Res Commun 1993; 192:1432-1438 with permission from Elsevier.

fibroin gene or the Ad2MLP is slow on relaxed DNA and accelerated by negative supercoiling of DNA. On the contrary, the preinitiation complex assembled rapidly on the *hsp70* gene irrespective of DNA topology. Tabuchi et al[23] have demonstrated that binding of TATA element binding protein (TBP) to the TATA element induces underwinding of the duplex DNA (approximately 0.5 linking difference per bound TBP molecule as shown in Fig. 3). The underwinding has been confirmed by crystal structure of a TBP-TATA element complex.[24,25] The change was facilitated by negative supercoiling of DNA on the fibroin promoter and the Ad2MLP but not on the *hsp70* promoter.[23] These data reveal that although supercoiling can affect both prokaryotic and eukaryotic transcription in vitro, the critical steps are different: open complex formation in most genes of prokaryotes vs TBP binding to the TATA element in most genes of eukaryotes. In transcription from the Ad2MLP, promoter clearance is also facilitated by negative supercoiling of DNA.[26] It is possible that clearance of the Ad2MLP goes up and then down with increasing negative supercoiling. Interestingly, the rate-limiting step in *hsp70* transcription is not preinitiation complex formation but restart of a paused RNA polymerase to productive elongation.[27] Probably *hsp70* transcription becomes independent of DNA topology so that it can be induced immediately upon heat shock.

Whether DNA supercoiling affects transcription in vivo is still elusive in eukaryotes. Although Dunaway and Ostrander have clearly shown that local domains of negative supercoiling activate the ribosomal RNA gene promoter in vivo,[28] other promoters have not been tested through a similar approach. Supercoiling factor (SCF) is a protein capable of generating negative supercoils in relaxed DNA in conjunction with topoisomerase II.[29] *D. melanogaster* SCF localizes to many interbands and puffs that are active sites of transcription[30] (Fig. 4), suggesting that SCF plays a role in formation of transcriptionally active chromatin. Recent study has shown that transcription of certain genes is compromised by targeting SCF with RNAi (Furuhashi and Hirose, unpublished). Because SWI/SNF-type chromatin remodeling factors require changes in DNA topology for their action,[31] SCF and topoisomerase II may facilitate chromatin remodeling through generation of negative supercoils.

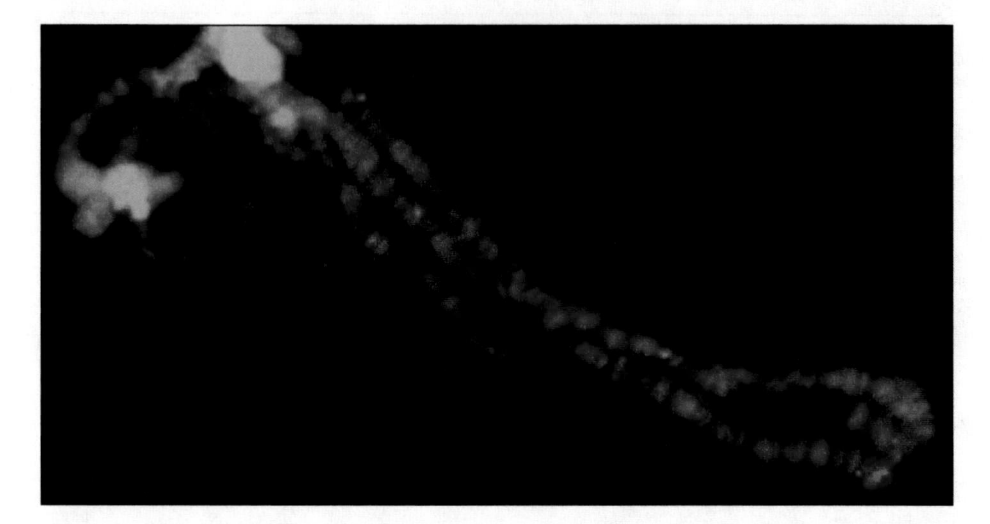

Figure 4. *Drosophila* SCF localizes to interbands and puffs on polytene chromosomes. Salivary gland polytene chromosomes were stained with antibody against SCF (red) and DAPI (blue). Reprinted from Kobayashi M, Aita N, Hayashi S et al. Mol Cell Biol 1998; 18:6373-6744 with permission from American Society for Microbiology.

Formation of unusual DNA structure (see Chapter 1 for details) such as Z-form is significantly facilitated by negative supercoiling.[32] Such unusual DNA structure can affect transcription. For example, Z-DNA in a promoter region has been suggested to participate in transcriptional activation in collaboration with a chromatin remodeling complex BAF.[33] On the contrary, Z-form within a transcribable region would inhibit transcription elongation. Finally positive supercoiling of DNA has been reported to diminish transcription in vivo.[34] These data indicate a possible role of DNA supercoiling in eukaryotic transcription in vivo.

Conclusion

DNA supercoiling can affect transcription in both prokaryotes and eukaryotes. While the critical step is open complex formation in most prokaryotic genes, TBP binding to the TATA element is affected in most eukaryotic genes. Furthermore, DNA supercoiling can affect transcription through chromatin remodeling in eukaryotes.

References

1. Jacob F, Monod J. Genetic regulatory mechanisms in the synthesis of proteins. J Mol Biol 1961; 3:318-356.
2. Wang JC. DNA topoisomerases. Ann Rev Biochem 1996; 65:635-692.
3. Sinden RR, Carlson JO, Pettijohn DE. Torsional tension in the DNA double helix measured with trimethylpsoralen in living *E. coli* cells: analogous measurements in insect and human cells. Cell 1980; 21:773-783.
4. Pruss GJ, Drlica K. DNA supercoiling and prokaryotic transcription. Cell 1989; 56:521-523.
5. Margolin P, Zumstein L, Sternglanz R et al. The *Escherichia coli supX* locus is *topA*, the structural gene for DNA topoisomerase I. Proc Natl Acad Sci USA 1985; 82:5437-5441.
6. Rudd KE, Menzzel R. His operons of *Escherichia coli* and *Salmonella typhimurium* are regulated by DNA supercoiling. Proc Natl Acad Sci USA 1987; 84:517-521.
7. Menzel R, Gellert M. Modulation of transcription by DNA supercoiling: a deletion analysis of the *Escherichia coli gyrA* and *gyrB* promoters. Proc Natl Acad Sci USA 1987; 84:4185-4189.
8. Drlica K. Control of bacterial DNA supercoiling. Mol Microbiol 1992; 6:425-433.
9. Kusano S. Ding Q, Fujita N et al. Promoter selectivity of *Escherichia coli* RNA polymerase $E\sigma^{70}$ and $E\sigma^{38}$ holoenzyme. J Biol Chem 1996; 271:1998-2004.

10. Mizushima T, Natori S, Sekimizu K. Relaxation of supercoiled DNA associated with induction of heat shock proteins in *Escherichia coli*. Mol Gen Genet 1993; 238:1-5.
11. Dorman CJ. Flexible response: DNA supercoiling, transcription and bacterial adaptation to environmental stress. Trends Microbiol 1996; 4:214-216.
12. Liu LF, Wang JC. Supercoiling of the DNA template during transcription. Proc Natl Acad Sci USA 1987; 84:7024-7027.
13. Pruss GJ, Drlica K. Topoisomerase I mutants: the gene on pBR322 that encodes resistance to tetracyclin affects plasmid DNA supercoiling. Proc Natl Acad Sci USA 1986; 83:8952-8956.
14. Lockshon D, Morris DR. Positively supercoiled plasmid DNA is produced by treatment of *Escherichia coli* with DNA gyrase inhibitors. Nucleic Acids Res 1983; 11:2999-3017.
15. Jupe ER, Sinden RR, Cartwright IL. Stably maintained microdomain of localized unrestrained supercoiling at a *Drosophila* heat shock gene locus. EMBO J 1993; 12:1067-1075.
16. Ljungman M, Hanawalt PC. Localyzed torsional tension in the DNA of human cells. Proc Natl Acad Sci USA 1992; 89:6055-6059.
17. Kramer PR, Sinden RR. Measurement of unrestrained negative supercoiling and topological domain size in living cells. Biochemistry 1997; 36:3151-3158.
18. Giaever GN, Wang JC. Supercoiling of intracellular DNA can occur in eukaryotic cells. Cell 1988; 55:849-856.
19. Matsumoto K, Hirose S. Visualization of unconstrained negative supercoils of DNA on polytene chromosomes of *Drosophila*. J Cell Sci 2004; 117:3797-3805.
20. Hirose S, Ohta T. DNA supercoiling and eukaryotic transcription-cause and effect. Cell Struct Funct 1990; 15:133-135.
21. Hirose S, Suzuki Y. In vitro transcription of eukaryotic genes is affected differently by the degree of DNA supercoiling. Proc Natl Acad Sci USA 1988; 85:718-722.
22. Mizutani M, Ura K, Hirose S. DNA superhelicity affects the formation of transcription preinitiation complex on eukaryotic genes differently. Nucleic Acids Res 1991; 19:2907-2911.
23. Tabuchi H, Handa H, Hirose S. Underwinding of DNA on binding of yeast TFIID to the TATA element. Biochem Biophys Res Commun 1993; 192:1432-1438.
24. Kim Y, Geiger JH, Hahn S et al. Crystal structure of a yeast TBP/TATA-box complex. Nature 1993; 365:512-520.
25. Kim JL, Nikolov DB, Burley SK. Co-crystal structure of TBP recongizing the minor groove of a TATA element. Nature 1993; 365:520-527.
26. Parvin JD, Sharp PA. DNA topology and a minimal set of basal factors for transcription by RNA polymerase II. Cell 1993; 73:533-540.
27. Rougvie AE, Lis JT. The RNA polymerase II molecule at the 5′ end of the uninduced *hsp70* gene of *D. melanogaster* is transcriptionally engaged. Cell 1988; 54:795-804.
28. Dunaway M, Ostrander EA. Local domains of supercoiling activate a eukaryotic promoter in vivo. Nature 1993; 361:746-748.
29. Ohta T, Hirose S. Purification of a DNA supercoiling factor from the posterior silk gland of *Bombyx mori*. Proc Natl Acad Sci USA 1990; 87:5307-5311.
30. Kobayashi M, Aita N, Hayashi S et al. DNA supercoiling factor localizes to puffs on polytene chromosomes in *Drosophila melanogaster*. Mol Cell Biol 1998; 18:6737-6744.
31. Gavin I, Horn PJ, Peterson CL. SWI/SNF chromatin remodeling requires changes in DNA topology. Mol Cell 2001; 7:97-104.
32. Nordheim A, Lafer EM, Peck LJ et al. Negatively supercoiled plamids contain left-handed Z-DNA segments as detected by specific antibody binding. Cell 1982; 31:309-318.
33. Liu R, Liu H, Chen X et al. Regulation of CSF1 promoter by the SWI/SNF-like BAF complex. Cell 2001; 105:309-318.
34. Gartenberg MR, Wang JC. Positive supercoiling of DNA greatly diminishes mRNA synthesis in yeast. Proc Natl Acad Sci USA 1992; 89:11461-11465.
35. Leng F, McMacken R. Potent stimulation of transcription-coupled DNA supercoiling by sequence-specific DNA-binding proteins. Proc Natl Acad Sci USA 2002; 99:9139-9144.

Part IV
DNA-Bending Proteins: Architectural Regulation of Transcription

Chapter 11

Gene Regulation by HMGA and HMGB Chromosomal Proteins and Related Architectural DNA-Binding Proteins

Andrew A. Travers

Abstract

The eukaryotic abundant high mobility group HMGA and HMGB proteins can act as architectural transcription factors by promoting the assembly of higher-order protein-DNA complexes which can either activate or repress gene expression. The structural organisation of both classes of protein is similar with either a single or repeated DNA binding domain preceding a short negatively charged C-terminal tail. In the HMGB class of proteins the HMG DNA-binding domain binds non-specifically and introduces a sharp bend into DNA whereas the AT-hook in the HMGA protein binds preferentially to A/T rich regions of DNA and stabilises a B-DNA structure. The acidic tails are hypothesised to facilitate the interaction of the proteins with nucleosomes by binding to the positively charged histone tails. Both classes of protein also interact with a large number of transcription factors that bind to specific DNA sequences.

Introduction

The eukaryotic nucleus contains three classes of abundant chromatin associated proteins – the High Mobility Group proteins of the HMGA, HMGB and HMGN classes (originally termed HMGI/Y, HMG1/2 and the HMG14/17; for recent nomenclature changes see ref. 1), so called because they were initially identified on the basis of their rapid migration through starch gels.[2] The HMGA and HMGB classes of chromosomal proteins in general share some common characteristics, notably a conserved acidic region and the ability to interact with several different transcription factors. A major role is to organise the structure of DNA-protein complexes in the context of chromatin.

Architectural DNA Binding Proteins

In both the eukaryotic nucleus and the bacterial nucleoid the trajectory of the DNA double helix is normally tightly constrained so that not only can the DNA be compacted without entanglement but also to provide an appropriate environment for the enzymatic machinery involved in DNA transcription, replication and recombination. This organisation is normally effected by abundant DNA binding proteins, termed architectural DNA-binding proteins,[3] that either induce DNA bending or facilitate the formation of multicomponent DNA-protein complexes. The term 'architectural' in this context implies that the protein is required for organising DNA but the proteins that fall within this definition are often otherwise functionally distinct and would include, for example, the histone octamer, abundant eukaryotic

DNA Conformation and Transcription, edited by Takashi Ohyama.

chromosomal proteins, such as the HMGA and HMGB proteins, abundant proteins associated with the prokaryotic nucleoid, such as FIS, H-NS, IHF and HU, as well as bona fide transcription factors exemplified by the TATA-binding protein (TBP). Some of these proteins have more than one architectural function. For example FIS can stabilise particular configurations of supercoiled DNA plasmids and also act to promote the assembly and activity of transcription, replication and recombination complexes. Many of the more generalised architectural proteins may be regarded as facilitators and are often not essential for viability while some more 'specialised' proteins, such as the histone octamer and TBP, are clearly essential.

Principles of Transcription Factor Induced Bending

The bending of DNA by transcription factors and by other protein complexes is a major component in the establishment of the overall morphology of protein-DNA complexes. This bending is usually a consequence of indirect readout, a mechanism by which the selectivity of binding is dependent not on making direct contacts between the aminoacids and bases, i.e., direct readout, but instead on the physicochemical properties of the DNA molecule itself.

Recognition of DNA by transcription factors often involves both direct and indirect readout. However, the principles of indirect readout are well illustrated by the histone octamer which, although not a transcription factor itself, completely lacks direct contacts between the aminoacid side chains and the bases of the bound DNA. The octamer binds 147 bp of DNA which are wrapped in a left-handed superhelix with a total curvature of approximate 10 radians.[4] This curvature contrasts with the stiffness of DNA in solution where the average persistence length (P), defined as the length over which the average deflection of the polymer axis caused by thermal agitation is one radian, is 140-150 bp,[5] i.e., the same length as that bound by the histone octamer. For DNA molecules that are not anisotropically curved the affinity of the DNA for the octamer is directly proportional to the flexibility (the inverse of the stiffness).[6] However the dependence of the binding energy on P is some 10-fold lower than the dependence of the bending energy in solution on P. This implies that the histone octamer increases the apparent flexibility substantially to compensate for the average increase in DNA curvature on binding.

How might this change in flexibility be effected? The histone core provides a DNA binding surface in the form of a positively charged ramp. On binding to this ramp the negative charges on one side of the DNA are neutralised. This asymmetric neutralisation, which can be mimicked in free DNA,[7] creates an imbalance in charge distribution on opposite sides of the double helix so that repulsion between the opposing sugar-phosphate backbones on the unneutralised side facilitates bending by increasing the width of the grooves. Concomitantly, the reduction in this repulsion on the inside of the bend permits greater freedom in the motions of the base-pairs, with a corresponding reduction in the width of the grooves. The greater flexibility of the motions between base-pairs is reflected in the periodic variation of twist and roll with groove width such that the ranges of values assumed for both are substantially larger than the corresponding ranges observed for DNA molecules free in solution.[8]

The correlation between flexibility and affinity for the histone octamer only applies strictly when a DNA molecule does not possess intrinsic anisotropic curvature. When it does the affinity may be relatively higher or lower. For example, the intrinsically curved TATA DNA sequence whose curvature is compatible with the surface of the histone octamer binds with an affinity that would normally be characteristic of a substantially more flexible isotropic binding site.[6,9] In this case binding is favoured by the lower entropic penalty on binding relative to an isotropically flexible molecule.[8] However if the intrinsic bend is too great and therefore less compatible with the protein binding surface the affinity is reduced relative to an isotropically flexible molecule.[10]

An extension of this principle of asymmetric alteration of the ionic environment of DNA is provided by the transcription factor TBP and the HMG-domain, found in HMGB proteins, a class of abundant chromosomal proteins and certain transcription factors such as SRY and LEF-1.[11] The HMGB proteins consist essentially of a small L-shaped protein domain with a cluster of hydrophobic residues on its inner surface and an extended unstructured basic region. When these proteins bind to DNA they produce a bend of 95-120° over about six base-pairs and decrease both the axial and torsional stiffness.[12] On the outer surface of the bend the hydrophobic 'wedge' towards the apex of the L binds in and widens the minor groove, concomitantly untwisting the DNA. This effect is believed to be facilitated by a local reduction in the dielectric constant which increases the repulsion between opposing sugar-phosphate backbones on the approach of the protein to DNA.[13,14] At the same time the basic region neutralises the phosphates bounding the major groove on the inside of the bend thus decreasing the repulsive forces and permitting the narrowing of the groove. Additionally the protein inserts, or intercalates, hydrophobic aminoacids into either a single base-step or into two base-steps that are themselves separated by a single base-step. The extent to which this intercalation increases or simply stabilises the induced bend is unclear. The bend induced by the intercalation contrasts with the smooth DNA bending induced by the histone octamer since the intercalation effectively introduces a kink in the DNA such that the stacking interactions between adjacent base-pairs are very substantially reduced.

In the TATA-binding protein this same principle of hydrophobic interactions predominates. Here two pairs of phenylalanine residues are intercalated at steps separated by 6 bases, kinking the DNA by ~45° at each intercalation site.[15,16] Between these pairs of phenylalanine residues a hydrophobic surface rests snugly within the minor groove. Again the minor groove is widened and untwisted. However, unlike the HMGB proteins there is no charge neutralisation on the opposing major groove face of the bent DNA and indeed the sharpness of the induced curvature is less than that for the HMGB proteins.

In other transcription factors there is substantial variation in the degree of induced bending. The *Escherichia coli* CAP (aka CRP) factor is a good example of mixed direct and indirect readout. This dimeric protein induces a bend of ~45° per monomer.[17] In this case the major bend occurs where the recognition helix of the helix-turn-helix motif binds in the major groove on the inside of the bend, concomitantly making direct contacts with the DNA bases and neutralising the sugar-phosphate backbone in the immediate vicinity.[18] Flanking the central recognition palindrome is a basic ramp which binds DNA and increases the overall DNA bend by indirect readout in a manner analogous to the histone octamer.

Biological Functions of DNA Bending

Although one of the principal roles of DNA bending in the living cell is to maintain the compaction of DNA, it also has important functions in transcriptional control and, in particular, in the assembly of regulatory complexes. A major consequence of introducing a tight bend into DNA is to bring DNA sequences which are far apart on a linear representation of a DNA molecule into close spatial proximity. This effect, which is also characteristic of plectonemically supercoiled DNA, is mediated in chromatin by the HMG-domain transcription factors, such as TCF-1, LEF-1 and SRY. In the case of TCF-1 acting at the enhancer of the *TCR* promoter, the bend induced by the factor brings together a normally unstable complex of the Ets-1 and PEBP2α DNA-binding proteins and ATF/CREB activator proteins to form a stable complex.[19] This example is probably a particular case of the more general phenomenon in which the DNA between a transcription factor and its target protein partner must be bent for protein-protein contacts to occur. The ease of bending will depend critically on the distance and the helical phase difference between the binding sites of the factor and its target. Normally unless one or both of the partner proteins are flexible contact will be facilitated when the binding sites are in helical phase, primarily because of the constraints on the torsional flexibility of DNA. However, at least in vitro, the constraints imposed by both torsional and axial rigidity can be overridden by the abundant DNA-bending proteins of the HMGB class. In the presence one of these proteins a requirement for an integral number of helical turns between binding sites is no longer crucial.[20] Furthermore the involvement of the HMGB protein in the formation of the complex need only be transient.

To what extent are variations in DNA flexibility reflected in genomic organisation? An excellent example of the dependence of biological function on DNA structure is provided by genome of the enteric bacterium *E. coli*. In this organism the strongest promoters for DNA transcription, often those directing the synthesis of rRNA and tRNA, are almost invariably associated with A/T rich, and hence flexible, DNA sequences extending upstream for 100-300 base-pairs from the transcription startpoint.[21] The activity of many of these promoters is strongly dependent on a high negative superhelical density stored in the DNA. This would in principle favour both DNA untwisting at sequences such as TATAAT close to startpoint[22] and also left-handed DNA wrapping around the protein complex responsible for initiating transcription. In many of these highly active promoters the DNA sequence also imparts curvature to the region, a feature that correlates both with the presence of multiple activating binding sites for the abundant DNA bending protein FIS (Factor for Inversion Stimulation).[21,23] These sites are often organised in helical phase such that the binding of FIS could constrain a negative superhelical loop. Indeed in the *rrnA* P1 regulatory region the presence of a far upstream FIS site centred at position -222 from the transcription startpoint results in the constraint of an additional supercoil in the initiation complex.[24] A primary function of this FIS-induced DNA looping is to promote the wrapping of DNA around the RNA polymerase prior to the initiation of transcription,[25] a phenomenon that has also been proposed for the activation of the *lac* promoter by the CAP DNA-bending transcription factor,[26] and consequently to facilitate the extended wrapping characteristic of the open complex.[27] In turn the FIS-induced constraint of negative superhelicity buffers this type of promoter against changes in the unconstrained superhelical density.[24,25,28] In other such promoters an alternative model proposes that the upstream activating sequence contains regions that are highly susceptible to DNA untwisting.[29] For both DNA wrapping and untwisting in the upstream region both models predict that the topological unwinding is transmitted to the 'TATA' sequence and promotes its untwisting.

DNA bending may also be required for the establishment of repressive regulatory complexes. Here, a DNA loop is often formed by the binding of an oligomeric repressor to two sites that are distant from each other along the DNA sequence. This loop, which can be as tightly bent as nucleosomal DNA, prevents the binding of RNA polymerase to the regulated promoter. Examples of this mode of regulation include repression by the AraC, LacI and GalR proteins.[30-33]

HMGA Proteins

The vertebrate HMGA proteins are small proteins of ~100-110 aminoacids and contain tandem copies, usually three, of a characteristic DNA-binding domain, the AT-hook, together with a C-terminal acidic region (Fig. 1).[34] The AT-hook is not restricted to the HMGA proteins as such as it is also found in a related *Drosophila* chromosomal protein, D1, which contains multiple copies of the motif,[35,36] in the motor subunits of various ATP-dependent chromatin remodelling complexes and in certain transcription factors that also contain a primary sequence-specific DNA-binding domain where it is assumed to act as an auxiliary DNA-binding element.[37]

The AT-hook is an unstructured short motif with the consensus sequence Arg/Lys-Pro-Arg-Gly-Arg-Gly-Pro-Arg/Lys[37,38] and selectively binds in the minor groove of A/T-rich regions of DNA.[39,40] The central Arg-Gly-Arg core adopts an extended conformation deep within the groove with the arginine side chains making extensive hydrophobic contacts along the base of the groove.[41] The proline residues change the trajectory of the backbone allowing the basic residues flanking the core mediate electrostatic and hydrophobic contacts with the DNA backbone. When bound to DNA the surface of the core motif contacting the DNA is concave and resembles that of the DNA binding drug netropsin, which has a similar selectivity for A/T-rich sequences.[41] For both netropsin and an AT-hook one consequence of DNA binding is a modest widening of the minor groove with a concomitant stabilisation of B-DNA structure. In some cases this widening results in a change in the direction of bending of DNA, particularly when that bending is dependent on a narrow minor groove width in A/T-rich sequences.[41,42]

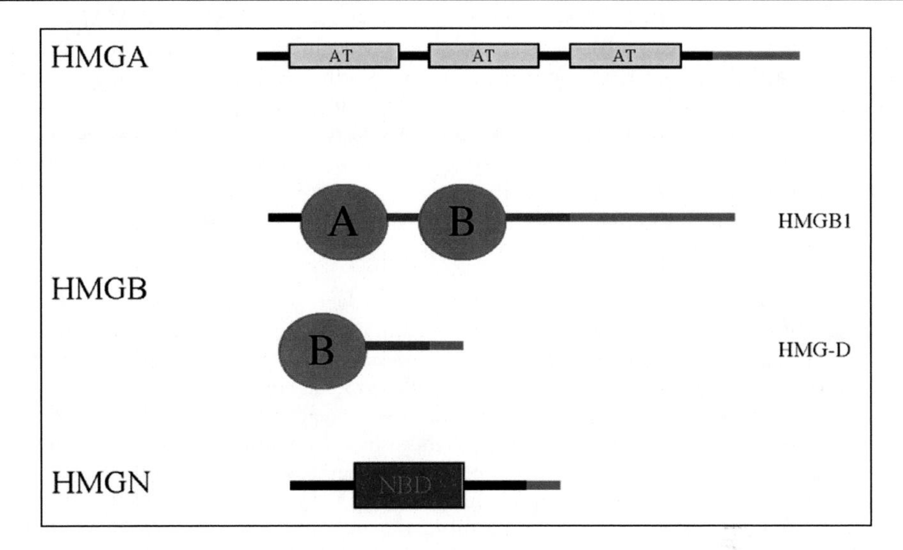

Figure 1. Comparison of the structural organisation of the HMG proteins. Regions of net negative charge outside the principal DNA binding or nucleosome binding domains are indicated in red and those of net positive charge in blue. The AT-hook DNA binding domain (AT) in HMGA proteins and the A- and B-type HMG domains in HMGB proteins are indicated. The nucleosome binding domain (NBD) in HMGN proteins is also shown.

Thus a small intrinsic bend of ~20° towards the minor groove in the IFN-β enhancer is reversed on binding HMGA1.[43] This could then facilitate recognition of the opposing major groove by transcription factors binding to specific sequences. Nevertheless although the HMGA proteins induce only small changes in DNA structure they bind tightly to DNA ligands with distorted or unusual features. These include supercoiled DNA, four-way junctions and base-unpaired regions of AT-rich DNA (reviewed in ref. 34). Strikingly in vitro these proteins can also introduce supercoils into relaxed DNA, possibly by stabilising cross-overs and thereby stabilising DNA loops.[44] This ability has been suggested as an explanation of the observation that in vitro HMGA1 represses the chick globin β^{A} gene promoter in the absence of the 3' enhancer but strongly activates transcription in its presence, regardless of whether or not the substrate is free DNA or is assembled into nucleosomes.[45] This stabilisation of loops could be mediated by the presence of multiple AT-hooks on each protein.

In mammals the HMGA proteins are encoded by two functional genes, *HMGA1* and *HMGA2*.[34] Alternative splicing of the transcripts of these genes increases the variety of protein products, of which the most abundant are HMGA1a (HMG-I) and HMGA1b (HMG-Y). These proteins appear to perform a variety of functions, of which the most studied are related to chromatin structure and to the facilitation or inhibition of transcription factor binding. In vitro HMGA proteins bind to nucleosomes, notably at the exit and entry points to the nucleosome core particle where they are in close proximity to histones H2A, H2B and H3.[46,47] The proteins can also bind to internal sites where they can induce local changes in the rotational setting of the wrapped DNA. The binding to the nucleosomal DNA is mediated by the AT-hooks but (by analogy to the HMGB class of proteins) it is conceivable that the acidic tail may also be involved in contacting the histone octamer and could perform a similar role to that of the HMGB proteins.

In mitotic chromosomes the HMGA proteins are associated with particular bands and these proteins are localised to the base of large chromatin loops in close proximity to scaffold-attachment regions (SARs).[48] As a consequence it has been suggested that the HMGA proteins are involved in the maintenance of the condensed mitotic chromosome structure in these regions.

Evidence supporting this view was adduced from the observation that synthetic 'MATH' proteins containing AT-hooks interfere with chromosome condensation during mitosis.[49] In contrast another model suggests that co-operative binding of HMGA molecules to a looped chromatin domain in interphase nuclei will facilitate the formation of an 'open' chromatin structure that is competent for transcription by competing with histone H1.[50] Although some experiments demonstrate that the MATH proteins can counteract the spreading of heterochromatin, as shown in particular by suppressing position-effect variegation in flies.[51] The mechanism by which this is accomplished remains to be established. However the expression of the HMGA proteins is strongly correlated with cell growth and is characteristically high in neoplastically transformed cells.[52]

HMGA1 has been shown to interact directly with a large variety of transcription factors including AT-1, ATF-3, NF-Y, IRF-1, SRF, NF-κB, p50, Tst-1/Oct-6 and c-Jun.[34] In some cases the protein regulates the formation of an enhanceosome. Thus at the virus-inducible β-interferon enhancer a complex containing both HMGA proteins and transcription factors forms and then acts to recruit RNA polymerase II and its associated general transcription factors.[53,54] In other cases HMGA proteins can block enhanceosome formation.[55] This modulation of transcription factor binding may be integrated with the regulation of chromatin organisation. Thus HMGA1a enhances the binding of the ATF-3 to a site at the edge of a nucleosome positioned on the HIV-1 promoter.[56] This combination of bound proteins can then recruit the remodelling complex hSWI/SNF. The interactions of HMGA with these factors can be modulated by covalent modifications including phosphorylation and methylation.[34]

HMGB Proteins

HMGB proteins are characterised by the HMG-box, a DNA-binding domain specific to eukaryotes. A major characteristic of this domain is to introduce a sharp bend into DNA (Fig. 2). Accordingly the domain also binds preferentially to a variety of distorted DNA structures, especially those in which the distortion itself induces a bend. These include negatively supercoiled DNA, small DNA circles, cruciforms, DNA bulges and cisplatin modified DNA.[10] The HMG-box domain is also found in several related types of protein, for example transcription factors such as SRY and LEF-1, and subunits of many chromatin remodelling complexes. All these proteins are predominantly nuclear and appear to act primarily as architectural facilitators in the manipulation of nucleoprotein complexes; for example, in the assembly of complexes involved in recombination and the initiation of transcription, as well as in the assembly and organisation of chromatin.

The archetypal HMGB proteins are highly abundant (~10-20 copies per nucleosome in the mammalian nucleus[57]) and often occur in two major forms, HMGB1 and HMGB2, originally termed HMG1 and HMG2, in vertebrates.[1] The two distinguishing features of these highly homologous proteins are two similar, but distinct, tandem HMG-box domains (A and B), and a long acidic C-terminal 'tail', consisting of ~30 (HMG1) or 20 (HMG2) acidic (aspartic and glutamic acid) residues, linked to the boxes by a short, predominantly basic linker (Fig. 1). However the most abundant HMG-box domain proteins in *Saccharomyces cerevisiae*, Nhp6ap and Nhp6bp (non-histone proteins 6A and 6B respectively), contain only a single HMG box, and lack an acidic tail. Likewise the two major HMG-box domain proteins in *Drosophila melanogaster*, HMG-D and HMG-Z, have only a single HMG box but, unlike the yeast proteins, contain a short C-terminal acidic tail in addition to a basic region (Fig. 1). These abundant proteins in yeast and *Drosophila* may be the general functional counterparts of HMGB1 and 2 in vertebrates.

The precise functions of the chromosomal HMGB proteins in vivo for a long time remained obscure. However there is now substantial evidence that they interact directly with both transcription factors and with the histone octamer. These interactions can affect transcription factor access to chromatin either directly or by promoting chromatin remodelling. In the latter case the proteins may facilitate repression or activation.

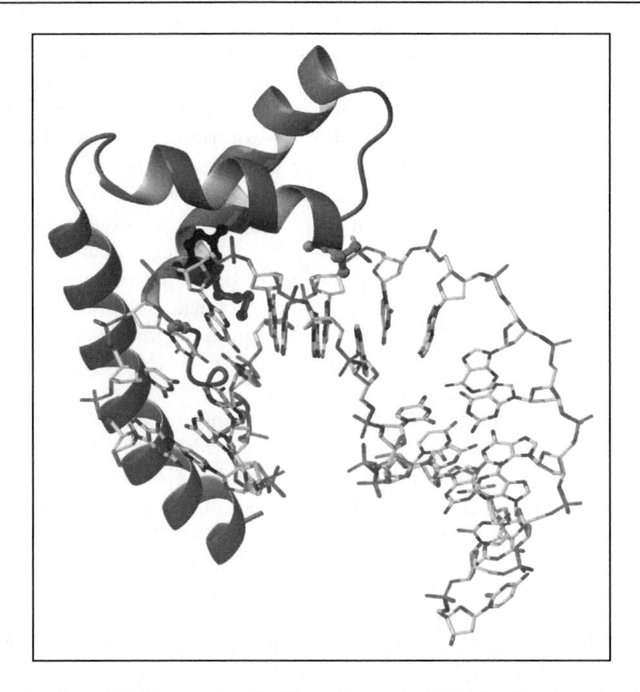

Figure 2. DNA bending by an HMG domain. The figure shows the DNA binding domain of the *Drosophila* High Mobility Group protein HMG-D binding to a short DNA fragment. The protein binds in the minor groove widening the groove and concomitantly stabilising a bend of ~100° in the DNA. This is achieved by inserting the sidechains of hydrophobic aminoacid residues (space-filling representation) between adjacent base-pairs at two locations separated by one base-step. The α-helices of the protein are depicted in red and yellow. The DNA structure shown contains a 'bulge' in which two adjacent bases on one of the strands are unpaired. Reproduced with permission from ref. 88.

There are two established cases in which the assembly of nucleoprotein complexes containing sequence-specific DNA-binding proteins is promoted by the DNA-bending properties of HMGB1 and 2, i.e., the proteins have a classical architectural role. First, in V(D)J recombination the lymphocyte-specific proteins RAG1 and RAG2 (human recombination activating genes 1 and 2) appear to recruit HMGB1 and 2 to the appropriate sites in chromatin[58-62] presumably by protein-protein contacts with the RAG1 homeodomain. Here they ensure the "12/23 rule". This requires that V(D)J recombination occurs only between specific recombination signal sequences (RSS). Each RSS is made up of a conserved heptamer and nonamer sequence separated by a non-conserved spacer of either 12 or 23 base pairs. HMGB1 (in concert with RAG1,2) facilitates recombination probably by bending the DNA between the two conserved sequences spaced by 23 bp and stabilising a nucleoprotein complex. The HMGB protein plays the dual role of bringing critical elements of the 23-RSS heptamer into the same phase as the 12-RSS to promote RAG binding and of assisting in the catalysis of 23-RSS cleavage. Recent footprinting experiments indicate that the HMGB1 (or HMGB2) protein is positioned 5' of the nonamer in 23-RSS complexes, interacting largely with the side of the duplex opposite the one contacting the RAG proteins.[63] A second instance in which an abundant HMGB protein may facilitate nucleoprotein complex assembly is in the formation of an enhanceosome containing the Epstein-Barr virus replication activator protein ZEBRA and HMGB1,[64] the two proteins bind cooperatively, HMGB1 binding to, and presumably bending, a specific DNA sequence between two ZEBRA recognition sites. Bending of DNA by HMGB1 and 2 has also been invoked to explain the essential role of these proteins in initiating DNA replication by loop formation at the MVM (minute virus of mice) parvovirus origin of replication.[65]

In vitro HMGB proteins can enhance the binding of various transcription factors (e.g. adenovirus MLTF, Oct-1 and 2, HoxD9, p53, steroid hormone receptors, Rel proteins, p73, Dof2 and the Epstein-Barr activator Rta) to their cognate DNA binding sites (reviewed in ref. 66). Similarly rat SSRP1 has been shown to facilitate the DNA binding of serum response factor[67] and human SSRP1 is associated with the γ isoform of p63 in vivo at the endogenous *MDM2* and *p21*$^{waf1/cip1}$ promoters.[68] In most of these cases, the interaction of the HMG protein with the transcription factor has been detected in vitro and could, in principle, serve as the mechanism for recruitment of HMGB1 or 2 to particular DNA sites. In some cases transfection experiments indicate functional interactions in vivo. Direct interactions between Nhp6p and the Gal4p and Tup1p transcription factors have also been inferred in vivo by a split-ubiquitin screen and confirmed by a pull-down assay.[69] Although the demonstrated interactions in vitro so far involve an HMGB protein and a single transcription factor, it is entirely possible that in vivo, in a natural regulatory context, the bending of DNA by HMGB1 and 2 could potentially allow the recruitment of a second transcription factor to the complex, in an analogous manner to the action of sequence-specific HMG-box transcription factors[1] such as LEF-1 in the enhanceosome at the T cell receptor alpha (TCRα).[70] HMGB1 may play a catalytic, chaperone role, since it does not appear to be stably incorporated into the final complex. Although a role for HMGB1 and HMGB2 induced DNA bending in the facilitated binding of transcription factors, while being entirely plausible, has not been directly established, it is strongly suggested by the ability of HU to substitute for the HMGB1-stimulated binding of the Epstein Barr virus transactivator Rta to its cognate binding sites.[71] A possible role for an HMGB1-induced change in DNA conformation in facilitation of transcription factor binding is also suggested by the observation that HMGB1 promotes binding of p53 to linear DNA but not to 66 bp DNA circles.[72] However, in this case the data do not distinguish between possible effects of DNA bending or untwisting.

The biological roles of the HMGB proteins have been studied using gene knock-outs. In mice the loss of HMGB1 but not of HMGB2 is lethal although in the former knock-out there are pleiotropic effects on glucose metabolism while in the latter spermatogenesis is impaired.[73,74] This suggests a functional redundancy between members of the HMGB1 and 2 family. A similar situation occurs with Nhp6ap and Nhp6bp in yeast.[75] However, the different phenotypes of the HMGB1 and HMGB2 null mice probably reflect specific roles for the two proteins in different tissues.[73,74] In *S. cerevisiae* the transcriptional effects of *NHP6* are not general but gene-specific. At the *CHA1* locus, loss of *NHP6* results both in an increase in the basal level of transcription and in a substantial decrease in the induced level.[76] This suggests an effect at the level of chromatin. The *CHA1* regulatory region contains a positioned nucleosome which occludes the TATA box under non-inducing conditions. On induction the TATA region becomes accessible.[77] However in the mutant strain, consistent with the increased basal level transcription, the chromatin structure of the TATA region in the uninduced state is similar to that in the induced wild-type strain. *NHP6* thus appears to be required for establishment of the organised chromatin structure characteristic of the uninduced state. The RSC remodelling complex is also required for this process[78] suggesting that RSC and Nhp6p may cooperate to remodel chromatin.

Further insights into how Nhp6ap and Nhp6bp function were provided by studies on the *HO* gene.[79] Loss of *NHP6* function can be suppressed by mutations that increase nucleosome accessibility and mobility, and enhanced by those with the opposite effect. Mutations both in the *SIN3* and *RPD3* genes, encoding components of a histone deacetylase complex, and in *SIN4*, partially restore wild-type function in cells lacking both Nhp6ap and Nhp6bp, while loss of the histone acetylase Gcn5p (also a component of the SAGA histone acetylase complex) in the same cells results in a more severe phenotype. Rpd3p and Gcn5p contribute to the dynamic balance between histone acetylation and deacetylation.[80] Both histone acetylation and the *sin* (SWI/SNF independence) phenotype are correlated with chromatin unfolding[81,82] and/or enhanced nucleosome accessibility[83] while histone deacetylation would be expected to favour folding. On this argument one role of the Nhp6p proteins would be to antagonise folding and possibly promote nucleosome accessibility.

The ability of the HMGB proteins to promote both transcription factor binding to their cognate sites and also chromatin remodelling implies that these activities could be coordinated to alter chromatin structure in the vicinity of a factor binding site. Like the HMGA proteins the abundant HMGB proteins bind to nucleosomes at sites close to the DNA exit and entry points. An insight into how HMGB proteins might alter the accessibility of nucleosomal DNA was provided by the observation that HMGB1 could facilitate the binding and subsequent remodelling function of the ACF remodelling complex in vitro.[84] Further observations showed that HMG-D, a *Drosophila* HMGB protein, when bound to nucleosome core particles increased the accessibility of nucleosomal DNA to restriction endonucleases at particular sites.[85] These sites were asymmetrically distributed, one site being located at one end of the bound DNA and the other in the vicinity of the nucleosome dyad. This effect required the acidic tail of the HMGB protein: without it the HMG-D reduced accessibility at all sites tested on the nucleosome. This result argues that certain HMGB proteins can alter the structure of nucleosomes and to do so presumably by interacting with an available basic region of the histone octamer. From the distribution of the sites with increased accessibility a prime candidate would be one (but not both) of the N-terminal tails of histone H3 or one of the C-terminal tails of histone H2A. It is important to note that the yeast Nhp6 proteins lack an acidic region and so could not interact with histones directly in this way. However they can associate with two other proteins, Pob3p and Spt16p, to form a complex, SPN, involved in chromatin remodelling.[86,87] Both these proteins contain extensive acidic regions and so, in principle, could substitute for the lack of an acidic region in Nhp6p.

Concluding Remarks

The abundant HMGA and HMGB chromosomal proteins share several common features. Both interact with nucleosomes, both can also bind a set of transcription factors, both are involved in enhanceosome formation and both can facilitate the recruitment of chromatin remodelling complexes. Interestingly the HMGN class of HMG proteins shares with the HMGA and HMGB classes the ability to interact with nucleosomes and also possesses a C-terminal region with a net negative charge.

References

1. Bustin M. Revised nomenclature for high mobility group (HMG) chromosomal proteins. Trends Biochem Sci 2001; 26:152-153.
2. Johns EW, Forrester S. Studies on nuclear proteins. The binding of extra acidic proteins to deoxyribonucleoprotein during the preparation of nuclear proteins. Eur J Biochem 1969; 8:547-551.
3. Grosschedl R, Giese K, Pagel J. HMG domain proteins: architectural elements in the assembly of nucleoprotein structures. Trends Genet 1994;10:94-100.
4. Luger K, Mäder AW, Richmond RK et al. Crystal structure of the nucleosome core particle at 2.8 Å resolution. Nature 1997; 389:251-260.
5. Hagerman PJ. Flexibility of DNA. Annu Rev Biophys Biophys Chem 1988; 17:265-286.
6. Travers A. The structural basis of DNA flexibility. Phil Trans R Soc Lond A 2004; in press.
7. Strauss JK, Maher LJ 3rd. DNA bending by asymmetric phosphate neutralization. Science 1994; 266:1829-1834.
8. Richmond TJ, Davey CA. The structure of DNA in the nucleosome core. Nature 2003; 423:145-150.
9. Widlund HR, Cao H, Simonsson S et al. Identification and characterization of genomic nucleosome-positioning sequences. J Mol Biol 1997; 267:807-817.
10. Anselmi C, Bocchinfuso G, De Santis P et al. Dual role of DNA intrinsic curvature and flexibility in determining nucleosome stability. J Mol Biol 1999; 286:1293-1301.
11. Scipioni A, Anselmi C, Zuccheri G et al. Sequence-dependent DNA curvature and flexibility from scanning force microscopy images. Biophys J 2002; 83:2408-2418.
12. Thomas JO, Travers AA. HMG1 and 2, and related architectural DNA-binding proteins. Trends Biochem Sci 2001; 26:167-174.

13. Travers AA. Reading the minor groove. Nat Struct Biol 1995; 2:615-618.
14. Elcock AH, McCammon JA. The low dielectric interior of proteins is sufficient to cause major structural changes in DNA on association. J Am Chem Soc 1996; 118:3787-3788.
15. Kim JL, Nikolov DB, Burley SK. Co-crystal structure of TBP recognizing the minor groove of a TATA element. Nature 1993; 365:520-527.
16. Kim Y, Geiger JH, Hahn S et al. Crystal structure of a yeast TBP/TATA-box complex. Nature 1993; 365:512-520.
17. Zinkel SS, Crothers DM. Comparative gel electrophoresis measurement of the DNA bend angle induced by the catabolite activator protein. Biopolymers 1990; 29:178-181.
18. Shultz SC, Shields GC, Steitz TA. Crystal structure of a CAP-DNA complex: the DNA is bent by 90°. Science 1991; 253:1001-1007.
19. Giese K, Cox J, Grosschedl R. The HMG domain of the lymphoid enhancer factor 1 bends DNA and facilitates assembly of functional nucleoprotein structures. Cell 1992; 69:185-195.
20. Ross ED, Hardwidge PR, Maher LJ, 3rd. HMG proteins and DNA flexibility in transcription activation. Mol Cell Biol 2001; 21:6598-6605.
21. Pedersen AG, Jensen LJ, Brunak S et al. A DNA structural atlas for *Escherichia coli.* J Mol Biol 2000; 299;907-930.
22. Drew HR, Weeks JR, Travers AA. Negative supercoiling induces spontaneous unwinding of a bacterial promoter. EMBO J 1985; 4:1025-1032.
23. Travers A, Muskhelishvili G. DNA microloops and microdomains: a general mechanism for transcription activation by torsional transmission. J Mol Biol 1998; 279:1027-1043.
24. Rochman M, Aviv M, Glaser G et al. Promoter protection by a transcription factor acting as a local topological homeostat. EMBO Rep 2002; 3:335-360.
25. Pemberton IK, Muskhelishvili G, Travers AA et al. FIS modulates the kinetics of successive interactions of RNA polymerase with the core and upstream regions of the *E. coli tyrT* promoter. J Mol Biol 2002; 318:651-663.
26. Buc H. Mechanism of activation of transcription by the complex formed between cyclic AMP and its receptor in *Escherichia coli.* Biochem Soc Trans 1986; 14:196-199.
27. Rivetti C, Guthold M., Bustamante C. Wrapping of DNA around the *E. coli* RNA polymerase open promoter complex. EMBO J 1999; 18:4464-4475.
28. Auner H, Buckle M, Deufel A et al. Mechanism of transcriptional activation by FIS: role of core promoter structure and DNA topology. J Mol Biol 2003; 331:331-344.
29. Hatfield GW, Benham CJ. DNA topology-mediated control of global gene expression in *Escherichia coli.* Annu Rev Genet 2002; 36:175-203.
30. Dunn TM, Hahn S, Ogden S et al. An operator at -280 base pairs that is required for repression of the *araBAD* operon promoter: addition of DNA helical turns between the operator and promoter cyclically hinders repression. Proc Natl Acad Sci USA 1984; 81:5017-5020.
31. Lyubchenko YL, Shlyakhtenko LS, Aki T et al. Atomic force microscopic demonstration of DNA looping by GalR and HU. Nucleic Acids Res 1997; 25:873-876.
32. Krämer H, Niemolle M, Amouyal M et al. *lac* repressor forms loops with linear DNA carrying two suitably placed *lac* operators. EMBO J 1987; 6:1481-1491.
33. Law SM, Bellomy GR, Schlax PJ et al. In vivo thermodynamic analysis of repression with and without looping in *lac* constructs. Estimates of free and local *lac* repressor concentrations and of physical properties of a region of supercoiled plasmid DNA in vivo. J Mol Biol 1993; 230:161-173.
34. Reeves R. Molecular biology of HMGA proteins: hubs of nuclear function. Gene 2001; 277:63-81.
35. Levinger L, Varshavsky A. Protein D1 preferentially binds A + T-rich DNA in vitro and is a component of *Drosophila melanogaster* nucleosomes containing A + T-rich satellite DNA. Proc Natl Acad Sci USA 1982; 79:7152-7156.
36. Ashley CT, Pendleton CG, Jennings WW et al. Isolation and sequencing of cDNA clones encoding *Drosophila* chromosomal protein D1. A repeating motif in proteins which recognize at DNA. J Biol Chem 1989; 264:8394-8401.
37. Aravind L, Landsman D. AT-hook motifs identified in a wide variety of DNA-binding proteins. Nucleic Acids Res 1998; 26:4413-4421.
38. Reeves R, Nissen MS. The AT-DNA-binding domain of mammalian high mobility group I chromosomal proteins: a novel peptide motif for recognizing DNA structure. J Biol Chem 1990; 265:8573-8582.
39. Churchill MEA, Travers AA. Protein motifs that recognize structural features of DNA. Trends Biochem Sci 1991; 16:92-97.
40. Solomon M, Strauss F, Varshavsky A. A mammalian high mobility group protein recognizes any stretch of six A-T base pairs in duplex DNA. Proc Natl Acad Sci USA 1986; 83:1276-1280.

41. Huth JR, Bewley CA, Nissen MS et al. The solution structure of an HMG-I(Y) DNA complex defines a new architectural minor groove binding motif. Nat Struct Biol 1997; 4:657-665.
42. Goodsell DS, Kopka ML, Dickerson RE. Refinement of netropsin bound to DNA: bias and feedback in electron density map interpretation. Biochemistry 1995; 34:4983-4993.
43. Falvo JV, Thanos D, Maniatis T. Reversal of intrinsic DNA bends in the IFN β gene enhancer by transcription factors and the architectural protein HMG I(Y). Cell 1995; 83:1101-1111.
44. Nissen MS, Reeves R. Changes in superhelicity are introduced into closed circular DNA by binding of high mobility group protein I/Y. J Biol Chem 1995; 270:4355-4360.
45. Bagga R, Michalowski S, Sabnis R et al. HMG I/Y regulates long-range enhancer-dependent transcription on DNA and chromatin by changes in DNA topology. Nucleic Acids Res 2000; 28:2541-2550.
46. Reeves R, Nissen MS. Interaction of high mobility group-I(Y) nonhistone proteins with nucleosome core particles. J Biol Chem 1993; 268:21137-21146.
47. Reeves R, Wolffe AP. Substrate structure influences binding of the non-histone protein HMG-I(Y) to free and nucleosomal DNA. Biochemistry 1996; 35:5063-5074.
48. Saitoh Y, Laemmli UK. Metaphase chromosome structure: bands arise from a differential folding path of the highly AT-rich scaffold. Cell 1994; 76:609-622.
49. Strick R, Laemmli UK. SARs are *cis* DNA elements of chromosome dynamics: synthesis of a SAR repressor protein. Cell 1995; 83:1137-1148.
50. Zhao K, Kas E, Gonzalez E et al. SAR-dependent mobilization of histone H1 by HMG-I/Y in vitro: HMG-I/Y is enriched in H1-depleted chromatin. EMBO J 1993; 12:3237-3247.
51. Girard F, Bello B, Laemmli UK et al. In vivo analysis of scaffold-associated regions in *Drosophila*: a synthetic high-affinity SAR binding protein suppresses position effect variegation. EMBO J 1998; 17:2079-2085.
52. Tallini G, Dal Cin P. HMGI(Y) and HMGI-C dysregulation: a common occurrence in human tumors. Adv Anat Pathol 1999; 6:237-246.
53. Agalioti T, Lomvardas S, Parekh B et al. Ordered recruitment of chromatin modifying and general transcription factors to the IFN-β promoter. Cell 2000; 103:667-678.
54. Kim TK, Maniatis T. The mechanism of transcriptional synergy of an in vitro assembled interferon-β enhanceosome. Mol Cell 1997; 1:119-129.
55. Klein-Hessling S, Schneider G, Heinfling A et al. HMG I(Y) interferes with the DNA binding of NF-AT factors and the induction of the interleukin 4 promoter in T cells. Proc Natl Acad Sci USA 1996; 93:15311-15316.
56. Henderson A, Holloway A, Reeves R et al. Recruitment of SWI/SNF to the human immunodeficiency virus type I promoter. Mol Cell Biol 2004; 24:389-397.
57. Duguet M, de Recondo AM. A deoxyribonucleic acid unwinding protein isolated from regenerating rat liver. Physical and functional properties. J Biol Chem 1978; 253:1660-1666.
58. Sawchuk DJ, Weis-Garcia F, Malik S et al. V(D)J recombination: modulation of RAG1 and RAG2 cleavage activity on 12/23 substrates by whole cell extract and DNA-bending proteins. J Exp Med 1997; 185:2025-2032.
59. van Gent DC, Hiom K, Paull TT et al. Stimulation of V(D)J cleavage by high mobility group proteins. EMBO J 1997; 16:2665-2670.
60. Kwon J, Imbalzano AN, Matthews A et al. Accessibility of nucleosomal DNA to V(D)J cleavage is modulated by RSS positioning and HMG1. Mol Cell 1998; 2:829-839.
61. West RB, Lieber MR. The RAG-HMG1 complex enforces the 12/23 rule of V(D)J recombination specifically at the double-hairpin formation step. Mol Cell Biol 1998; 18:6408-6415.
62. Aidinis V, Bonaldi T, Beltrame M et al. The RAG1 homeodomain recruits HMGB1 and HMGB2 to facilitate recombination signal sequence binding and to enhance the intrinsic DNA-bending activity of RAG1-RAG2. Mol Cell Biol 1999; 19:6532-6542.
63. Swanson PC. Fine structure and activity of discrete RAG-HMG complexes on V(D)J recombination signals. Mol Cell Biol 2002; 22:1340-1351.
64. Ellwood KB, Yen YM, Johnson RC et al. Mechanism for specificity by HMG-1 in enhanceosome assembly. Mol Cell Biol 2000; 20:4359-4370.
65. Cotmore SF, Christensen J, Tattersall P. Two widely spaced initiator binding sites create an HMG1-dependent parvovirus rolling-hairpin replication origin. J Virol 2000; 74:1332-1341.
66. Travers AA, Thomas JO. Chromosomal HMG-box proteins. In: Zlatanova J, Leuba SH, eds. Chromatin structure and dynamics: state-of-the-art. New Comprehensive Biochemistry. Amsterdam: Elsevier Science, 2004:103-134.
67. Spencer JA, Baron MH, Olson EN. Cooperative transcriptional activation by serum response factor and the high mobility group protein SSRP1. J Biol Chem 1999; 274:15686-15693.

68. Zeng SX, Dai MS, Keller DM et al. SSRP1 functions as a co-activator of the transcriptional activator p63. EMBO J 2002; 21:5487-5497.
69. Laser H, Bongards C, Schüller J et al. A new screen for protein interactions reveals that the *Saccharomyces cerevisiae* high mobility group proteins Nhp6A/B are involved in the regulation of the GAL1 promoter. Proc Natl Acad Sci USA 2000; 97:13732-13737.
70. Giese K, Kingsley C, Kirshner JR et al. Assembly and function of a TCRα enhancer complex is dependent on LEF-1-induced DNA bending and multiple protein-protein interactions. Genes Dev 1995; 9:995-1008.
71. Mitsouras M, Wong B, Arayata C et al. The DNA architectural protein HMGB1 displays two distinct modes of action that promote enhanceosome assembly. Mol Cell Biol 2002; 22:4390-4401.
72. McKinney K, Prives C. Efficient specfic DNA binding by p53 requires both its central and C-terminal domains as revealed by studies with High-Mobility Group 1 protein. Mol Cell Biol 2002; 22:6797-6808.
73. Calogero S, Grassi F, Aguzzi A et al. The lack of chromosomal protein HMGB1 does not disrupt cell growth but causes hypoglycaemia in newborn mice. Nat Genet 1999; 22:276-279.
74. Ronfani L, Ferraguti M, Croci L et al. Reduced fertility and spermatogenesis defects in mice lacking chromosomal protein Hmgb2. Development 2001; 128:1265-1273.
75. Costigan C, Kolodrubetz D, Snyder M. NHP6A and NHP6B, which encode HMG1-like proteins, are candidates for downstream components of the yeast SLT2 mitogen-activated protein kinase pathway. Mol Cell Biol 1994; 14:2391-2403.
76. Moreira JMA, Holmberg S. Chromatin-mediated transcriptional regulation by the yeast architectural factors NHP6A and NHP6B. EMBO J 2000; 19:6804-6813.
77. Moreira JM, Holmberg S. Nucleosome structure of the yeast *CHA1* promoter: analysis of activation-dependent chromatin remodeling of an RNA-polymerase-II-transcribed gene in TBP and RNA pol II mutants defective in vivo in response to acidic activators. EMBO J 1998; 17:6028-6038.
78. Moreira JM, Holmberg S. Transcriptional repression of the yeast *CHA1* gene requires the chromatin-remodeling complex RSC. EMBO J 1999; 18:2836-2844.
79. Yu Y, Eriksson P, Stillman DJ. Architectural factors and the SAGA complex function in parallel pathways to activate transcription. Mol Cell Biol 2000; 20:2350-2357.
80. Verdone L, Wu J, van Riper K et al. Hyperacetylation of chromatin at the *ADH2* promoter allows Adr1 to bind in repressed conditions. EMBO J 2002; 21:1101-1111.
81. Tse C, Sera T, Wolffe AP et al. Disruption of higher-order folding by core histone acetylation dramatically enhances transcription of nucleosomal arrays by RNA polymerase III. Mol Cell Biol 1998; 18:4629-4638.
82. Horn PJ, Crowley KA, Carruthers LM et al. The SIN domain of the histone octamer is essential for intramolecular folding of nucleosomal arrays. Nat Struct Biol 2002; 9:167-171.
83. Anderson JD, Lowary PT, Widom J. Effects of histone acetylation on the equilibrium accessibility of nucleosomal DNA target sites. J Mol Biol 2001; 307:977-985.
84. Bonaldi T, Längst G, Strohner R et al. The DNA chaperone HMGB1 facilitates ACF/CHRAC-dependent nucleosome sliding. EMBO J 2002; 21:6865-6873.
85. Ragab A, Travers A. HMG-D and histone H1 alter the local accessibility of nucleosomal DNA. Nucleic Acids Res 2003 31:7083-7089.
86. Formosa T, Eriksson P, Wittmeyer J et al. Spt16-Pob3 and the HMG protein Nhp6 combine to form the nucleosome-binding factor SPN. EMBO J 2001; 20:3506-3517.
87. Brewster NK, Johnston GC, Singer RA. A bipartite yeast SSRP1 analog comprised of Pob3 and Nhp6 proteins modulates transcription. Mol Cell Biol 2001; 21:3491-3502.
88. Cerdan R, Payet D, Yang JC, Travers AA, Neuhaus D. HMG-D complexed to a bulge DNA: an NMR model. Protein Sci 2001;10:504-518.

CHAPTER 12

Molecular Mechanisms of Male Sex Determination:

The Enigma of SRY

Michael A. Weiss

Abstract

The human testis-determining gene *Sry*, a single-copy gene on the short arm of the Y chromosome, encodes a high-mobility-group (HMG) box, a DNA-bending motif conserved among architectural transcription factors. The SRY-DNA complex exhibits a dramatic reorganization of the double helix. Although *Sry*-related *Sox* genes are of broad interest in relation to development, the mechanistic role of SRY in gene regulation has remained enigmatic. It is not known whether the HMG box is the sole functional domain of the protein. Additional unresolved issues include identification of target genes and interacting proteins. Although sex-reversal mutations commonly impair DNA binding, this correlation is not rigorous and does not exclude alternative regulatory mechanisms, such as possible SRY-directed RNA splicing. New studies of transgenic XX mice expressing chimeric SRY proteins suggest a powerful methodology to investigate structure-function relationships. Progress may benefit from genetic, genomic- and proteomic-based technologies to delineate the downstream pathway of SRY.

Introduction

Sexual dimorphism provides a model of a genetic switch between alternative programs of development.[1] The male phenotype in eutherian mammals is determined by *Sry* (sex-determining region of the Y chromosome;[2-4] genes and DNA sites are herein indicated in italics as above, and proteins in capital letters), a gene on the short arm of the Y chromosome. Assignment of SRY as the testis-determining factor (TDF) is supported by studies of transgenic murine models[5] and human intersex abnormalities.[6-10] Because SRY was identified in the absence of a priori biochemical information, its molecular activity was inferred by sequence homology: SRY contains a sequence-specific high mobility group (HMG)-box,[11] a conserved motif of DNA binding and DNA bending.[12] The HMG box is of broad interest in relation to human development. Its binding to DNA induces a dramatic reorganization of DNA structure,[13,14] which is proposed to contribute to assembly of specific transcriptional preinitiation complexes.[15-17] Although the association between SRY-induced DNA bending and sex determination is highly suggestive, how SRY functions at the molecular level to initiate testicular differentiation remains enigmatic. Does SRY operate via specific DNA bending, and if so, what is the relationship between DNA architecture and gene regulation? If this association is spurious or incomplete, what alternative mechanisms may be operating?

DNA Conformation and Transcription, edited by Takashi Ohyama.

In this chapter I review the structure of SRY and discuss current questions related to possible biochemical mechanisms of testicular differentiation. SRY belongs to a subfamily of related HMG boxes, designated SOX. *Sox* genes are classified in seven families (designated A-G) based on extent of homology (> 80% within a family). SOX proteins share similar DNA-binding and DNA-bending properties.[18,19] Of particular interest is SOX9, proposed to function downstream of SRY and site of mutations associated with campomelic dysplasia and sex reversal.[20,21] Section I focuses on the structures of specific and non-specific HMG boxes and their respective DNA complexes. Section II highlights similarities and differences between human and murine SRY in relation to studies of engineered sex reversal in transgenic mice. Design of chimeric murine *Sry* transgenes[22,23] promises an important new approach toward structure-function relationships in vivo. In Section III current hypotheses and approaches toward the functional delineation of a pathway of testicular differentiation are described.

Structure of the HMG Box and Protein-DNA Complexes

The HMG box defines a superfamily of eukaryotic DNA-binding proteins.[12] This ca. 80-residue domain, originally described in non-histone chromosomal proteins HMG1 and HMG2, exhibits an L-shaped structure[17] (Fig. 1A). Three α-helices and an N-terminal β-strand pack to form major- and minor wings.[24-27] The structure presents an angular surface as a template for DNA bending[13,14,28-30] (asterisk in Fig. 1A). A side view illustrates its flat architecture (Fig. 1B). Two classes of HMG boxes are distinguished by their DNA-binding properties. Whereas HMG1 and related proteins typically contain two or more HMG boxes that recognize distorted DNA structures with weak or absent sequence specificity, specific architectural transcription factors contain one HMG box that recognizes both distorted DNA structures and specific DNA sequences.[17] Domains of either class dock within a widened minor groove to direct bending of an underwound double helix. Such architectural distortion can enhance binding of unrelated DNA-binding motifs to flanking target sites.[13,14,28-30] The extent of DNA bending varies among HMG boxes, but in each case the protein binds on the outside of the DNA bend to compress the major groove.

NMR and crystallographic analyses of non-specific HMG boxes[24-27] demonstrate that, unlike conventional globular domains, the two wings of the HMG box contain discrete hydrophobic cores. The primary core, located between helix 1, helix 2, and the proximal portion of helix 3, stabilizes the confluence of the major wing (lower panel of Fig. 2). A "mini-core" occurs in the minor wing between α-helix 3 and the N-terminal β-strand (upper panel of Fig. 2). Both wings contribute to the motif's angular DNA-binding surface.[13,14,28-30] Structures of non-specific DNA complexes[28-31] are remarkable for non-polar contacts between the protein and an expanded, underwound and bent minor groove. Structures of bound and free HMG boxes are similar (for reviews, see refs. 32 and 33).

Structures of specific DNA complexes containing the HMG boxes of SRY and lymphoid enhancer factor-1 (LEF-1) have been determined by NMR[13,14,34] (Fig. 3A, B). The bound HMG boxes strongly resemble non-specific HMG boxes. Comparison of specific and non-specific complexes has provided insight into the origins of sequence specificity, which seems to reflect sequence changes at only a handful of positions.[29] SRY and LEF-1 complexes exhibit overall similarities as well as key differences. Although differing in bend angle (54° [SRY] and 110° [LEF-1]), each exhibits a single side-chain "cantilever" at corresponding positions: partial intercalation by Ile (SRY; position 13 of the HMG box consensus, lower panel of Fig. 2) or Met (LEF-1) similarly disrupts base stacking but not base pairing.[13,14,35,36] These "cantilever" side chains are shown in black in Figure 3. No additional sites of insertion (as defined in non-specific complexes[28-30]) are observed. Similar structural features have recently been observed in ternary DNA complexes containing the SOX2 HMG box and a POU domain.[37,38]

SRY and LEF-1 each contain a basic region C-terminal to the HMG box (red segments in Fig. 3). Although conserved among specific HMG boxes, basic tails are not generally present in non-specific domains.[12] Although not well ordered in the free domains (dashed line in Fig.

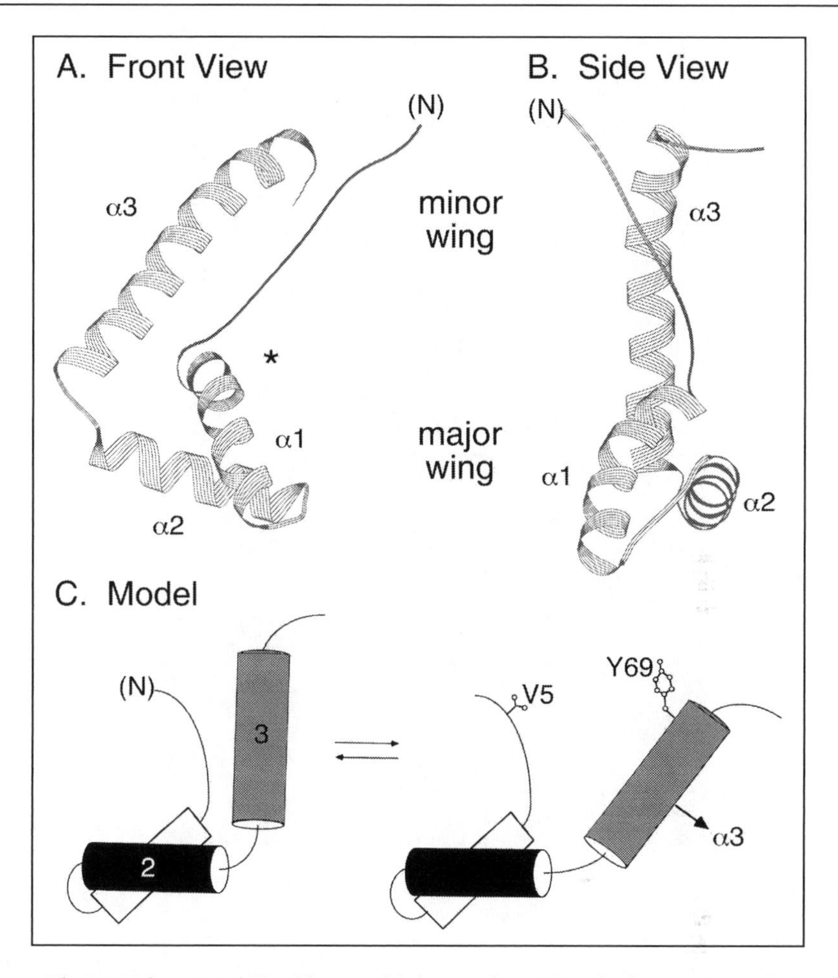

Figure 1. The HMG box. A and B) Ribbon model showing front (A) and side views (B). α-Helices 1 and 2 form major wing; helix 3 and the N-terminal β-strand form minor wing. Asterisk indicates position of cantilever side chain. C) Proposed model of SRY in which the free minor wing exists in equilibrium between open and closed structures. V5 and Y69 (right-hand panel; HMG-box consensus numbering) stabilize minor wing (see Fig. 2). In DNA complex an interface forms between helix 3 and the N-terminal β-strand. Figure is reprinted with permission from Weiss MA. Floppy SOX: Mutual induced fit in HMB (high-mobility group) box-DNA recognition. Mol Endocrinol 2001; 15:353-362.

4A), the tails play critical functional roles. Comprising residues 70-85 in an extended HMG box consensus sequence (Fig. 4B), this region also contains a nuclear localization signal (NLS; an additional NLS is present near the N-terminus of the HMG box[39,40]) and sites of clinical mutations.[41-44] R133 (position 78 in the HMG-box consensus; residue in the asterisked box in Fig. 4B) is required for function of the C-terminal NLS; its substitution by tryptophan impairs nuclear import, but not DNA binding or bending.[40,45] Although SRY and LEF-1 contain analogous basic tails, the two proteins exhibit different patterns of basic residues, prolines and glycines (Fig. 4B). These distinct patterns reflect structural differences between the tails in respective protein-DNA complexes[14,34] (Fig. 3). Truncation of the LEF-1 tail impairs DNA bending (from 130° to 52°) and impairs specific DNA affinity by at least 100-fold.[46-49]

Although the structure of free SRY or LEF-1 has not been determined, insights have been obtained from studies of SRY-related SOX HMG boxes.[19] Solution structures of SOX4 and

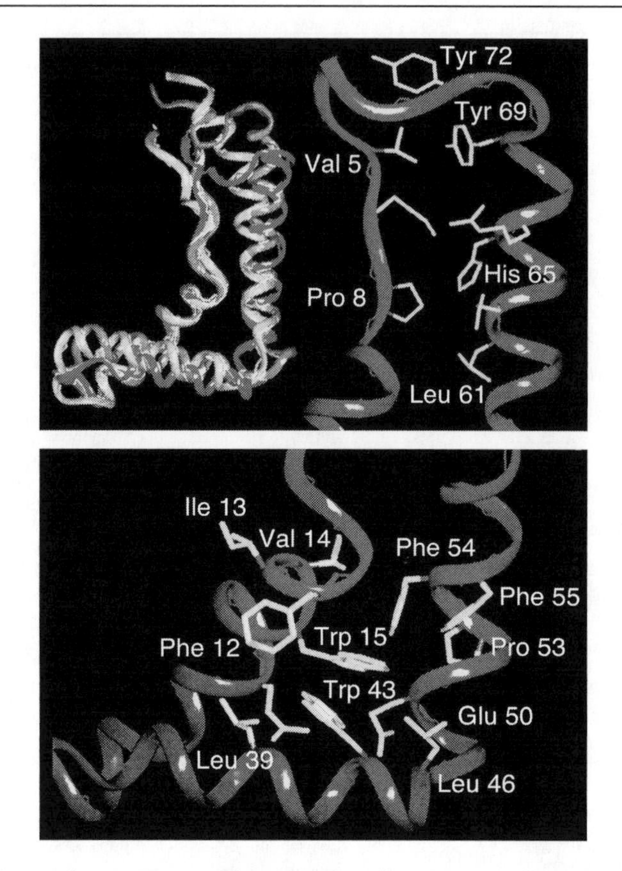

Figure 2. Side-chain packing in SRY HMG box. Upper panel) Comparison of bound SRY box[13] (dark grey) and non-specific boxes (HMG-D; and HMG-1B) at left. Side-chain packing in minor wing of bound SRY is shown at right. Lower panel) Major wing of SRY in DNA complex. Numbering refers to the HMG-box consensus. Figure is reprinted with permission from Weiss MA. Floppy SOX: Mutual induced fit in HMG (high-mobility group) box-DNA recognition. Mol Endocrinol 2001; 15:353-362.

SOX5 in the absence of DNA have been found to exhibit a novel combination of order and disorder. The three canonical α-helices are present and locally well ordered.[50,51] Whereas the major wing is well defined, however, the minor wing is not (Fig. 1C and Fig. 5). NMR models of SOX4[50] exhibit no fixed relationship between α-helix 3 and the major wing (α-helices 1 and 2); the N-terminal strand is disordered and detached. The presence of ordered α-helical segments with imprecise tertiary relationship is reminiscent of a molten globule[52] in equilibrium between open and closed states (Fig. 1C). Flexibility of the minor wing may enable SRY and SOX proteins to accommodate a broad range of sequence-dependent DNA bend angles.[53]

Insights from Comparison of Human and Murine SRY

Evidence that SRY is sufficient to initiate male organogenesis in an otherwise female genetic background was provided by studies of *Sry* transgenes in XX mice:[5] a 14 kb genomic segment of the murine Y chromosome containing the murine *Sry* gene (*mSry*) open reading frame (*orf*) and flanking control elements (Fig. 6A) could direct male somatic differentiation.[5] Surprisingly, an analogous 24 kb segment of the human Y chromosome containing the human *Sry* gene (*hSry*) could not function similarly in transgenic mice. This result was unexpected in light of the broad conservation of organogenetic pathways among mammals. However, human and

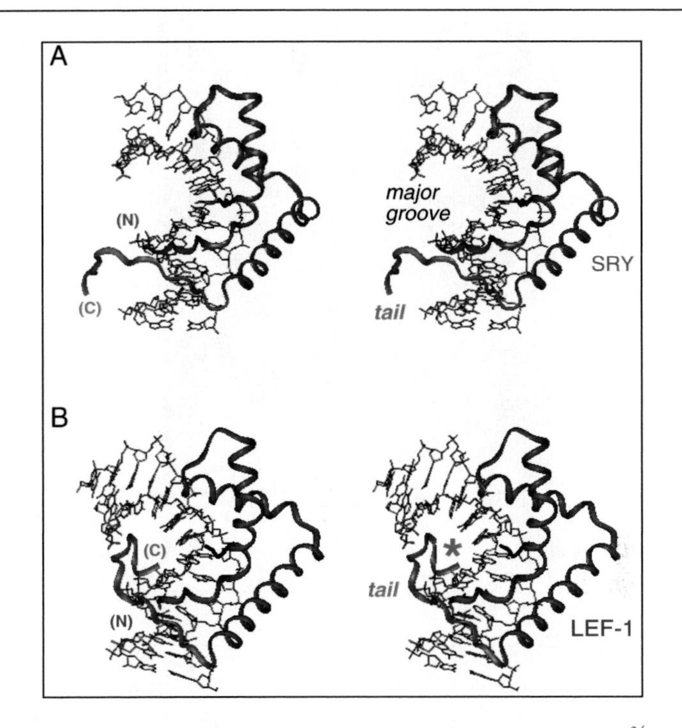

Figure 3. Structures of HMG box/DNA complexes (stereo pairs). A) SRY-DNA complex[34] (protein databank accession code 1J46). Ribbon model of protein is shown in green (HMG box) and red (tail; residues 70-85 in the HMG-box consensus). DNA is shown as gray sticks. B) LEF-1-DNA complex[14] (protein databank accession code 2LEF). Ribbon model of protein is shown in blue (HMG box) and red (tail; residues 70-85). DNA is shown as gray sticks. Asterisk (right panel) indicates binding of LEF-1 tail within compressed major groove. Cantilever side chains I13 (SRY) and M13 (LEF-1; M11 in protein data bank [PDB] entry) are shown as black sticks.

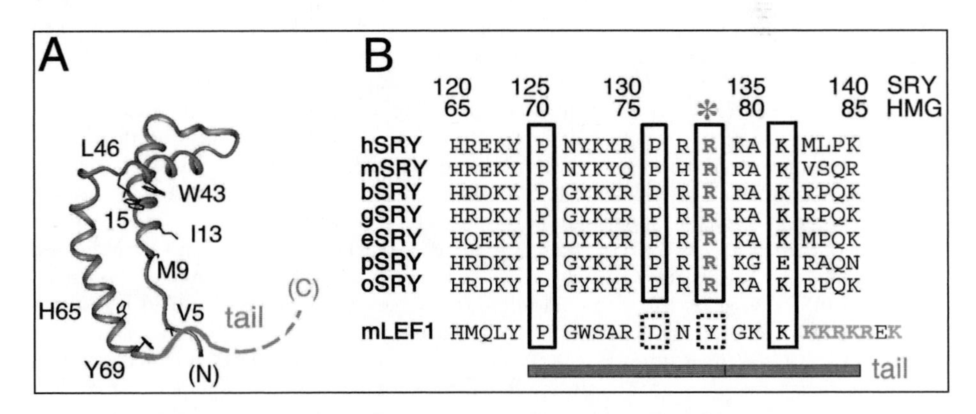

Figure 4. Tail sequences and mode of DNA binding. A) Ribbon model of SRY HMG box with selected side chains. Position of disordered tail is shown as dashed line. B) C-terminal sequences among mammalian SRY alleles (upper group) and other sequence-specific HMG boxes (lower group). Bar indicates position of tail; left-hand segment (residues 70-78 in the HMG box) delimits truncated tail. Asterisk indicates R133 (site of sex-reversal mutation).[40] Boxes indicate residues of functional interest. Dashed boxes indicate divergent residues in LEF-1/T-cell factor-1 (TCF-1) family of HMG boxes. C-terminal basic side chains are highlighted in light grey (mLEF1). Sequences are designated by species: hSRY, human; mSRY, murine; bSRY, bovine; gSRY, goat; eSRY, equine; pSRY, porcine and oSRY, sheep.

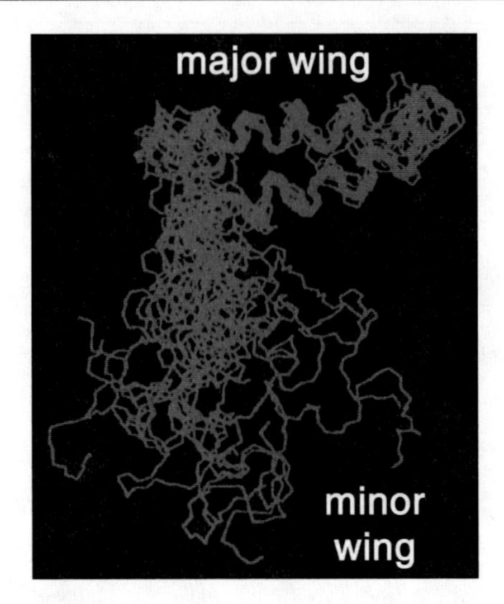

Figure 5. Molten structure of SOX-4 HMG box: well-ordered major wing and disordered minor wing.[50] Figure is reprinted with permission from Weiss MA. Floppy SOX: Mutual induced fit in HMG (migh mobility group) box-DNA recognition. Mol Endocrinol 2001; 15:353-362.

murine SRY (hSRY and mSRY) exhibit marked sequence divergence both within and extrinsic to the HMG box.[54] Substitutions occur at 28 of 85 amino-acid positions in the HMG box, for example, spanning its DNA-binding surface (including the cantilever side chains), and hydrophobic core. Further, rodent *Sry* coding regions contain a CAG DNA repeat encoding a novel glutamine-rich domain C-terminal to the HMG box[55,56] (Fig. 6A). This repeat can function as a transcriptional activation domain in model systems.[57] Failure of hSRY-directed male gene expression in the original studies of transgenic XX mice could have reflected impaired DNA bending or less stringent sequence specificity,[58] absence of a glutamine-rich repeat in hSRY, or divergence of *cis*-acting control elements between human and murine genomic segments.

To distinguish between these possibilities, an innovative approach employing chimeric transgenes has been developed.[22,23] This approach utilizes the 14 kb *mSry orf* (Fig. 6, top) to retain appropriate *cis*-acting regulatory elements to direct expression of *mSry* (panel A) in the differentiating gonadal ridge. In this context "domain swap" of DNA segments encoding the mSRY HMG box by DNA segments encoding the boxes of mSOX3 or mSOX9 was observed not to impair *Sry*-directed male sex determination (Fig. 6B and 6C), indicating their functional equivalence in the context of mSRY.[22] Swap of murine and human HMG boxes and amino-terminal non-box sequences (Fig. 6D) likewise yields a chimeric protein able to direct testicular differentiation.[23] An identical phenotype was observed following replacement of the 5′-portion of the *mSry orf* by the *hSry* coding region, including its stop codon (Fig. 6E); induction of sex reversal in XX transgenic mice by hSRY lacking the divergent C-terminal non-box sequences of mSRY suggests that the glutamine-rich domain (dark grey segment in Fig. 6E) is not required for male-specific gene regulation.[23] We note in passing that this conclusion is not in accord with transgenic studies of truncation mutants of *mSry* transgenes in which an intact glutamine-rich domain was apparently required for XX sex reversal.[59] It could not be excluded, however, that such truncations led to proteolyic instability or aggregation of the variant proteins, thus preventing SRY-directed gene regulation. It remains formally possible that mSRY functions through its glutamine-rich domain whereas hSRY functions through recruitment of an unrelated set of interacting proteins,[60] coincidentally leading to identical phenotypes.

Conservation of DNA bend angles among primate SRY-DNA complexes has motivated the hypothesis that a precise DNA bend angle is required to initiate the male program.[61] The functional equivalence of mSRY and hSRY in sex-reversed XX mice,[23] as demonstrated by

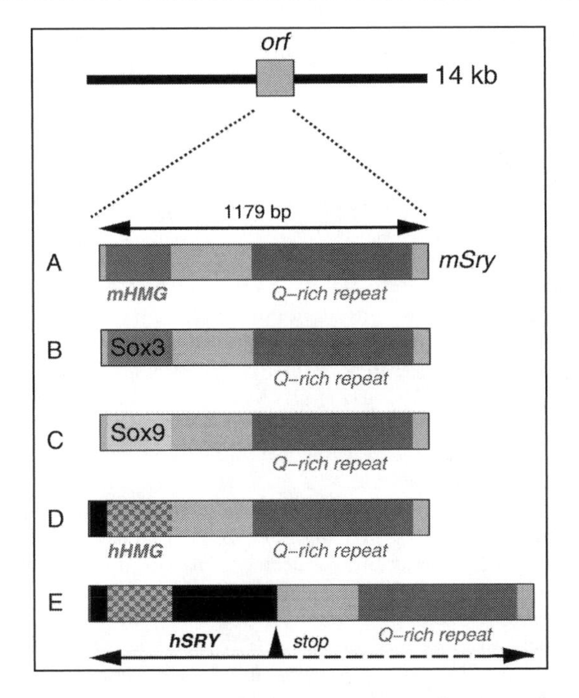

Figure 6. Transgenic studies of *Sry*-directed sex reversal in XX mice. Male somatic phenotype is induced by 14 kb genomic fragment of Y chromosome containing *Sry* open reading frame[5] (*orf*; top panel). Domain structure of mSRY is delineated in (A): N-terminal HMG box and C-terminal glutamine-rich region due to CAG repeat. B and C) Chimeric *mSry* constructions encoding "domain swap" of mSRY HMG box by those of mSOX3 (B) or mSOX9[22] (C). In chimeric construction (D) the N-terminal portion of *mSry* HMG box is replaced by that of *hSry*, including coding regions for N-terminal non-box sequences (black) and HMG box[23] (checkerboard). Chimeric construction (E) contains intact *hSry orf*, including coding region for C-terminal non-box sequences (black) and 3′ stop codon (arrowhead), instead of N-terminal segment of *mSry orf*. The glutamine-rich domain of mSRY is thus not expressed.[23] Chimeric constructions in each panel function in XX transgenic mice to induce testicular differentiation and male somatic development.[22,23] Sequences of SOX3 and SOX9 HMG boxes are identical in human and murine genes. Figure is reprinted with permission from Philips NB et al. *Sry*-directed sex rversal in transgenic mice is robust with respect to enhanced DNA bending: Comparison of human and murine HMG boxes. Biochemistry 2004; 43:7066-7084.

chimeric transgenes, may bear on this. Although hSRY and mSRY each induce sharp bends in DNA, Grosschedl and coworkers reported that a human SRY domain bent a consensus SRY target site (5′-ATTGTT-3′ and complement) 25-30° less sharply than did a murine domain.[58] These findings rationalized the seeming inability of hSRY to cause sex reversal in XX transgenic mice[5] but will require reassessment in light of the new generation of transgenic studies described above.[23] In fact, a survey of multiple biochemical studies suggests that the relative extent of DNA bending by human and murine SRY is uncertain as a broad range of bend angles have been inferred from permutation gel electrophoresis. This range may be due in part to differences among studies in protein constructions, DNA target sites, conditions of electrophoresis, length of DNA probes and method of interpretation.

The Biochemical Basis for the Genetic Function of *Sry*

How SRY regulates testicular differentiation is not well understood. The absence of identified target genes poses a major barrier to future advances. The presumption that SRY functions as an architectural transcription factor is nonetheless supported by a wealth of indirect observations.

i. The SRY HMG box is conserved as a specific DNA-bending motif[11,62] whereas sequences N- and C-terminal to the HMG box exhibit marked divergence among mammalian *Sry* alleles.[54]
ii. Almost all point mutations causing human sex reversal occur in the HMG box and impair specific DNA binding[4,62] or DNA bending.[34,63]
iii. Studies of chimeric murine and human *Sry* transgenes in XX mice indicate that divergent sequences C-terminal to the HMG box (murine or human; see Fig. 6 above) are equally compatible with testis determination.[23]
iv. Nuclear localization of human SRY is critical as demonstrated by a sex-reversal variant (R133W) that is specifically impaired in nuclear localization.[40,45]

Together, these observations suggest—but do not establish—that the DNA bending contributes to assembly of sex-specific transcriptional complexes.[15-17]

A central role for DNA bending is in accord with studies of other specific HMG-box proteins. Examples of target genes and *cis*-acting DNA control sites are well characterized in the case of SOX2, SOX9 and LEF-1,[64-69] for example, and so the simplest hypothesis would be that homologous protein motifs operate through homologous mechanisms. Nonetheless, nature does not always respect such economy: structural motifs may and often do diverge in function and may be utilized in diverse pathways. It is therefore possible that non-DNA-mediated activities of SRY in the nucleus may play ancillary or even central roles in its function. A possible role in RNA splicing, for example, has been described.[70,71] One may also imagine that the SRY HMG box could participate in protein-protein interactions to regulate the activities of other DNA- or RNA-binding factors. Definitive evidence in favor of or excluding these possibilities will require biochemical reconstitution of SRY-dependent regulatory complexes.

We anticipate that genetic insight will precede and provide guidance for biochemical studies. Participation of additional factors in the pathway of male development is implied by the absence of SRY mutations or deletions in the majority of patients with 46, XY gonadal dysgenesis.[72] In most such cases the responsible gene or genes are unknown. Further, the existence of SRY-interacting proteins is suggested by the observations of inherited SRY mutations in patients with gonadal dysgenesis and their (male and fertile) fathers. Although stochastic effects on gene expression and protein stability cannot be excluded as a mechanism of phenotypic variability, variable penetrance is likely to reflect autosomal polymorphisms in other genes required for testicular differentiation. This would be in accord with studies of male sex determination in mouse strains *Mus domesticus* and *Mus musculus*, which have implicated at least three autosomal genes in testicular differentiation.[73,74] Even if SRY operates through the induction of a specific nucleoprotein architecture,[13,15,16,36,63,75] it is not known how such architectures affect transcription. Binding of SRY, for example, may facilitate binding of other proteins to neighboring sites or competitively displace such factors.[73,76-78]

Anomalous Sex-Reversal Mutations

Because of the provisional nature of the DNA-bending hypothesis, it is worthwhile to summarize observations that may be inconsistent. Two mutations in the N-terminal non-HMG-box region of SRY (S18N and R30I) have been identified in patients with partial gonadal dysgenesis. In each case the same mutation was noted in normal male members of the proband's family.[79,80] The uncertain molecular basis of these sex-reversal phenotype may reflect present limitations in knowledge rather than evidence against the DNA-bending hypothesis. The R30I mutation, for example, may perturb a protein kinase A (PKA) phosphorylation site that has been shown to enhance DNA binding affinity in vitro.[81] Involvement of C-terminal non-box sequences is suggested by the a case report of 46, XY sex reversal associated with an SRY mutation causing deletion of the C-terminal 41 residues.[82] Because the deletion spares the HMG box, the variant protein is predicted to retain native DNA-binding and DNA-bending properties. Interestingly, whereas protein sequences outside of the SRY HMG box are generally divergent,[3] *Sry* alleles encode a conserved putative PDZ-binding peptide sequence at the extreme C-terminus, which is deleted in the above patient. Indeed, a candidate human SRY-interacting

PDZ protein (*S*RY *I*nteracting *P*rotein 1; SIP1) has been identified by the yeast two-hybrid assay.[83] Based on this and other considerations, participation of SIPs in SRY-directed gene regulation is proposed.[4,73]

Although mutations in the HMG box of SRY generally impair DNA binding, variants with near-native DNA-binding properties have also been described. Variant proteins with native-like DNA-binding properties are proposed to exhibit structural perturbations in DNA architecture,[34,63,84] decreased stability leading to accelerated degradation in vivo,[13] or impaired nuclear localization.[40] The following mutations are of special interest. (i) *V60L and V60A*. Located at position 5 of the HMG box, the native valine packs in the mini-core of the minor wing (see Fig. 2A). V60L is inherited[9] whereas V60A is uncharacterized.[85] Although an initial report indicated that V60L blocked detectable DNA-binding activity[9] (a finding seemingly at odds with its presence in a male father), a subsequent study found essentially native DNA binding and induced DNA structure.[86] The V60A variant has not been characterized. (ii) *M64I and M64R.* Located at position 9 of the HMG box, the native residue projects from the N-terminal β-strand at the edge of the DNA.[13,34] M64I has been reported to impair specific DNA bending but only modestly affect DNA-binding affinity.[63] The structural basis of the bending defect has been elucidated at atomic resolution.[34] Although specific DNA binding is reduced by about two-fold, it is not clear that this decrement is in itself sufficient to block testicular differentiation. M64R impairs specific DNA binding by approximately five fold with complete loss of DNA bending.[4] Abolition of DNA bending by a point mutation is unusual and suggests a fundamentally different mode of DNA binding. (iii) *M78T.* Located at position 23 of the HMG box, the native side chain packs at the back of the protein-DNA complex (magenta in Fig. 7). It is not known whether the mutation is de novo or familial. The mutation impairs specific DNA binding by less than two fold with little change in DNA bending. It is possible that M78T impairs binding of an SRY-interacting factor or accelerates intracellular degredation. (iv) *F109S.* Located at position 54 of the HMG box, the native side chain seals the back surface of the hydrophobic core[34] (magenta in Fig. 7). The mutation is inherited and also found in several unaffected male members of the family.[84] The mutation does not significantly perturb specific DNA binding or DNA bending[63] but is predicted to accelerate proteolytic degredation in vivo.[13]

DNA Bend Angle and Transcriptional Regulation

A central role for DNA bending in SRY-mediated gene regulation provides an attractive mechanism in light of conservation of DNA bend angles among primate SRY complexes.[61] Further, a decrement in DNA bending similar in magnitude to that reported between mSRY and hSRY was characterized in a variant hSRY-DNA complex associated with de novo sex reversal.[63] We propose that sharp DNA bending above a critical threshold is necessary to direct male-specific transcriptional regulation (Fig. 8A). This model envisages that SRY-directed DNA bending within a permissive range of angles facilitates assembly of DNA-multiprotein preinitiation complexes[77,78] ("enhanceosomes" and "repressosomes"; giving rise to architectural gene regulation).[15,17] The notion of a threshold in DNA bending giving rise to a range of functional DNA bend angles is supported by studies of transcriptional activation in *Escherichia coli.*[87-89] This model also posits corresponding thresholds in specific DNA affinity and kinetic lifetime of the multiprotein-DNA assembly.[90]

Decreased DNA bending below a critical threshold may impair transcriptional regulation. Functional analysis of nucleotide substitutions in the LEF-1-responsive enhancer element TCRα in T cells has shown that variant target sites bent to 90° exhibit impaired transcriptional activation uncorrelated with effects of the substitutions on protein binding.[91] Similar studies of amino-acid substitutions in the HMG box of SOX2 and nucleotide substitutions in its 5′-TTTGTTT-3′ (and complement) target site demonstrated that spatially precise DNA bending is essential to its transcriptional regulation of a target gene *fgf4* (fibroblast growth factor 4).[68] Variant SOX2-binding site 5′-TTTGGTT-3′ (and complement) exhibits a DNA bend

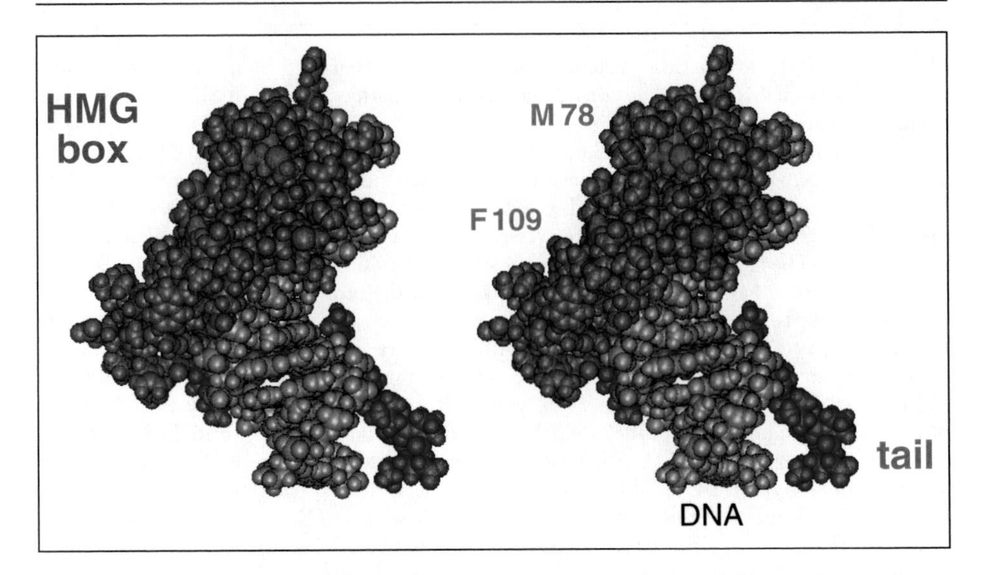

Figure 7. Space-filling model of SRY-DNA complex highlighting positions of M78 and F109 (magenta; residues 23 and 54 in HMG-box consensus) on "back" surface. The HMG box is otherwise shown in blue, C-terminal basic tail in red (residues 73-85), and bend DNA site in gray (12 bp). Coordinates obtained from ref. 34.

angle of 42° rather than 80° as in the native complex with no change in phase orientation. Although SOX2 binds well to the variant site, transactivation of the *fgf4* enhancer is blocked.[68]

Binding and bending of DNA control sites by SRY may enable the recruitment of additional factors, which might be non-sex specific. Such recruitment can be mediated by protein-protein interactions (model I in Fig. 8B), either between SRY and other DNA-binding proteins or accessory proteins. It is possible that a subset of sex-reversal mutations not observed in the protein-DNA interface (e.g., see Fig. 7) may weaken such interactions. Cooperative recruitment of other DNA-binding proteins may also be indirectly mediated by changes in DNA structure. Assembly of such complexes could regulate transcriptional initiation or induce an altered chromatin structure, leading in turn to recruitment of trithorax-group-, polycomb-group- or higher-order architectural proteins involved in the long-range regulation of gene expression. An example of cooperative recruitment is provided by SOX and POU domains.[37,38] Unlike SRY, SOX proteins often exhibit conservation outside of the HMG box, including within classical domains of transcriptional activation or repression.[18,19] Almost all SRY proteins lack discrete transactivation- or repression domains (possibly excepting the glutamine-rich repeat of rodent alleles; see Fig. 6). Mutations in or truncation of the transactivation domains of SOX9 is associated with campomelic dysplasia and sex reversal, presumably due to a block in the pathway downstream of SRY.[20,21]

A major barrier to further progress is posed by the present absence of known target genes. A variety of indirect evidence raises the possibility that *Sox9* is the only critical target gene (for review, see ref. 92). Although SOX9 is downstream of SRY and also required for testis termination, whether or how SRY might regulate its expression, nuclear localization, or activity is not understood. New technologies in genomics and proteomics offer great promise to surmount this barrier. Identification of target genes will enable characterization of DNA control elements, if present, and so permit testing of the DNA-bending hypothesis. Comparison of human and murine pathways should continue to be fruitful. The diversity of HMG boxes functionally tolerated in chimeric *mSry* transgenes contrasts with the subtle biochemical deficits observed or inferred in a subset of sex-reversal human SRY variants.[63,80,82]

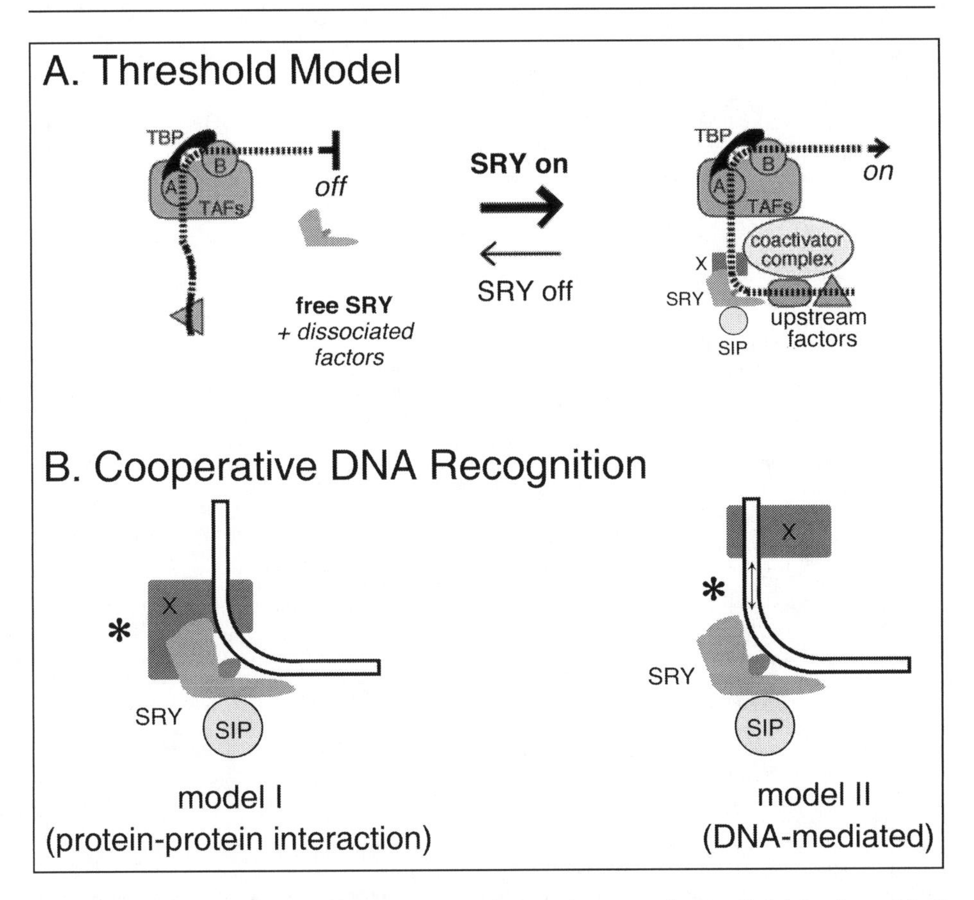

Figure 8. Proposed mechanism of SRY-directed architectural gene regulation. A) Schematic model of SRY-directed assembly of a male-specific transcriptional preinitiation complex through sharp DNA bending. Regulation requires a threshold in DNA bending, DNA affinity, and kinetic stability. Right, activated transcription occurs in the presence of bound SRY as specific DNA bend permits assembly of stable activator-coactivator-basal preinitiation complex[77] (enhanceosome). Left, activated transcription is off in absence of bound SRY due to disassembly of DNA-multiprotein complex and dissociation of activator-coactivator complexes. Putative factor X (at right) is proposed to bind cooperatively with SRY to an adjoining DNA site. B) Enlargement of proposed SRY multi-protein/DNA complex. Cooperative binding of factor X, a putative sequence-specific DNA-binding protein, is proposed to bind at a DNA site adjoining that of SRY. Cooperativity may be mediated either via protein-protein interactions (model I; e.g., by an SRY-interacting protein at left) or via changes in local DNA structure induced by SRY (model II at right). Figure is reprinted with permission from Philips NB et al. *Sry*-directed sex rversal in transgenic mice is robust with respect to enhanced DNA bending: Comparison of human and murine HMG boxes. Biochemistry 2004; 43:7066-7084.

Conclusions

The male phenotype in mammals is determined by SRY. Critical questions are: Does SRY function as an architectural transcription factor? Is its HMG box the only functional element? Is *Sox9* a direct target, and if so, the only functional target? Integrating the SRY-mediated switch into a broader genetic framework of organogenesis would have important implications for the pathogenesis and therapy of diverse human diseases.

References

1. Lovell-Badge R. Sex determining gene expression during embryogenesis. Philos Trans R Soc Lond B Biol Sci 1993;339:159-164.
2. Sinclair AH, Berta P, Palmer MS et al. A gene from the human sex-determining region encodes a protein with homology to a conserved DNA-binding motif. Nature 1990; 346:240-244.
3. Goodfellow PN, Lovell-Badge R. SRY and sex determination in mammals. Annu Rev Genet 1993; 27:71-92.
4. Mitchell CL, Harley VR. Biochemical defects in eight SRY missense mutations causing XY gonadal dysgenesis. Molec Genet Metab 2002; 77:217-225.
5. Koopman P, Gubbay J, Vivian N et al. Male development of chromosomally female mice transgenic for *Sry*. Nature 1991; 351:117-121.
6. Berta P, Hawkins JR, Sinclair AH et al. Genetic evidence equating SRY and the testis-determining factor. Nature 1990; 348:448-450.
7. McElreavy K, Vilain E, Abbas N et al. XY sex reversal associated with a deletion 5' to the SRY "HMG box" in the testis-determining region. Proc Natl Acad Sci USA 1992; 89:11016-11020.
8. Hawkins JR, Taylor A, Berta P et al. Mutational analysis of SRY: nonsense and missense mutations in XY sex reversal. Hum Genet 1992; 88:471-474.
9. Harley VR, Jackson DI, Hextall PJ et al. DNA binding activity of recombinant SRY from normal males and XY females. Science 1992; 255:453-456.
10. Vilain E, Jaubert F, Fellous M et al. Pathology of 46, XY pure gonadal dysgenesis absence of testis differentiation associated with mutations in the testis-determining factor. Differentiation 1993; 52:151-159.
11. Gubbay J, Collignon J, Koopman P et al. A gene mapping to the sex-determining region of the mouse Y chromosome is a member of a novel family of embryonically expressed genes. Nature 1990; 346:245-250.
12. Ner SS. HMGs everywhere. Curr Biol 1992; 2:208-210.
13. Werner HM, Huth JR, Gronenborn AM et al. Molecular basis of human 46X,Y sex reversal revealed from the three-dimensional solution structure of the human SRY-DNA complex. Cell 1995; 81:705-714.
14. Love JJ, Li X, Case DA et al. Structural basis for DNA bending by the architectural transcription factor LEF-1. Nature 1995; 376:791-795.
15. Grosschedl R. Higher-order nucleoprotein complexes in transcription: analogies with site-specific recombination. Curr Opin Cell Biol 1995; 7:362-370.
16. Kornberg RD, Lorch Y. Interplay between chromatin structure and transcription. Curr Opin Cell Biol 1995; 7:371-375.
17. Bewley CA, Gronenborn AM, Clore GM. Minor groove-binding architectural proteins: structure, function, and DNA recognition. Ann Rev Biophys Biomol Struct 1998; 27:105-131.
18. Pevny LH, Lovell-Badge R. *Sox* genes find their feet. Curr Opin Genet Dev 1997; 7:338-344.
19. Wegner M. From head to toes: the multiple facets of Sox protein. Nucleic Acids Res 1999; 6:1409-1420.
20. Foster JW, Dominguez-Steglich MA, Guioli S et al. Campomelic dysplasia and autosomal sex reversal caused by mutations in an *SRY*-related gene. Nature 1994; 372:525-530.
21. Wagner T, Wirth J, Meyer J et al. Autosomal sex reversal and campomelic dysplasia are caused by mutations in and around the *SRY*-related gene *SOX9*. Cell 1994; 79:1111-1120.
22. Bergstrom DE, Young M, Albrecht KH et al. Related function of mouse SOX3, SOX9, and SRY HMG domains assayed by male sex determination. Genesis 2000; 28:111-124.
23. Lovell-Badge R, Canning C, Sekido R. Sex-determining genes in mice: building pathways. In: Chadwick D, Good J, eds. The Genetics and Biology of Sex Determination. West Sussex: John Wiley & Sons Ltd., 2002:4-22.
24. Read CM, Cary PD, Crane-Robinson C et al. Solution structure of a DNA-binding domain from HMG1. Nucleic Acids Res 1993; 21:3427-3436.
25. Weir HM, Kraulis PJ, Hill CS et al. Structure of the HMG box motif in the B-domain of HMG1. EMBO J 1993; 12:311-319.
26. Jones DN, Searles MA, Shaw GL et al. The solution structure and dynamics of the DNA-binding domain of HMG-D from *Drosophila melanogaster*. Structure 1994; 2:609-627.
27. Hardman CH, Broadhurst RW, Raine AR et al. Structure of the A-domain of HMG1 and its interaction with DNA as studied by heteronuclear three- and four-dimensional NMR spectroscopy. Biochemistry 1995; 34:16596-16607.
28. Ohndorf UM, Rould MA, He Q et al. Basis for recognition of cisplatin-modified DNA by high-mobility-group proteins. Nature 1999; 399:708-712.

29. Murphy FV, Sweet RM, Churchill ME. The structure of a chromosomal high mobility group protein-DNA complex reveals sequence-neutral mechanisms important for non-sequence-specific DNA recognition. EMBO J 1999; 18:6610-6618.
30. Allain FH, Yen YM, Masse JE et al. Solution structure of the HMG protein NHP6A and its interaction with DNA reveals the structural determinants for non-sequence-specific binding. EMBO J 1999; 18:2563-2579.
31. Masse JE, Wong B, Yen Y-M et al.The *S. cerevisiae* architectural HMGB protein NHP6A complexed with DNA: DNA and protein conformational changes upon binding. J Mol Biol 2002; 323:263-284.
32. Murphy FV 4th, Churchill ME. Nonsequence-specific DNA recognition: a structural perspective. Structure Fold Des 2000; 8:R83-89.
33. Travers A. Recognition of distorted DNA structures by HMG domains. Curr Opin Struct Biol 2000; 10:102-109.
34. Murphy EC, Zhurkin VB, Louis JM et al. Structural basis for SRY-dependent 46-X,Y sex reversal: modulation of DNA bending by a naturally occurring point mutation. J Mol Biol 2001; 312:481-499.
35. King CY, Weiss MA. The SRY high-mobility-group box recognizes DNA by partial intercalation in the minor groove: a topological mechanism of sequence specificity. Proc Natl Acad Sci USA 1993; 90:11990-11994.
36. Haqq CM, King CY, Ukiyama E et al. Molecular basis of mammalian sexual determination: activation of Mullerian inhibiting substance gene expression by SRY. Science 1994; 266:1494-1500.
37. Remenyi A, Lins K, Nissen LJ et al. Crystal structure of a POU/HMG/DNA ternary complex suggests differential assembly of Oct4 and Sox2 on two enhancers. Genes Dev 2003; 17:2048-2059.
38. Williams DCJ, Cai M, Clore GM. Molecular basis for synergistic transcriptional activation by Oct1 and Sox2 revealed from the solution structure of the 42-kDa Oct1.Sox2.*Hoxb1*-DNA ternary transcription factor complex. J Biol Chem 2004; 279:1449-1457.
39. Sudbeck P, Scherer G. Two independent nuclear localization signals are present in the DNA-binding high-mobility group domains of SRY and SOX9. J Biol Chem 1997; 272:27848-27852.
40. Li B, Zhang W, Chan G et al. Human sex reversal due to impaired nuclear localization of SRY. A clinical correlation. J Biol Chem 2001; 276:46480-46484.
41. Poulat F, Soullier S, Goze C et al. Description and functional implications of a novel mutation in the sex-determining gene *SRY*. Hum Mutat 1994; 3:200-204.
42. Schmitt-Ney M, Thiele H, Kaltwasser P et al. Two novel SRY missense mutations reducing DNA binding identified in XY females and their mosaic fathers. Am J Hum Genet 1995; 56:862-869.
43. Veitia R, Ion A, Barbaux S et al. Mutations and sequence variants in the testis-determining region of the Y chromosome in individuals with a 46, XY female phenotype. Hum Genet 1997; 99:648-652.
44. Lundberg Y, Ritzén M, Harlin J et al. Novel Missense mutation (P131R) in the HMG box of *SRY* in XY sex reversal. Hum Mutat 1998; Suppl. 1:S328.
45. Harley VR, Layfield S, Mitchell CL et al. Defective importin β recognition and nuclear import of the sex-determining factor SRY are associated with XY sex-reversing mutations. Proc Natl Acad Sci USA 2003; 100:7045-7050.
46. Giese K, Amsterdam A, Grosschedl R. DNA-binding properties of the HMG domain of the lymphoid-specific transcriptional regulator LEF-1. Genes Dev 1991; 5:2567-2578.
47. Carlsson P, Waterman ML, Jones KA. The hLEF/TCF-1α HMG protein contains a context-dependent transcriptional activation domain that induces the TCRα enhancer in T cells. Genes Dev 1993; 7:2418-2430.
48. Read CM, Cary PD, Preston NS et al. The DNA sequence specificity of HMG boxes lies in the minor wing of the structure. EMBO J 1994; 13:5639-5646.
49. Lnenicek-Allen M, Read CM, Crane-Robinson C. The DNA bend angle and binding affinity of an HMG box increased by the presence of short terminal arms. Nucleic Acids Res 1996; 24:1047-1051.
50. van Houte LP, Chuprina VP, van der Wetering M et al. Solution structure of the sequence-specific HMG box of the lymphocyte transcriptional activator Sox-4. J Biol Chem 1995; 270:30516-30524.
51. Crane-Robinson C, Read CM, Cary PD et al. The energetics of HMG box interactions with DNA. Thermodynamic description of the box from mouse Sox-5. J Mol Biol 1998; 281:705-717.
52. Redfield C, Smith RA, Dobson CM. Structural characterization of a highly-ordered 'molten globule' at low pH. Nat Struct Biol 1994; 1:23-29.
53. Weiss MA. Floppy SOX: mutual induced fit in HMG (High-Mobility Group) Box-DNA recognition. Mol Endocrin 2001; 15:353-362.
54. Whitfield LS, Lovell-Badge R, Goodfellow PN. Rapid sequence evolution of the mammalian sex-determining gene *SRY*. Nature 1993; 364:713-715.

55. Coward P, Nagai K, Chen D et al. Polymorphism of a CAG trinucleotide repeat within *Sry* correlates with B6.Y^{Dom} sex reversal. Nat Genet 1994; 6:245-250.
56. Gubbay J, Vivian N, Economou A et al. Inverted repeat structure of the *Sry* locus in mice. Proc Natl Acad Sci USA 1992; 89:7953-7957.
57. Dubin RA, Ostrer H. Sry is a transcriptional activator. Mol Endocrinol 1994; 8:1182-1192.
58. Giese K, Pagel J, Grosschedl R. Distinct DNA-binding properties of the high mobility group domain of murine and human SRY sex-determining factors. Proc Natl Acad Sci USA 1994; 91:3368-3372.
59. Bowles J, Cooper L, Berkman J et al. SRY requires a CAG repeat domain for male determination in *Mus musculus*. Nat Genet 1999; 22:405-408.
60. Poulat F, Girard F, Chevron MP et al. Nuclear localization of the testis determining gene product SRY. J Cell Biol 1995; 128:737-748.
61. Pontiggia A, Whitfield S, Goodfellow PN et al. Evolutionary conservation in the DNA-binding and -bending properties of HMG-boxes from SRY proteins of primates. Gene 1995; 154:277-280.
62. Ferrari S, Harley VR, Pontiggia A et al. SRY, like HMG1, recognizes sharp angles in DNA. EMBO J 1992; 11:4497-4506.
63. Pontiggia A, Rimini R, Harley VR et al. Sex-reversing mutations affect the architecture of SRY-DNA complexes. EMBO J 1994; 13:6115-6124.
64. Giese K, Grosschedl R. LEF-1 contains an activation domain that stimulates transcription only in a specific context of factor-binding sites. EMBO J 1993; 12:4667-4676.
65. Yuan H, Corbi N, Basilico C et al. Developmental-specific activity of the *FGF-4* enhancer requires the synergistic action of Sox2 and Oct-3. Genes Dev 1995; 9:2635-2645.
66. Kamachi Y, Sockanathan S, Liu Q et al. Involvement of SOX proteins in lens-specific activation of crystallin genes. EMBO J 1995; 14:3510-3519.
67. Bell DM, Leung KK, Wheatley SC et al. SOX9 directly regulates the type-II collagen gene. Nat Genet 1997; 16:174-178.
68. Scaffidi P, Bianchi ME. Spatially precise DNA bending is an essential activity of the sox2 transcription factor. J Biol Chem 2001; 276:47296-47302.
69. Zhang P, Jimenez SA, Stokes DG. Regulation of human *COL9A1* gene expression. Activation of the proximal promoter region by SOX9. J Biol Chem 2003; 278:117-123.
70. Ohe K, Lalli E, Sassone-Corsi P. A direct role of SRY and SOX proteins in pre-mRNA splicing. Proc Natl Acad Sci USA 2002; 99:1146-1151.
71. Lalli E, Ohe K, Latorre E et al. Sexy splicing: regulatory interplays governing sex determination from *Drosophila* to mammals. J Cell Sci 2003; 116:441-445.
72. Pivnick EK, Wachtel S, Woods D et al. Mutations in the conserved domain of SRY are uncommon in XY gonadal dysgenesis. Hum Genet 1992; 90:308-310.
73. Lau Y-FC, Zhang J. Sry interactive proteins: implication for the mechanisms of sex determination. Cytogenet Cell Genet 1998; 80:128-132.
74. Eicher EM, Washburn LL, Schork NJ et al. Sex-determining genes on mouse autosomes identified by linkage analysis of C57BL/6J-Y^{POS} sex reversal. Nat Genet 1996; 14:206-209.
75. Giese K, Cox J, Grosschedl R. The HMG domain of lymphoid enhancer factor 1 bends DNA and facilitates assembly of functional nucleoprotein structures. Cell 1992; 69:185-195.
76. Graves JAM. Two uses for old SOX. Nat Genet 1997; 16:114-115.
77. Maniatis T, Falvo JV, Kim TH et al. Structure and function of the interferon-β enhanceosome. Cold Spring Harb Symp Quant Biol 1998; 63:609-620.
78. Adhya S, Geanacopoulos M, Lewis DE et al. Transcription regulation by repressosome and by RNA polymerase contact. Cold Spring Harb Symp Quant Biol 1998; 63:1-9.
79. Domenice S, Yumie Nishi M, Correia Billerbeck AE et al. A novel missense mutation (S18N) in the 5' non-HMG box region of the *SRY* gene in a patient with partial gonadal dysgenesis and his normal male relatives. Hum Genet 1998; 102:213-215.
80. Assumpcao JG, Benedetti CE, Maciel-Guerra AT et al. Novel mutations affecting SRY DNA-binding activity: the HMB box N65H associated with 46, XY pure gonadal dysgenesis and the familial non-HMG box R301 associated with variable phenotypes. J Mol Med 2002; 80:782-790.
81. Desclozeaux M, Poulat F, de Santa Barbara P et al. Phosphorylation of an N-terminal motif enhances DNA-binding activity of the human SRY protein. J Biol Chem 1998; 273:7988-7995.
82. Tajima T, Nakae J, Shinohara N et al. A novel mutation localized in the 3' non-HMG box region of the *SRY* gene in 46, XY gonadal dysgenesis. Hum Mol Genet 1994; 3:1187-1189.
83. Poulat F, Barbara PS, Desclozeaux M et al. The human testis determining factor SRY binds a nuclear factor containing PDZ protein interaction domains. J Biol Chem 1997; 272:7167-7172.
84. Jäger RJ, Harley VR, Pfeiffer RA et al. A familial mutation in the testis-determining gene *SRY* shared by both sexes. Hum Genet 1992; 90:350-355.

85. Hiort O. True hermaphroditism with 46, XY karyotype and a point mutation in the *SRY* gene. J Pediatrics 1995; 126:1022.
86. Benevides JM, Chan G, Lu XJ et al. Protein-directed DNA structure. I. Raman spectroscopy of a high- mobility-group box with application to human sex reversal. Biochemistry 2000; 39:537-547.
87. Rice PA, Yang S, Mizuuchi K et al. Crystal structure of an IGF-DNA complex: a protein-induced DNA U-turn. Cell 1996; 87:1295-1306.
88. Parekh BS, Hatfield GW. Transcriptional activation by protein-induced DNA bending: evidence for a DNA structural transmission model. Proc Natl Acad Sci USA 1996; 93:1173-1177.
89. Engelhorn M, Geiselmann J. Maximal transcriptional activation by the IHF protein of *Escherichia coli* depends on optimal DNA bending by the activator. J Mol Microbiol 1998; 30:431-441.
90. Ukiyama E, Jancso-Radek A, Li B et al. SRY and architectural gene regulation: the kinetic stability of a bent protein-DNA complex can regulate its transcriptional potency. Mol Endocrinol 2001; 15:363-377.
91. Giese K, Pagel J, Grosschedl R. Functional analysis of DNA bending and unwinding by the high mobility group domain of LEF-1. Proc Natl Acad Sci USA 1997; 94:12845-12850.
92. Harley VR, Clarkson MJ, Argentaro A. The molecular action and regulation of the testis-determining factors, SRY (sex-determining region on the Y chromosome) and SOX9 [SRY-related high-mobility group (HMG) box 9]. Endocr Rev 2003; 24:466-487.

Part V
Chromatin Infrastructure in Transcription: Roles of DNA Conformation and Properties

CHAPTER 13

The Role of Unusual DNA Structures in Chromatin Organization for Transcription

Takashi Ohyama

Abstract

The structural and mechanical properties of DNA influence nucleosome positioning and the manner in which DNA is organized in chromatin. Curved DNA structures, poly(dA•dT) sequences, and Z-DNA-forming sequences frequently occur near transcription start sites. Many reports have indicated that curved DNA structures play an important role in the formation, stability and positioning of nucleosomes, and consequently in DNA packaging in nuclei. Curved DNA structures and poly(dA•dT) sequences can increase the accessibility of target DNA elements of activators in chromatin to facilitate initiation of transcription. Z-DNA seems to be implicated in gene activation coupled with chromatin remodeling, and eukaryotes may use triplex DNA and cruciform structures to manipulate chromatin structure in a site-specific manner.

Introduction

DNA is highly compacted in a nucleus. In humans, the genomic DNA measures about a meter if unraveled. Thus, it follows that in the nucleus of a somatic cell, about $1x10^{-5}$ m in diameter, our chromosomal DNA must be compacted in length by as much as 200,000-fold. Biologically important DNA regions, such as the origins of replication, regulatory regions of transcription, and recombination loci must also be compacted. A narrow fiber of DNA is first folded into nucleosomes, the most fundamental unit of chromatin. It is generally thought that if nucleosomes assemble over a promoter region, they block initiation of transcription, because they inhibit access and/or assembly of transcription factors.

Histone modifications and ATP-dependent chromatin remodeling play a central role to suppress or amplify the inherently repressive effects of chromatin.[1-14] There are several mechanisms to explain how these phenomena participate in gene regulation. One model of transcription initiation is as follows: (i) a transcription factor (activator) binds to its target sequence in chromatin; (ii) the activator recruits a remodeling complex by direct protein-protein interaction; (iii) the complex alters the structure of the surrounding nucleosomes; (iv) the altered chromatin structure allows general transcription factors and RNA polymerase to bind to the promoter; (v) transcription starts.[4] Activators can bind to their target DNA elements, even when the target is adjacent to nucleosomes, or actually within a nucleosome.[15-19]

What chromatin structure is it that activators can bind? At present, we cannot clearly answer this question. This problem is still a "missing link" in the transcription cascade. The structural and mechanical properties of DNA have been often argued in relation to their effects on the nucleosome positioning and their effects on the way DNA is organized in chromatin. In

DNA Conformation and Transcription, edited by Takashi Ohyama.

Table 1. Repetitive sequences that contain a curved DNA structure in the repeating unit

Organism	Repetitive Sequence	Reference(s)
Monkey	Satellite	49
Cow (*Bos taurus*)	Satellite	51
Rat (*Rattus norvegicus*)	Satellite	49,51
Mouse	Satellite	47,49
Chicken (*Gallus gallus*)	Satellite	51
White dove (*Columba risoria*)	Satellite	51
Pigeon (*Columba livia*)	Satellite	51
Komodo dragon (*Varanus komodoensis*)	Satellite	51
Monitor lizard (*Varanus dumereliaddi*)	Satellite	51
Boa constrictor (*Boa constrictor*)	Satellite	51
Frog (*Xenopus laevis*)	Satellite	50
Oatmeal nematode (*Panagrellus redivirus*)	Satellite	51
Shrimp (*Artemia franciscana*)	*AluI* family	48
Tobacco	Highly repetitive DNA sequence family	52

this chapter, I focus on whether unusual DNA structures affect the packaging of genomic DNA into chromatin, and on how transcription is initiated.

Curved DNA and DNA Packaging in Chromatin

Both intrinsic DNA curvature and anisotropic DNA bendability (flexibility) influence the formation, stability and positioning of nucleosomes.[20-42] Thus, they may play an important role in the packaging of transcriptional control regions into chromatin.[43,44] This section focuses on the role of intrinsic DNA curvature, and considers how the packaged DNA keeps *cis* elements accessible to transcription factors. The role of DNA bendability is described in the next chapter.

DNA curvature seems to have general significance for DNA packaging. Because DNA has to be bent to fit closely around a histone core, it seems thermodynamically favorable to form nucleosomes on DNA sequences that are already appropriately curved. In fact, it has been experimentally shown that nucleosomes often preferentially associate with curved DNA fragments.[26,28-30,35,37,38,41] For example, Widlund et al constructed a library of nucleosome core DNA from the mouse genome, and screened those sequences that form the most stable nucleosomes.[37] The identified fragments contained phased runs of three or more consecutive adenines (or thymines), and showed retarded migration in non-denaturing polyacrylamide gel electrophoresis. Thus, curved DNA structure was found to be the most common feature among the screened fragments.

Curved DNA structures may also stabilize chromatin through their interaction with histone N-terminal tail domains that are the major sites of histone modifications such as acetylation, methylation and phosphorylation.[1,5,8,10,12,13] These modifications are implicated in transcription activation and gene silencing. Interactions between N-terminal domains and intrinsic DNA curvature could influence nucleosome positioning and stability.[45,46] On rigid, intrinsically curved DNA sequences, interactions between DNA and the histone tails stabilizes the formation of nucleosomes by ca. 250 cal/mol.[46]

The next question is whether curved DNA structures occur frequently in eukaryotic genomes. There are some clear answers to this. For example, repetitive DNA sequences, including satellite DNAs, very often contain one or more curved DNA structures (Table 1).[47-52] Curved DNA structures may be a common feature shared by all satellites, which are universally

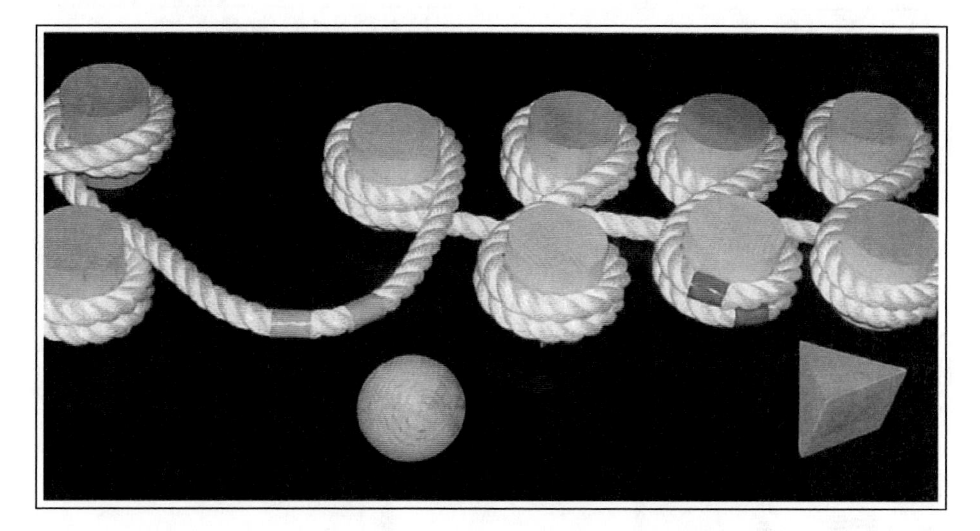

Figure 1. Target DNA elements of transcription activators could become accessible in chromatin, either by making the region free of nucleosomes, or by exposing it toward the environment on the nucleosome. In the figure, the rope represents the DNA, and wooden cylinders represent histone octamers. Taped sections represent activator target DNA elements. The sphere and pyramid represent activators.

associated with regions of constitutive heterochromatin and comprise anywhere from a few percent to > 50% of mammalian genomes.[53,54] If the hypothesis is correct, then curved DNA structures must contribute significantly to genome packaging. Some satellites, however, do not show the electrophoretic retardation characteristic of curved DNA structures. A fragment from bovine satellite I DNA is one such example. It behaves normally in non-denaturing polyacrylamide gels. Interestingly, an "unseen DNA curvature" was found in the fragment, in which another structural property that causes rapid migration had suppressed the effect of the curved DNA.[55] Interestingly, repeatedly occurring curved DNA sites are not restricted to satellite DNA, but are also reported for human ε-, Gγ-Aγ-ψβ-, δ-, and β-globin, *c-myc*, and immunoglobulin heavy chain μ loci, and in mouse β^{major}-globin locus.[56-59] Considering that most findings of naturally occurring curved DNA structures have been based on detection of retarded migration, the finding of "unseen curved DNA" strongly suggests that there are many more curved DNA loci on eukaryotic genomes than expected.

Now, let's consider promoter packaging into chromatin. Positioning of nucleosomes on a DNA sequence plays an important role in controlling the access of specific DNA-binding proteins to regulatory DNA elements.[60-65] Curved DNA often occurs in transcriptional control regions irrespective of the promoter type (Chapter 5). Thus, curved DNA may regulate positioning of nucleosomes in these regions, so as to allow the binding of activators.[44] Logically, the target DNA elements could become accessible by one of two mechanisms: either by positioning the target on a nucleosome and exposing it toward the environment; or by making it free of nucleosomes (Fig. 1).

An example of the first mechanism is the nucleosome structure formed on the long terminal repeat of the mouse mammary tumor virus (MMTV-LTR). In this case, four glucocorticoid receptor recognition elements (GREs) are located on the surface of a positioned nucleosome, and the major grooves of two GREs are exposed towards the environment.[60] These sites can be recognized by the receptor (a zinc finger protein), which initiates transcription. By what mechanism are these two GREs exposed on the surface of the nucleosome? An early study implicated curved DNA.[66] It was subsequently suggested that this curved DNA has a left-handed curved trajectory of its helical axis.[44]

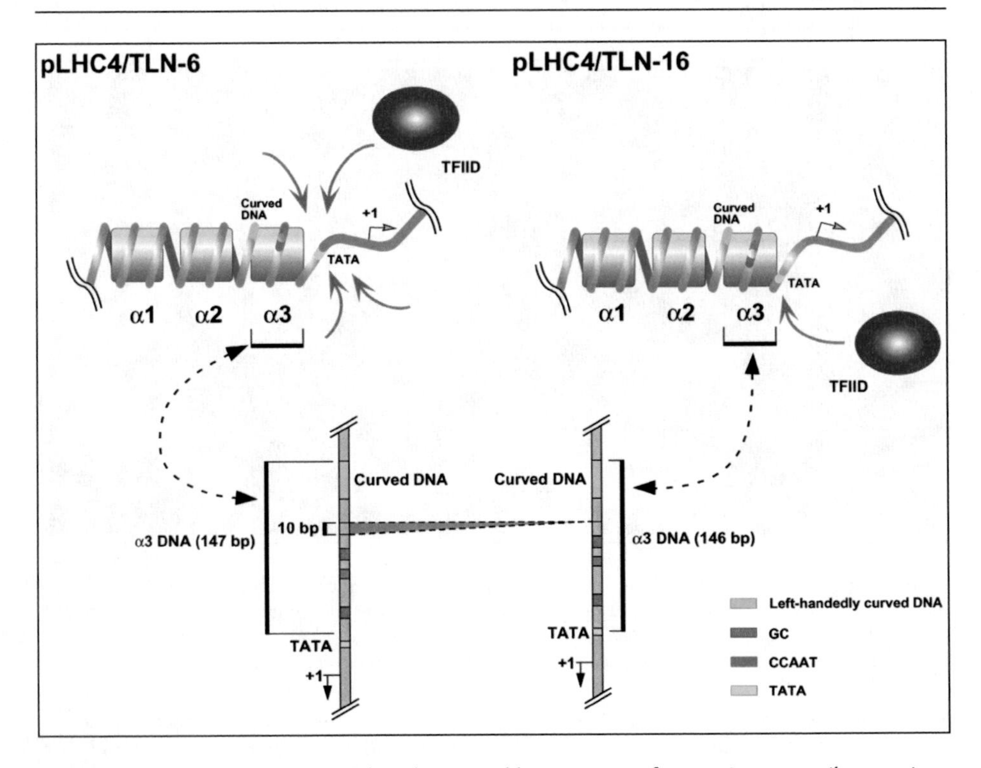

Figure 2. Synthetic curved DNA with a close resemblance to part of a negative supercoil can activate transcription by modulating local chromatin structure. The figure shows an example using the HSV *tk* promoter as a test system. This curved DNA can attract histone octamers. When it is linked to the core promoter at a specific rotational phase and distance, it can position the TATA box in the linker DNA region (pLHC4/TLN-6), or at the edge of the nucleosome (pLHC4/TLN-16) with its minor groove facing outwards. Both structures enhance accessibility of the TATA box and transcription is activated, although the first structure is more active than the second. The symbols α1, α2 and α3 indicate nucleosomes formed on the promoter region. Reproduced with permission from ref. 65, ©2003 Oxford University Press.

If the helical axis adopts a left-handed curved trajectory, it will resemble the negatively supercoiled DNA seen on a nucleosome, and thus it may recruit core histones easily. Furthermore, if a *cis*-DNA element is involved in, or is located near, the curved DNA structure, rotational setting of the element on the histone core (or even in linker DNA region in some cases) would be restricted by the DNA curvature. When its recognition site is displayed toward the environment by the curvature, the recognition step would be facilitated. Recently, this hypothesis was substantiated by using synthetic DNA segments with different conformations.[65] When left-handed curved DNA was linked to the herpes simplex virus thymidine kinase (HSV *tk*) promoter at a specific rotational phase and distance, in COS-7 cells, it activated the promoter approximately 10-fold. Mechanistically, the curved DNA attracted a histone core and the TATA box was thereby left in the linker DNA with its minor groove facing outwards (Fig. 2). Neither planar DNA curvature, nor right-handedly curved DNA, nor straight DNA, had this effect.

On the other hand, when a given DNA is dissimilar to the negative supercoil, it would make the region free of nucleosomes (the second mechanism). The adenylate kinase gene promoter of *Saccharomyces cerevisiae*, which has a curved DNA of this type, seems to be an example of this. This promoter was shown to be free of nucleosomes.[67] In the yeast *GAL1* promoter and *GAL80* promoter, curved DNA may make the UAS (upstream activation sequence) escape from being incorporated into nucleosomes. The underlying mechanisms are, however, different. Under inactivated (non-inducing) conditions, the *GAL1* promoter, which has two strong DNA curvatures in the upstream of the TATA box,[68] is incorporated into a nucleosome, referred to as nucleosome B. The bend centers lie within the terminal 20 bp or so on each end of the 147 bp sequence bound to the nucleosome B.[68] The UAS_G is located in the non-nucleosomal region just upstream of the nucleosome B. In this case, high nucleosome-forming ability of the two curved DNA structures seems to be used to make the UAS free of nucleosomes. In contrast, in the *GAL80* promoter, a single intrinsic DNA curvature which is located close to UAS_{GAL80} seems to exclude nucleosome formation on the UAS.[68] It can be imagined that similarity or dissimilarity between a given DNA curvature and the negatively supercoiled DNA seen on a nucleosome determines how easily the curved DNA can be incorporated into nucleosomes.

Besides the mechanism described above, curved DNA may also alter nucleosome structures to make target DNA elements accessible on the surface of nucleosomes. An interesting result was obtained in an experiment using DNA fragments composed of a synthetic DNA bending sequence (the repeated $(A/T)_3NN(G/C)_3NN$ motifs; TG-motifs) and the binding site for the nuclear factor 1 (NF-1) with an A_5 tract on both sides.[69] The TG-motifs are anisotropically flexible and have a high nucleosome-forming ability.[27] When nucleosomes were reconstituted on the fragments, the NF-1 binding affinity was higher when the flanking A-tracts were out-of-phase with the TG-motifs, than when they were in-phase. An altered nucleosome structure was also formed on a poly(dA•dT) sequence,[17] which is described in the next section.

Poly(dA•dT) Sequences and Nucleosome Positioning

DNA sequences of $(dA•dT)_n$ also frequently occur in eukaryotic genomes. They are rigid and adopt a unique DNA conformation that has a narrow minor groove.[70-72] In *Homo sapiens, Caenorhabditis elegans, Arabidopsis thaliana* and *Saccharomyces cerevisiae,* there are more poly(dA•dT) sequences present than would be expected if the DNA sequence were random, while in *Escherichia coli* and *Mycobacterium tuberculosis,* no difference is observed between actual and expected occurrences.[73] In promoter regions, $(dA•dT)_n$-rich sequences, where several $(dA•dT)_n$ sequences are connected by other short sequences, have frequently been found. For example, in yeast, promoters of the genes *HIS3, PET56, DED1, CBS2, ARG4, URA3* and *ADH2* contain or are flanked by them.[74-78] The $(dA•dT)_n$-rich sequences act as upstream promoter elements in *HIS3, PET56, DED1, ARG4,* and *URA3.*[74,75,77] In the rest of this section, the relationships between poly(dA•dT) sequences, nucleosome formation, and transcription are considered further.

In Vitro Reconstitution of Nucleosomes on Poly(dA•dT) Sequences

It is not yet clear whether poly(dA•dT) sequences always impede nucleosome formation. Earlier studies showed that long poly(dA•dT) sequences resisted nucleosome formation.[79-81] It was also shown that nucleosome formation over one member of a young *Alu* subfamily, which had recently transposed immediately downstream of a $T_{14}A_{11}$ stretch in the human neurofibromatosis type 1 gene locus, was impeded by the stretch.[82] On the other hand, human genomic DNA fragments containing long $(dA•dT)_n$ tracts (e.g., n=32, 34, or 41) were successfully incorporated into nucleosome cores.[83,84] Furthermore, nucleosomes were reconstituted successfully on the yeast *DED1* promoter, containing a T_6 tract, two T_5 tracts and a T_9 tract.[85,86] In this case, the characteristic T-tract conformation was lost upon folding into nucleosomes, indicating that the structural constraints in a nucleosome dominate over the intrinsic conformation of the T-tract. Higher temperatures apparently favor reconstitution.[87] The length of poly(dA•dT), the number of the poly(dA•dT) sequences, the DNA sequences surrounding poly(dA•dT), and conditions used for reconstitution, all seem to determine whether poly(dA•dT) sequences can form nucleosomes.

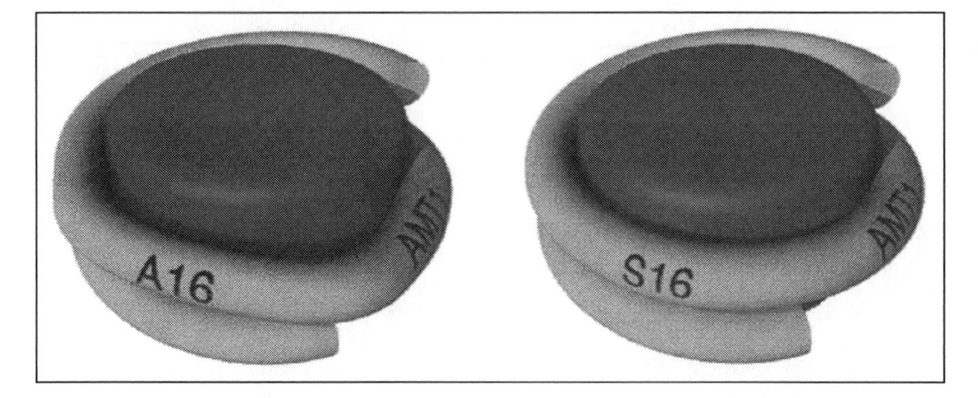

Figure 3. A putative nucleosome structure formed on the poly(dA•dT)-containing wild-type *AMT1* promoter (left). Also shown is a putative nucleosome structure formed on a mutant promoter (right), which carried "normal" DNA instead of the poly(dA•dT) sequence. A16: the $(dA•dT)_{16}$ tract; S16 ("normal" DNA): a random sequence predicted to assume a B-form DNA structure; AMT1: Amt1-binding site. In the wild-type, compared to the mutant, the Amt1-binding site is more accessible owing to the adjacent homopolymeric (dA•dT) tract. Reproduced with permission from ref. 17, ©1996 Elsevier.

Influence of Poly(dA•dT) Sequences on Nucleosome Formation in Vivo

Some promoters carrying one or more poly(dA•dT) sequences are not packaged into stable nucleosomes in vivo,[88-90] while others are packaged.[78,91] Shimizu et al found that in the yeast minichromosome, $A_{15}TATA_{16}$ and A_{34} tracts disrupt nucleosome formation, whereas a shorter A_5TATA_4 tract is incorporated into the positioned nucleosome.[92] They also reported that the longer A-tracts retained their unique DNA conformation in vivo. Using in vivo UV photofootprinting and DNA repair by photolyase, Suter et al demonstrated that in yeast, poly(dA•dT) sequences in promoters such as *HIS3*, *URA3* and *ILV1* were not folded in nucleosomes.[89] Like the report by Shimizu et al, this group also suggested that poly(dA•dT) sequences maintain their characteristic DNA structure in vivo. Interestingly, in the *Candida glabrata AMT1* gene (encoding copper-metalloregulatory transcription factor), nucleosome formation was allowed but the poly(dA•dT) sequence influenced the resulting nucleosome structure. The promoter harbors a $(dA•dT)_{16}$ sequence slightly upstream of a metal response element (MRE). These two sequences are packaged into a positioned nucleosome that exhibits the $(dA•dT)_{16}$-dependent localized distortion (Fig. 3). This nucleosome makes the MRE accessible.[17,93]

Functional Significance of Poly(dA•dT) Sequences in Transcription

Poly(dA•dT) sequences seem to make target DNA elements in chromatin more accessible, which is essentially the same as the proposed effect of curved DNA. To do this, they either prevent nucleosome formation, or change nucleosome structures. As described above, in some cases, the poly(dA•dT) sequences are incorporated into nucleosomes, with either the original conformation, or with an altered conformation,[17,86,92,93] while in other cases they are not incorporated.[89,92] Although it is not clear what determines this difference, the lengths of poly(dA•dT) sequences and a slight difference in the conformational and/or mechanical properties of the poly(dA•dT)-containing sequences may be key parameters. In addition, to establish non-nucleosomal regions, or to form distorted nucleosomes, assistance of some factors (e.g., poly(dA•dT)-binding proteins or histone modifying enzymes) may be required. In this sense, the HMG-I(Y) family of "high mobility group" proteins may be important. HMG proteins organize the structure of DNA-protein complexes in the context of chromatin (Chapter 11). HMG-I(Y) can preferentially bind to certain types of poly(dA•dT) sequences on the surface of nucleosomes and alter the local setting of DNA on the nucleosomes.[94] Thus, poly(dA•dT) sequences may also function as a signal to introduce structural changes into nucleosomes.

Chromatin and Z-DNA, Triple-Stranded DNA, and Cruciform DNA

The bulk of the eukaryotic genome is believed not to be torsionally stressed, even though it is negatively supercoiled, because such supercoilings are largely accommodated by the DNA writhing in nucleosomes. Unconstrained negative supercoils, however, can be still generated. For example, they are generated behind an RNA polymerase transcribing a DNA template (Chapter 10).[95] The negative supercoils stabilize non-B DNA structures such as Z-DNA, triplex DNA and cruciform DNA. DNA elements with sequences suitable for the formation of Z-DNA are found at various positions in genomes. An early study estimated that the human genome contains approximately 100,000 copies of potential Z-DNA-forming sequences.[96] Interestingly, similar to curved DNA and poly(dA•dT) sequences, Z-DNA-forming sequences occur more frequently near transcription start sites.[97] Do they function to position nucleosomes or to inhibit nucleosome formation? It is thought that the actual Z-DNA structure lies in non-nucleosomal regions in chromatin.[98] However, we do not yet know whether Z-DNA can regulate nucleosome position.

An interesting study has been reported recently. Z-DNA seems to be implicated in gene activation coupled with chromatin remodeling (Fig. 4). The promoter of the human colony-stimulating factor 1 (CSF1) gene is flanked by TG repeats (Z-DNA forming sequence), which were converted to Z-DNA upon activation by the SWI/SNF-like BRG1-associated factor (BAF) complex in vivo. Furthermore, the in vitro data showed that the BAF complex facilitates Z-DNA formation in a nucleosomal template.[99] These data suggest that at the *CSF1* promoter, BAF-induced Z-DNA formation stabilizes an open chromatin structure. This illustrates why promoters sometimes contain, or are flanked by, Z-DNA forming sequences.

Triple-stranded DNA seems unable to be accommodated within nucleosomes.[100] This conclusion is strengthened by the report by Espinas et al, who performed in vitro assembly of mono-nucleosomes onto 180 bp DNA fragments containing $(GA\bullet TC)_{22}$, or onto 190 bp fragments with $(GA\bullet TC)_{10}$.[101] Although the repeated sequences themselves had no influence on nucleosome positioning, nucleosome assembly was strongly inhibited when the triple-stranded DNA was formed at the $(GA\bullet TC)_n$ site. On the other hand, triplex formation was difficult when the $(GA\bullet TC)_n$ site was incorporated into a nucleosome. Thus, nucleosome assembly and triplex formation are presumably competing processes. In conclusion, triplexes seem unable to determine the position of nucleosomes by recruiting histone cores.

Cruciform structures are located mainly on internucleosomal DNA,[102] perhaps because they cannot associate with histone cores[103] and as a result, they could induce an alternative positioning of nucleosomes. The cruciform structures could also act over a distance to destabilize adjacent nucleosomes.[104] Thus, cruciforms are probably not used to recruit histone octamers to form positioned nucleosomes. However, eukaryotes may use triplex structures and cruciforms to form open chromatin structures.

Linker histones are probably implicated in transcriptional regulation.[105,106] However, the interaction between unusual DNA structures and linker histones has not been studied adequately. Interestingly, H1 seems to bind preferentially to curved DNA structures that are flanked with specific sequences.[107] It is evident that more information is needed, on the interaction between linker histones and DNA of various conformations.

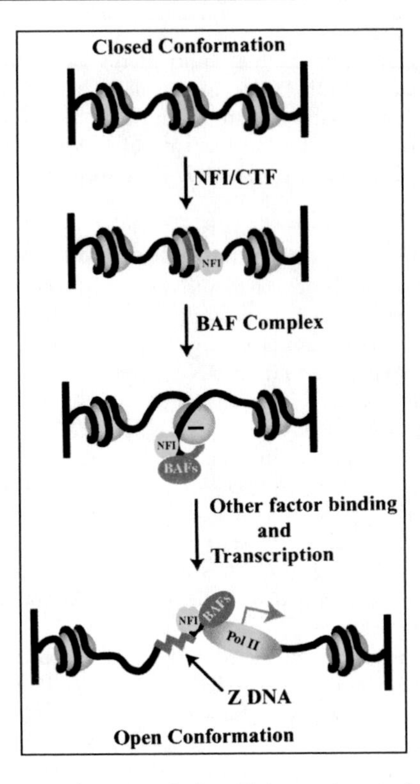

Figure 4. A model depicting NFI- and Z-DNA-facilitated chromatin remodeling by the BAF complex, at the *CSF1* promoter. Prior binding of NFI/CTF to its target site in the *CSF1* promoter is required for the recruitment of the BAF complex. Activation of the promoter by the BAF-complex requires Z-DNA-forming sequences, which are converted to Z-DNA conformation upon activation, removing the nucleosome. Reproduced with permission from ref. 99, ©2001 Elsevier.

Conclusion

Curved DNA and poly(dA•dT) structures can enhance the accessibility of *cis*-DNA elements in chromatin by exposing them to the milieu while on the nucleosome (curved DNA), or by preventing nucleosome formation (both curved DNA and poly(dA•dT)), or in some cases by forming altered nucleosomal structures (poly(dA•dT)). Z-DNA seems to be implicated in gene activation coupled with chromatin remodeling, and triplex DNA and cruciform structures may be used to form open chromatin structures.

Acknowledgements

The author would like to acknowledge the contributions of Jun-ichi Nishikawa, Yoshiro Fukue, and Junko Ohyama. The studies reported from my laboratory were supported in part by Grants-in-Aid for Scientific Research from the Ministry of Education, Science, Sports and Culture of Japan.

References

1. Imhof A, Wolffe AP. Transcription: gene control by targeted histone acetylation. Curr Biol 1998; 8:R422-424.
2. Workman JL, Kingston RE. Alteration of nucleosome structure as a mechanism of transcriptional regulation. Annu Rev Biochem 1998; 67:545-579.
3. Kornberg RD, Lorch Y. Twenty-five years of the nucleosome, fundamental particle of the eukaryote chromosome. Cell 1999; 98:285-294.
4. Aalfs JD, Kingston RE. What does 'chromatin remodeling' mean? Trends Biochem Sci 2000; 25:548-555.
5. Turner BM. Histone acetylation and an epigenetic code. Bioessays 2000; 22:836-845.
6. Vignali M, Hassan AH, Neely KE et al. ATP-dependent chromatin-remodeling complexes. Mol Cell Biol 2000; 20:1899-1910.
7. Wu J, Grunstein M. 25 years after the nucleosome model: chromatin modifications. Trends Biochem Sci 2000; 25:619-623.
8. Jenuwein T. Re-SET-ting heterochromatin by histone methyltransferases. Trends Cell Biol 2001; 11:266-273.
9. Becker PB, Hörz W. ATP-dependent nucleosome remodeling. Annu Rev Biochem 2002; 271:247-273.
10. Geiman TM, Robertson KD. Chromatin remodeling, histone modifications, and DNA methylation-how does it all fit together? J Cell Biochem 2002; 87:117-125.
11. Narlikar GJ, Fan HY, Kingston RE. Cooperation between complexes that regulate chromatin structure and transcription. Cell 2002; 108:475-487.
12. Turner BM. Cellular memory and the histone code. Cell 2002;111:285-291.
13. Carrozza MJ, Utley RT, Workman JL et al. The diverse functions of histone acetyltransferase complexes. Trends Genet 2003; 19:321-329.
14. Lusser A, Kadonaga JT. Chromatin remodeling by ATP-dependent molecular machines. Bioessays 2003; 25:1192-1200.
15. Almer A, Rudolph H, Hinnen A et al. Removal of positioned nucleosomes from the yeast PHO5 promoter upon PHO5 induction releases additional upstream activating DNA elements. EMBO J 1986; 5:2689-2696.
16. Archer TK, Cordingley MG, Wolford RG et al. Transcription factor access is mediated by accurately positioned nucleosomes on the mouse mammary tumor virus promoter. Mol Cell Biol 1991;11:688-698.
17. Zhu Z, Thiele DJ. A specialized nucleosome modulates transcription factor access to a *C. glabrata* metal responsive promoter. Cell 1996; 87:459-470.
18. Wolffe AP. Chromatin: Structure and Function. 3rd ed. London: Academic press, 1998.
19. Onishi Y, Kiyama R. Interaction of NF-E2 in the human β-globin locus control region before chromatin remodeling. J Biol Chem 2003; 278:8163-8171.
20. Zhurkin VB, Lysov YP, Ivanov VI. Anisotropic flexibility of DNA and the nucleosomal structure. Nucleic Acids Res 1979; 6:1081-1096.
21. Trifonov EN, Sussman JL. The pitch of chromatin DNA is reflected in its nucleotide sequence. Proc Natl Acad Sci USA 1980; 77:3816-3820.
22. Dickerson RE, Kopka ML, Pjura P. A random-walk model for helix bending in B-DNA. Proc Natl Acad Sci USA 1983; 80:7099-7103.
23. Drew HR, Travers AA. DNA bending and its relation to nucleosome positioning. J Mol Biol 1985; 186:773-790.
24. Zhurkin VB. Sequence-dependent bending of DNA and phasing of nucleosomes. J Biomol Struct Dyn 1985; 2:785-804.
25. Satchwell SC, Drew HR, Travers AA. Sequence periodicities in chicken nucleosome core DNA. J Mol Biol 1986; 191:659-675.
26. Pennings S, Muyldermans S, Meersseman G et al. Formation, stability and core histone positioning of nucleosomes reassembled on bent and other nucleosome-derived DNA. J Mol Biol 1989; 207:183-192.
27. Shrader TE, Crothers DM. Artificial nucleosome positioning sequences. Proc Natl Acad Sci USA 1989; 86:7418-7422.
28. Wolffe AP, Drew HR. Initiation of transcription on nucleosomal templates. Proc Natl Acad Sci USA 1989; 86:9817-9821.
29. Costanzo G, di Mauro E, Salina G et al. Attraction, phasing and neighbour effects of histone octamers on curved DNA. J Mol Biol 1990; 216:363-374.

30. Shrader TE, Crothers DM. Effects of DNA sequence and histone-histone interactions on nucleosome placement. J Mol Biol 1990; 216:69-84.
31. Ioshikhes I, Bolshoy A, Trifonov EN. Preferred positions of AA and TT dinucleotides in aligned nucleosomal DNA sequences. J Biomol Struct Dyn 1992; 9:1111-1117.
32. Patterton H-G, Simpson RT. Modified curved DNA that could allow local DNA underwinding at the nucleosomal pseudodyad fails to position a nucleosome in vivo. Nucleic Acids Res 1995; 23:4170-4179.
33. Sivolob AV, Khrapunov SN. Translational positioning of nucleosomes on DNA: the role of sequence-dependent isotropic DNA bending stiffness. J Mol Biol 1995; 247:918-931.
34. Baldi P, Brunak S, Chauvin Y et al. Naturally occurring nucleosome positioning signals in human exons and introns. J Mol Biol 1996; 263:503-510.
35. De Santis P, Kropp B, Leoni L et al. Influence of DNA superstructural features and histones aminoterminal domains on mononucleosome and dinucleosome positioning. Biophys Chem 1996; 62:47-61.
36. Ioshikhes I, Bolshoy A, Derenshteyn K et al. Nucleosome DNA sequence pattern revealed by multiple alignment of experimentally mapped sequences. J Mol Biol 1996; 262:129-139.
37. Widlund HR, Cao H, Simonsson S et al. Identification and characterization of genomic nucleosome-positioning sequences. J Mol Biol 1997; 267:807-817.
38. Fitzgerald DJ, Anderson JN. Unique translational positioning of nucleosomes on synthetic DNAs. Nucleic Acids Res 1998; 26:2526-2535.
39. Olson WK, Gorin AA, Lu XJ et al. DNA sequence-dependent deformability deduced from protein-DNA crystal complexes. Proc Natl Acad Sci USA 1998; 95:11163-11168.
40. Anselmi C, Bocchinfuso G, De Santis P et al. Dual role of DNA intrinsic curvature and flexibility in determining nucleosome stability. J Mol Biol 1999; 286:1293-1301.
41. Widlund HR, Kuduvalli PN, Bengtsson M et al. Nucleosome structural features and intrinsic properties of the TATAAACGCC repeat sequence. J Biol Chem 1999; 274:31847-31852.
42. Roychoudhury M, Sitlani A, Lapham J et al. Global structure and mechanical properties of a 10-bp nucleosome positioning motif. Proc Natl Acad Sci USA 2000; 97:13608-13613.
43. Pedersen AG, Baldi P, Chauvin Y et al. DNA structure in human RNA polymerase II promoters. J Mol Biol 1998; 281:663-673.
44. Ohyama T. Intrinsic DNA bends: an organizer of local chromatin structure for transcription. Bioessays 2001; 23:708-715.
45. Kropp B, Leoni L, Sampaolese B et al. Influence of DNA superstructural features and histone amino-terminal domains on nucleosome positioning. FEBS Lett 1995; 364:17-22.
46. Widlund HR, Vitolo JM, Thiriet C et al. DNA sequence-dependent contributions of core histone tails to nucleosome stability: differential effects of acetylation and proteolytic tail removal. Biochemistry 2000; 39:3835-3841.
47. Radic MZ, Lundgren K, Hamkalo BA. Curvature of mouse satellite DNA and condensation of heterochromatin. Cell 1987; 50:1101-1108.
48. Benfante R, Landsberger N, Tubiello G et al. Sequence-directed curvature of repetitive *AluI* DNA in constitutive heterochromatin of *Artemia franciscana*. Nucleic Acids Res 1989; 17:8273-8282.
49. Martínez-Balbás A, Rodríguez-Campos A, García-Ramírez M et al. Satellite DNAs contain sequences that induce curvature. Biochemistry 1990; 29:2342-2348.
50. Pasero P, Sjakste N, Blettry C et al. Long-range organization and sequence-directed curvature of *Xenopus laevis* satellite 1 DNA. Nucleic Acids Res 1993; 21:4703-4710.
51. Fitzgerald DJ, Dryden GL, Bronson EC et al. Conserved patterns of bending in satellite and nucleosome positioning DNA. J Biol Chem 1994; 269:21303-21314.
52. Kralovics R, Fajkus J, Kovarík A et al. DNA curvature of the tobacco GRS repetitive sequence family and its relation to nucleosome positioning. J Biomol Struct Dyn 1995; 12:1103-1119.
53. John B, Miklos GLG. Functional aspects of satellite DNA and heterochromatin. Int Rev Cytol 1979; 58:1-114.
54. Singer MF. Highly repeated sequences in mammalian genomes. Int Rev Cytol 1982; 76:67-112.
55. Ohyama T, Tsujibayashi H, Tagashira H et al. Suppression of electrophoretic anomaly of bent DNA segments by the structural property that causes rapid migration. Nucleic Acids Res 1998; 26:4811-4817.
56. Wada-Kiyama Y, Kiyama R. Periodicity of DNA bend sites in human ε-globin gene region. Possibility of sequence-directed nucleosome phasing. J Biol Chem 1994; 269: 22238-22244.
57. Wada-Kiyama Y, Kiyama R. Conservation and periodicity of DNA bend sites in the human β -globin gene locus. J Biol Chem 1995; 270: 12439-12445.
58. Wada-Kiyama Y, Kiyama R. An intrachromosomal repeating unit based on DNA bending. Mol Cell Biol 1996; 16:5664-5673.

59. Ohki R, Hirota M, Oishi M et al. Conservation and continuity of periodic bent DNA in genomic rearrangements between the c-*myc* and immunoglobulin heavy chain μ loci. Nucleic Acids Res 1998; 26: 3026-3033.
60. Piña B, Brüggemeier U, Beato M. Nucleosome positioning modulates accessibility of regulatory proteins to the mouse mammary tumor virus promoter. Cell 1990; 60: 719-731.
61. Schild C, Claret F-X, Wahli W et al. A nucleosome-dependent static loop potentiates estrogen-regulated transcription from the *Xenopus* vitellogenin B1 promoter in vitro. EMBO J 1993; 12:423-433.
62. Imbalzano AN, Kwon H, Green MR et al. Facilitated binding of TATA-binding protein to nucleosomal DNA. Nature 1994; 370:481-485.
63. Godde JS, Nakatani Y, Wolffe AP. The amino-terminal tails of the core histones and the translational position of the TATA box determine TBP/TFIIA association with nucleosomal DNA. Nucleic Acids Res 1995; 23:4557-4564.
64. Wong J, Li Q, Levi B-Z et al. Structural and functional features of a specific nucleosome containing a recognition element for the thyroid hormone receptor. EMBO J 1997; 16:7130-7145.
65. Nishikawa J, Amano M, Fukue Y et al. Left-handedly curved DNA regulates accessibility to *cis*-DNA elements in chromatin. Nucleic Acids Res 2003; 31: 6651-6662.
66. Piña B, Barettino D, Truss M et al. Structural features of a regulatory nucleosome. J Mol Biol 1990; 216:975-990.
67. Angermayr M, Oechsner U, Gregor K et al. Transcription initiation in vivo without classical transactivators: DNA kinks flanking the core promoter of the housekeeping yeast adenylate kinase gene, *AKY2*, position nucleosomes and constitutively activate transcription. Nucleic Acids Res 2002; 30:4199-4207.
68. Bash RC, Vargason JM, Cornejo S et al. Intrinsically bent DNA in the promoter regions of the yeast *GAL1-10* and *GAL80* genes. J Biol Chem 2001; 276:861-866.
69. Blomquist P, Belikov S, Wrange Ö. Increased nuclear factor 1 binding to its nucleosomal site mediated by sequence-dependent DNA structure. Nucleic Acids Res 1999; 27:517-525.
70. Alexeev DG, Lipanov AA, Skuratovskii IY. Poly(dA)•poly(dT) is a B-type double helix with a distinctively narrow minor groove. Nature 1987; 325:821-823.
71. Nelson HC, Finch JT, Luisi BF et al. The structure of an oligo(dA)•oligo(dT) tract and its biological implications. Nature 1987; 330:221-226.
72. Park HS, Arnott S, Chandrasekaran R et al. Structure of the α-form of poly[d(A)] •poly[d(T)] and related polynucleotide duplexes. J Mol Biol 1987; 197:513-523.
73. Dechering KJ, Cuelenaere K, Konings RN et al. Distinct frequency-distributions of homopolymeric DNA tracts in different genomes. Nucleic Acids Res 1998; 26:4056-4062.
74. Struhl K. Naturally occurring poly(dA-dT) sequences are upstream promoter elements for constitutive transcription in yeast. Proc Natl Acad Sci USA 1985; 82:8419-8423.
75. Roy A, Exinger F, Losson R. *cis*- and *trans*-acting regulatory elements of the yeast *URA3* promoter. Mol Cell Biol 1990; 10:5257-5270.
76. Schlapp T, Rödel G. Transcription of two divergently transcribed yeast genes initiates at a common oligo(dA-dT) tract. Mol Gen Genet 1990; 223:438-442.
77. Thiry-Blaise LM, Loppes R. Deletion analysis of the *ARG4* promoter of *Saccharomyces cerevisiae*: a poly(dAdT) stretch involved in gene transcription. Mol Gen Genet 1990; 223:474-480.
78. Verdone L, Camilloni G, Di Mauro E et al. Chromatin remodeling during *Saccharomyces cerevisiae ADH2* gene activation. Mol Cell Biol 1996; 16:1978-1988.
79. Simpson RT, Künzler P. Chromatin and core particles formed from the inner histones and synthetic polydeoxyribonucleotides of defined sequence. Nucleic Acids Res 1979; 6:1387-1415.
80. Rhodes D. Nucleosome cores reconstituted from poly (dA-dT) and the octamer of histones. Nucleic Acids Res 1979; 6:1805-1816.
81. Kunkel GR, Martinson HG. Nucleosomes will not form on double-stranded RNA or over poly(dA)•poly(dT) tracts in recombinant DNA. Nucleic Acids Res 1981; 9:6869-6888.
82. Englander EW, Howard BH. A naturally occurring $T_{14}A_{11}$ tract blocks nucleosome formation over the human neurofibromatosis type 1 (NF1)-Alu element. J Biol Chem 1996; 271:5819-5823.
83. Fox KR. Wrapping of genomic polydA•polydT tracts around nucleosome core particles. Nucleic Acids Res 1992; 20:1235-1242.
84. Brown PM, Fox KR. DNA triple-helix formation on nucleosome-bound poly(dA)•poly(dT) tracts. Biochem J 1998; 333:259-267.
85. Losa R, Omari S, Thoma F. Poly(dA)•poly(dT) rich sequences are not sufficient to exclude nucleosome formation in a constitutive yeast promoter. Nucleic Acids Res 1990; 18:3495-3502.
86. Schieferstein U, Thoma F. Modulation of cyclobutane pyrimidine dimer formation in a positioned nucleosome containing poly(dA•dT) tracts. Biochemistry 1996; 35:7705-7714.

87. Puhl HL, Behe MJ. Poly(dA)•poly(dT) forms very stable nucleosomes at higher temperatures. J Mol Biol 1995; 245:559-567.
88. Lascaris RF, de Groot E, Hoen PB et al. Different roles for Abf1p and a T-rich promoter element in nucleosome organization of the yeast *RPS28A* gene. Nucleic Acids Res 2000; 28:1390-1396.
89. Suter B, Schnappauf G, Thoma F. Poly(dA•dT) sequences exist as rigid DNA structures in nucleosome-free yeast promoters in vivo. Nucleic Acids Res 2000; 28:4083-4089.
90. Iyer V, Struhl K. Poly(dA:dT), a ubiquitous promoter element that stimulates transcription via its intrinsic DNA structure. EMBO J 1995; 14:2570-2579.
91. Rubbi L, Camilloni G, Caserta M et al. Chromatin structure of the *Saccharomyces cerevisiae* DNA topoisomerase I promoter in different growth phases. Biochem J 1997; 328:401-407.
92. Shimizu M, Mori T, Sakurai T et al. Destabilization of nucleosomes by an unusual DNA conformation adopted by poly(dA)•poly(dT) tracts in vivo. EMBO J 2000; 19:3358-3365.
93. Koch KA, Thiele DJ. Functional analysis of a homopolymeric (dA-dT) element that provides nucleosomal access to yeast and mammalian transcription factors. J Biol Chem 1999; 274:23752-23760.
94. Reeves R, Wolffe AP. Substrate structure influences binding of the non-histone protein HMG-I(Y) to free nucleosomal DNA. Biochemistry 1996; 35:5063-5074.
95. Liu LF, Wang JC. Supercoiling of the DNA template during transcription. Proc Natl Acad Sci USA 1987; 84:7024-7027.
96. Hamada H, Kakunaga T. Potential Z-DNA forming sequences are highly dispersed in the human genome. Nature 1982; 298:396-398.
97. Schroth GP, Chou PJ, Ho PS. Mapping Z-DNA in the human genome. Computer-aided mapping reveals a nonrandom distribution of potential Z-DNA-forming sequences in human genes. J Biol Chem 1992; 267:11846-11855.
98. van Holde K, Zlatanova J. Unusual DNA structures, chromatin and transcription. Bioessays 1994; 16:59-68.
99. Liu R, Liu H, Chen X et al. Regulation of CSF1 promoter by the SWI/SNF-like BAF complex. Cell 2001; 106:309-318.
100. Westin L, Blomquist P, Milligan JF et al. Triple helix DNA alters nucleosomal histone-DNA interactions and acts as a nucleosome barrier. Nucleic Acids Res 1995; 23:2184-2191.
101. Espinás ML, Jiménez-García E, Martínez-Balbás Á et al. Formation of triple-stranded DNA at $d(GA{\bullet}TC)_n$ sequences prevents nucleosome assembly and is hindered by nucleosomes. J Biol Chem 1996; 271:31807-31812.
102. Battistoni A, Leoni L, Sampaolese B et al. Kinetic persistence of cruciform structures in reconstituted minichromosomes. Biochim Biophys Acta 1988; 950:161-171.
103. Nobile C, Nickol J, Martin RG. Nucleosome phasing on a DNA fragment from the replication origin of simian virus 40 and rephasing upon cruciform formation of the DNA. Mol Cell Biol 1986; 6:2916-2922.
104. Kotani H, Kmiec EB. DNA cruciforms facilitate in vitro strand transfer on nucleosomal templates. Mol Gen Genet 1994; 243:681-690.
105. Zlatanova J. Histone H1 and the regulation of transcription of eukaryotic genes. Trends Biochem Sci 1990; 15: 273-276.
106. Alami R, Fan Y, Pack S et al. Mammalian linker-histone subtypes differentially affect gene expression in vivo. Proc Natl Acad Sci USA 2003; 100:5920-5925.
107. Yaneva J, Schroth GP, van Holde KE et al. High-affinity binding sites for histone H1 in plasmid DNA. Proc Natl Acad Sci USA 1995; 92:7060-7064.

Chapter 14

DNA Bendability and Nucleosome Positioning in Transcriptional Regulation

Mensur Dlakić, David W. Ussery and Søren Brunak

Abstract

The placement of nucleosomes along genomic DNA is determined by signals that can be specific or degenerate at the level of sequence; the latter signals are harder to find using conventional methods. In recent years, the development of sophisticated machine learning techniques that can extract subtle phased signals has improved our ability to distinguish between various classes of nucleosome-positioning sequences. Our knowledge of the structural mechanics of free DNA also has reached the point where it can be fruitfully incorporated into predictive models. More importantly, the accumulation of high-resolution structures with proteins bound to DNA, and those of nucleosomes in particular, has provided important clues about the role of DNA bending and flexibility in nucleosome positioning.

Introduction

Eukaryotic DNA is compacted and organized in nucleosome arrays that make up chromosomes.[1] The nucleosome core, a basic unit of chromatin, contains two copies each of the core histones H2A, H2B, H3 and H4, and about 146 base pairs of DNA wrapped around the protein octamer.[2] Further compaction of DNA by the linker histone H1 is achieved in higher-order structures assembled from repetitive nucleosome cores and linker DNA. Although nucleosomes show no clear binding preference for particular DNA sequences, they are not randomly distributed on DNA. Translational and rotational positioning of nucleosomes along the DNA molecule are in part determined by signal sequences that are often degenerate, and sometimes have periodicity corresponding to the helical repeat of DNA.[3-9]

Structural properties of DNA, such as intrinsic bending and flexibility, play important roles in DNA recognition by sequence-specific DNA-binding proteins.[10-13] A similar role has been proposed for these DNA features in nucleosome positioning,[14-16] and verified using a variety of experimental and theoretical approaches.[5,6,8,17-24] There is a statistical preference for rotational positioning of DNA around the histone octamer such that AAA•TTT and AAT•ATT trinucleotides have the DNA minor groove facing the octamer, while the minor groove of GGC•GCC and AGC•GCT faces away from protein.[5,17] These observations were confirmed by designing artificial nucleosome positioning sequences that form nucleosomes significantly better than bulk nucleosomal DNA.[6] In addition to static bending properties of DNA, the variation in sequence-dependent bendability[19] can also impart strong rotational and translational orientation to nucleosomal DNA.[8,9]

DNA Conformation and Transcription, edited by Takashi Ohyama. ©2005 Eurekah.com and Springer Science+Business Media.

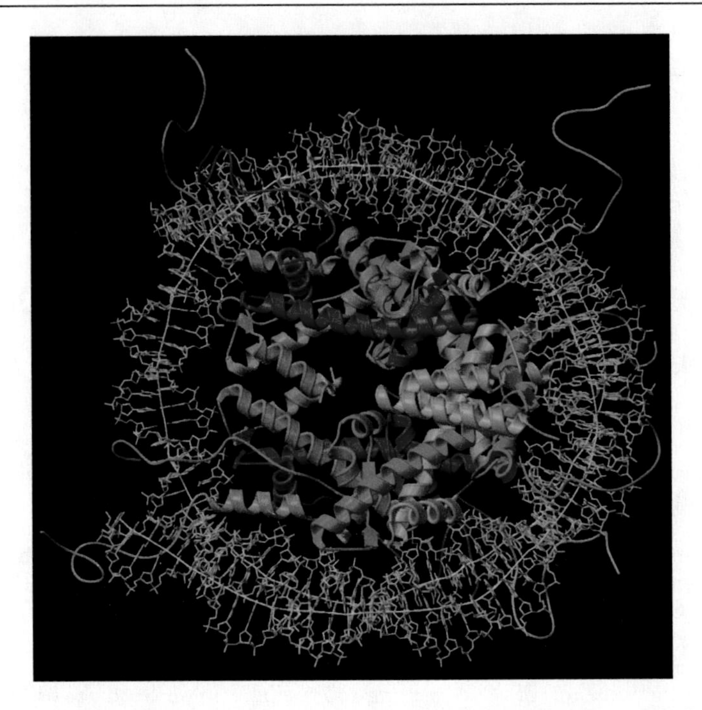

Figure 1. Nucleosome core particle. Ribbon representations of core histones (H2A: green; H2B: light green;) H3: cyan; H4: red) bound to DNA (yellow). DNA axis is shown as a continuous white line going through centers of base pairs. The first 20 residues of the histone H3 tails were omitted for clarity.

Comparison of Nucleosomal DNA with B-DNA Oligonucleotides

The recent flurry of nucleosome core structures from Richmond and Luger labs[2,25-27] has revealed a wealth of structural details about protein-DNA interactions (Fig. 1). These efforts culminated with a 1.9 Å structure that provided a high-resolution view of the DNA conformation as well.[28,29] The resolution of this structure is comparable to high-resolution structures of oligonucleotides, providing a solid basis for statistical analyses aimed at extracting general principles of conformational variability in nucleosomal DNA. We therefore set out to determine to what degree the nucleosomal DNA angular parameters and deformability compare with those of B-DNA oligonucleotides (Table 1), and whether these differences can be rationalized in terms of nucleosome positioning. Many sequence-dependent characteristics of DNA are shared between the two data sets, clearly showing that intrinsic conformational properties of the double helix are utilized by the histones much the same way as in specific protein-DNA complexes.[13] On the other hand, certain features of nucleosomal DNA are not found as major trends in B-DNA oligonucleotides.

Sequence-dependent deformability of DNA is reflected in the dispersion of base-pair parameters. The inspection of Table 1 reveals that nucleosomal DNA shows greater flexibility than B-DNA in terms of roll and tilt angles, and only slightly lower flexibility for twist (standard deviations from mean values are shown as subscripts in Table 1). Higher flexibility of DNA angular parameters was also observed when comparing specific protein-DNA complexes to B-DNA.[13] Although reduced variability of B-DNA is likely caused in part by crystal packing effects, it is clear that the entropy of protein-DNA complex formation enables, and possibly requires, larger conformational flexibility of DNA.

DNA duplex bends towards the minor or major grooves (roll) much more easily than in a direction along the longer base-pair axis (tilt);[14,30] this bending anisotropy holds true in

Table 1. Comparison of structural parameters between nucleosomal DNA and B-DNA oligomers

	Nucleosome 147				B-DNA			
	Roll	Tilt	Twist	R_{Corr}	Roll	Tilt	Twist	R_{Corr}
All	$1.82_{8.61}$	$0.00_{4.11}$	$34.68_{4.94}$	-0.73	$0.15_{5.54}$	$0.00_{3.26}$	$35.66_{5.94}$	-0.48
RR	$2.09_{8.16}$	$-0.21_{3.79}$	$34.46_{3.84}$	-0.57	$1.04_{4.17}$	$-1.26_{2.90}$	$35.47_{4.97}$	-0.46
RY	$2.40_{6.63}$	$0.00_{3.58}$	$31.94_{4.15}$	-0.78	$-3.42_{5.73}$	$0.00_{3.46}$	$34.59_{4.87}$	-0.58
YR	$0.81_{10.91}$	$0.00_{5.08}$	$37.87_{5.47}$	-0.92	$2.64_{5.79}$	$0.00_{3.24}$	$37.42_{8.29}$	-0.78
AA	$4.20_{7.07}$	$-0.60_{3.61}$	$34.31_{4.24}$	-0.57	$0.00_{3.96}$	$-0.93_{3.02}$	$35.25_{3.85}$	-0.32
AG	$1.09_{8.63}$	$0.51_{3.98}$	$34.31_{3.60}$	-0.54	$4.55_{3.09}$	$-2.25_{2.88}$	$29.64_{4.34}$	-0.37
GA	$1.35_{5.32}$	$-2.69_{3.86}$	$35.98_{3.57}$	-0.86	$-0.06_{4.00}$	$-1.01_{2.80}$	$39.63_{2.88}$	-0.37
GG	$6.59_{7.57}$	$0.18_{3.92}$	$33.08_{5.06}$	-0.80	$2.79_{3.09}$	$-1.74_{2.23}$	$35.41_{3.21}$	-0.20
AT	$0.35_{4.19}$	$0.00_{2.98}$	$31.67_{3.30}$	-0.81	$-0.77_{3.48}$	$0.00_{2.32}$	$31.21_{3.58}$	-0.72
AC	$7.58_{4.63}$	$0.41_{3.74}$	$29.68_{2.81}$	-0.51	$-0.43_{4.88}$	$-0.97_{3.71}$	$33.00_{5.42}$	-0.23
GC	$-2.80_{7.44}$	$0.00_{4.55}$	$36.40_{4.25}$	-0.80	$-6.22_{5.93}$	$0.00_{4.02}$	$37.47_{3.62}$	-0.40
TA	$4.31_{10.43}$	$0.00_{4.10}$	$35.64_{5.01}$	-0.89	$-0.08_{4.86}$	$0.00_{3.67}$	$43.43_{5.91}$	-0.80
TG	$-0.20_{11.02}$	$-0.53_{5.38}$	$38.50_{5.53}$	-0.93	$1.02_{6.08}$	$-0.22_{2.96}$	$39.45_{8.73}$	-0.80
CG					$6.34_{3.91}$	$0.00_{3.36}$	$31.16_{4.00}$	-0.27

The coordinate files of B-DNA oligomers (without bound drugs or intercalating molecules, bulges, abasic sites, nicks or any base modifications other than methylation and bromination) were obtained using the search facility of the Nucleic Acid Database.[54] When multiple molecules per asymmetric unit were present and were not related by crystallographic symmetry, we treated them as independent entries for the calculation purposes. The nucleic acid parameters were determined using a local calculation scheme of CURVES.[55] Similar to earlier analyses, terminal dinucleotides were excluded as not being representative of true sequence-dependent features of DNA.[35,56] Nucleosome core structure (accession code 1KX5) was obtained from Protein Data Bank.[57] Numbers represent mean values of angular parameters for a given dinucleotide, with standard deviations shown in subscript. Only ten unique dinucleotides are shown. Complementary dinucleotides have identical values with the changed sign for tilt. R_{Corr} is a correlation coefficient (between -1 and 1) that measures the degree to which twist and roll angles are linearly related. (If there is perfect linear relationship with positive slope between the two variables, the correlation coefficient is 1; in case of a perfect linear relationship with negative slope between the two variables, the correlation coefficient is -1). CG dinucleotide was not present in the nucleosome core structure.

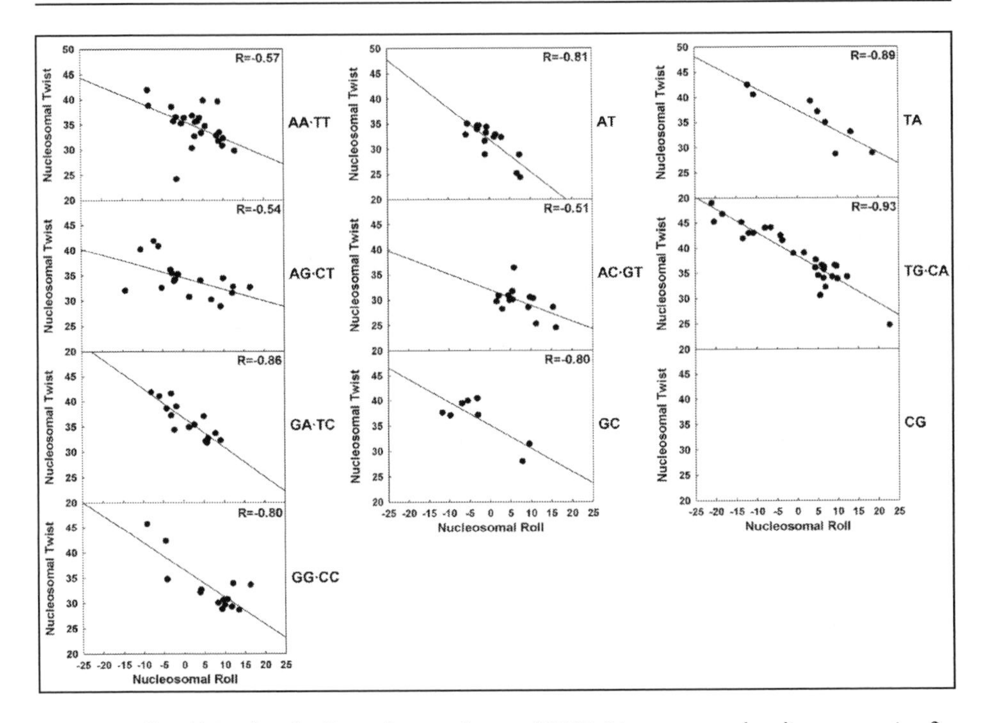

Figure 2. Roll vs. Twist plots for dimers from nucleosomal DNA. Lines represent best linear regression fits through the data points. R-values in the upper right corner are coefficients of correlation. Empty plot for CG indicates that for that dinucleotide no entries were present in the structure.

protein-DNA complexes as well.[13] In nucleosomal DNA, there is even greater preference for roll over tilt, in terms of both mean values and the dispersion (Table 1). The roll in nucleosomal DNA contributes to smooth bending into either groove and to kinking into the minor groove.[12,29] The latter feature is never seen in B-DNA oligonucleotides and is rarely seen in protein-DNA complexes.[30] Kinking into minor groove is observed almost exclusively at CA•TG steps (roll angles are in the range -12° to -21°), in marked contrast with protein-DNA complexes where CA•TG step has mostly positive roll.[30] However, CA•TG steps have a preference for low-roll/high-twist in a YCAR context (Y stands for pyrimidine and R for purine),[31] and most of nucleosomal CA•TG steps with negative roll values are indeed preceded by a pyrimidine and followed by a purine.

Pyrimidine-purine (YR) dimers are the most easily deformed steps both in B-DNA crystals[15,32,33] and protein-DNA complexes.[13,30] YR dimers in nucleosomal DNA are even more flexible as judged by standard deviations of roll angles,[29] living up to their billing as "flexible hinges" that fit the DNA duplex to the protein surface.[13] Though both TA and CA•TG steps have the highest flexibility of all 10 unique dimers in terms of twist and roll values (Table 1), most of negative roll angles occur at CA•TG steps, while TA steps have mostly positive roll angles.[34]

Greater flexibility of nucleosomal DNA is achieved by concerted changes in roll and twist angles. A strong anti-correlation between twist and roll has been observed in earlier analyses,[13,32,35] and is also present in our subset of B-DNA oligomers shown in Table 1. Overall, all dimers in nucleosomal DNA have excellent correlation between twist and roll values (Fig. 2), as is the case for specific protein-DNA complexes,[12,13] which supports the notion that sequence-specific constraints of the sugar phosphate backbone contribute primarily to the conformational variability of protein-bound DNA. In B-DNA oligomers, where crystal packing forces induce subtle but noticeable variations in DNA deformability,[36] the correlation between twist and roll angles is less prominent (Table 1). It is important to note that YR steps have the most significant twist-roll correlation, consistent with their highest flexibility. In particular, roll angles in CA•TG steps cover the span of more than 40° (from -21° to +23°). This remarkable variability enables them to be positioned either on the "inside" or on the "outside" of the histone octamer.

DNA Bending and Flexibility Are Utilized for Nucleosome Positioning

The analysis above shows that nucleosome positioning, for the most part, takes advantage of the intrinsic structural mechanics of the double helix. When DNA is bound by specific transcription factors, most of the free energy contribution comes from specific interactions between protein side-chains and DNA bases. In this case the protein has to accommodate structural properties of only a limited number of nucleotides, and YR dimers have been selected over the course of evolution as most frequent sequence elements to "fit" DNA around the protein because of their unique conformational properties.[13,15,30] When wrapping DNA around its surface, a sequence-specific DNA-binding protein needs to solve a "local" optimization problem, as its binding site will typically be short. This is achieved efficiently by utilizing only part of the conformational space where YR dimers have positive roll angles.[13,30] In contrast, core histones have to wrap tightly a longer piece of DNA regardless of its sequence. This is a "global" optimization problem and can be solved only by exploiting a wider range of conformational variability of DNA. Part of the solution is similar to sequence-specific DNA-binding proteins in a sense that YR dimers in nucleosomal DNA are most flexible and roll is preferred over tilt. However, nucleosomal DNA has certain characteristics that are only partially employed in B-DNA oligomers and specific protein-DNA complexes: (1) higher overall flexibility of all dinucleotides; (2) extremely tight coupling of twist and roll angles; (3) negative roll angles in CA•TG steps.

Integrating the results from the present analysis with earlier theoretical and experimental data, we propose that nucleosomes are positioned by a combination of static and dynamic signals encoded in DNA. Statistical analysis of nucleosomal DNA cleavage[17] showed that GGC•GCC elements strongly prefer to have the minor groove facing "outside" (positive roll angles). The same trend is seen for the GG•CC step in our analysis, as well as AC•GT, AA•TT and TA steps. On the other hand, the GC step, and to a lesser degree the AT step, show stronger preference for negative roll values. We propose that sequence elements with stronger bending preferences set the initial frame for nucleosome positioning by assuming their preferred conformation. When these sequences are positioned in such a way that satisfies most of their rotational preferences, the rest of DNA is "molded" around the histone octamer by exploiting the conformational variability of DNA. The CA•TG step is most useful in this regard as it can conform both to "inside" and "outside" positions; other DNA sequences are also capable of adopting alternative conformations depending on the context.[37] This concept implies that strong nucleosome positioning can be achieved by sequences with properly phased rotational signals, by sequences that contain many flexible elements, and by those that contain favorable combinations of the two. Indeed, all three types of nucleosome positioning sequences have been observed experimentally.[6,20,22]

Preference of Nucleosomal Positioning in Exon and Intron DNA

Phased structural features can be "hidden signals" for nucleosomal positioning, which are not apparent from looking at the sequence. To extract meaningful information from the noisy data, we used machine learning methods which learn the theory automatically from the data by inference and model fitting. For this purpose, we adopted probabilistic models known as hidden Markov models (HMMs); an HMM is composed of connected states that emit observable outputs. Using this technique, we have found a pattern in human DNA which codes for proteins.[8] There is a periodicity of about 10 bp for human exon DNA, whilst intron DNA

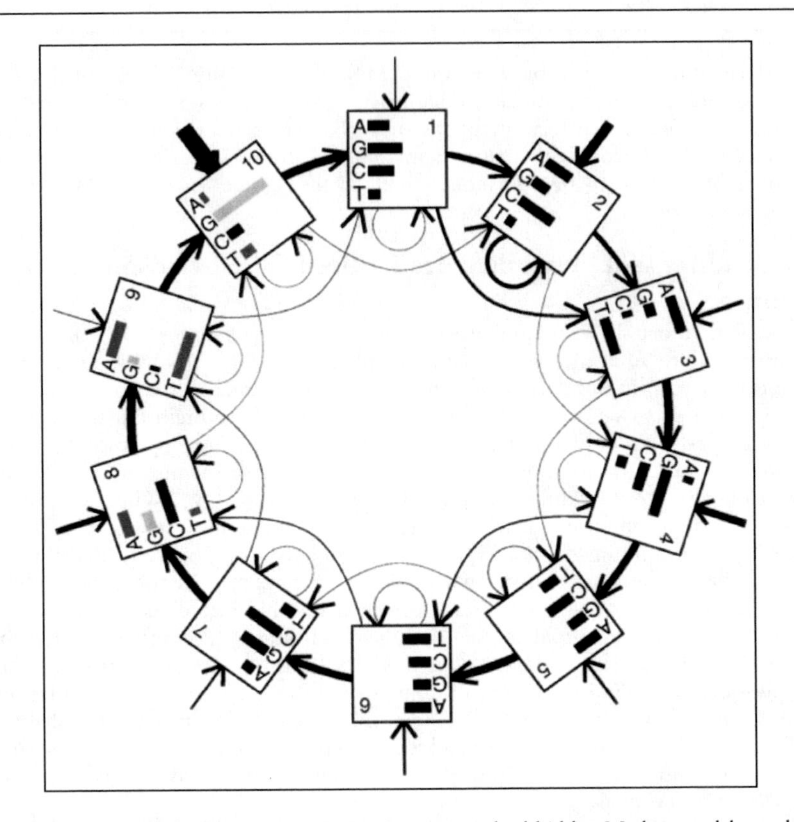

Figure 3. Circular HMM for human exon DNA. A 10 state wheel hidden Markov model reveals DNA periodicity, as described in ref. 8. The HMM is completely defined by the set of 10 states, the alphabet of 4 symbols, a probability transition, and a probability emission matrix. The model is intended to describe a stochastic system that evolves from state to state, while randomly emitting symbols from the alphabet. In the wheel architecture the thickness of the external arrows shows the probability of starting in the corresponding state. Emission probabilities are represented by bars inside boxes. (This figure was originally Figure 1A in ref. 8, and is used with permission from Elsevier.)

contains a somewhat weaker periodicity signal. However, it is unlikely that the periodicity in the exon DNA is coming from α-helix encoding sequences; the functional saturation of the genetic code precludes its involvement in nucleosomal positioning.[8,38] Figure 3 displays a wheel shaped HMM architecture for human exon DNA (in this case of length 10 nucleotides), where sequences can enter the wheel at any point. The thickness of the arrows from "outside" represents the probability of starting from the corresponding state. After training, the emission parameters in the wheel model showed a clearly recognizable periodic pattern [^T][AT]G (non-T, A or T, G) in states 8, 9 and 10. This wheel with a model of ten nucleotides provided the best fit from several different numbers of states. Furthermore, using the wheel model to estimate the average negative log-likelihood per nucleotide, values specific for various types of exons, introns and intergenic regions were also computed. The ranking of these also strongly indicate that the above described periodic pattern is strongest in exons. The period in the alignments (average distance between state 9 nucleotides) is in the order of 10.1-10.2 bp.[8] This value is the periodicity of DNA wrapped around nucleosomes, as discussed above.

By looking for periodicity signals in prokaryotic genomes, at the DNA structural level, we found evidence for horizontal DNA transfer from an Archaea to the bacterium *Thermotoga maritime.*[39] Is it possible that such signals exist in DNA that is more compacted, than in

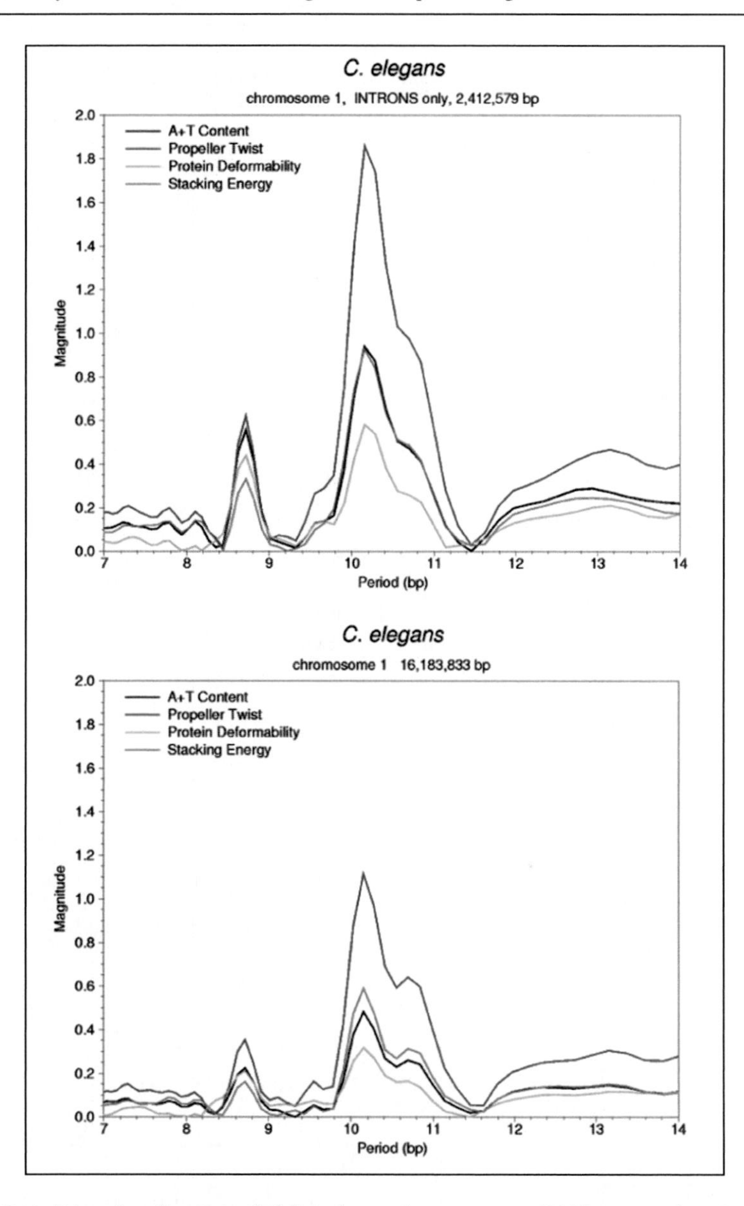

Figure 4. Periodicity plots for *Caenorhabditis elegans* chromosome 1. DNA structural periodicity was calculated for four different measures, as described previously.[49] The four panels correspond to the whole chromosome, intron-containing DNA, exon-containing DNA, and the *E. coli* K-12 chromosome. Figure continued on next page.

regions of highly expressed genes? One possibility is that there might be a difference between DNA which contains protein-coding regions vs. the rest of the DNA. A good test of this is chromosome 1 from *Caenorhabditis elegans* (~16 Mbp), which has a coding density of about 25%-that is, about 25% (~4 Mbp) of the chromosome is transcribed into mRNA. Furthermore, this "coding" region is divided into two almost equal fractions of intron and exon DNA (~2Mbp each). Figure 4 shows the periodicity plot for the whole chromosome, as well as for the

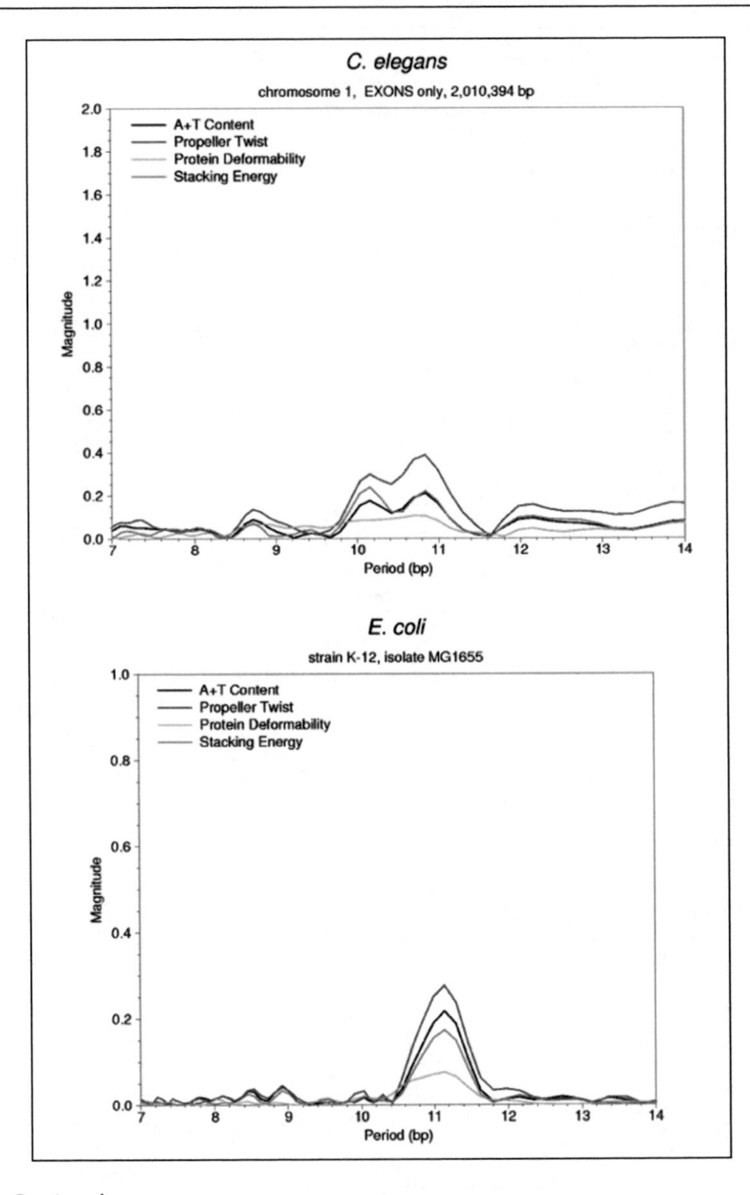

Figure 4. Continued.

exon and intron containing DNA. It is quite clear that, in contrast to the results for human DNA, the dominant contribution to the periodicity is from the DNA containing introns, with the exon DNA having a less strong periodicity. For comparison, the periodicity plot for the 4.6 Mbp chromosome of *Escherichia coli* K-12, which mostly contains coding DNA, is also shown in Figure 4. Thus, it seems likely, at least for *C. elegans* chromosome 1, that the non-coding DNA is more likely to be wrapped tightly in nucleosome complexes, whilst the coding DNA might have a greater chance to exist in a more open conformation. The strong peak of around 10.2 nucleotides in *C. elegans* has been previously found using several different methods, and has been localized to introns and intergenic regions.[40,41] Repeats of various sizes have been found in other chromosomes.[42] It is likely that these periodic repeats are reflections of global chromatin properties.

Nucleosome Positioning and Transcriptional Regulation

The idea has been around for more than 20 years—phased nucleosomes can result in more compact chromatin structures, and these regions are less available for the transcriptional machinery.[43] The size of an RNA polymerase molecule is about the same size as a nucleosome octamer with DNA wrapped around it, as shown in Figure 1. Thus if these nucleosomes are tightly bound together, in some sort of higher order structure, then there is little chance that the RNA polymerase enzyme can get access to the promoter region, and hence transcription will be repressed.

Although this idea is simple, in actuality there are complications. The degree of compaction of the chromatin depends on many things, including modification of the histone protein tails[44,45] as well as DNA methylation,[46] and there are many pathways which can regulate these modifications.[47,48] In this chapter, we are focusing on the structure of DNA, and whether there are some general properties of the double helix which can affect the ability of certain sequences to condense.

Based on the sequences of DNA wrapped around trimmed nucleosomal cores, a trinucleotide model was developed for the preference of certain trinucleotides to be in phase with the helical repeat. Depending on where these trinucleotides are located, they will have either the major or minor groove facing away from the histone octamer.[17] Trinucleotides were given a number based on the frequency of occurrence; those favoring the major groove were assigned a negative value, whilst if the minor groove faced away from the nucleosome, they were given a positive number. By taking the absolute value of the numbers, one has a measure of the relative propensity of a sequence to be positioned in nucleosomes, which we call "position preference".[9] In human promoters, we have found that several trinucleotides known to have high propensity for major groove compression occur much more frequently in the regions downstream of the transcriptional start point, whilst the upstream regions contain more low-bendability triplets. Within the region downstream of the start point, we find a periodic pattern in sequence and bendability, which is in phase with the DNA helical pitch. The periodic bendability profile shows bending peaks roughly at every 10 bp with stronger bending at 20 bp intervals. These observations suggest that DNA in the region downstream of the transcriptional start point is able to wrap around"--otein in a manner reminiscent of DNA in a nucleosome. This notion was further supported by the finding that the periodic bendability is caused mainly by the complementary triplet pairs CAG•CTG and GGC•GCC, which previously have been found to correlate with nucleosome positioning.

When these values are calculated for individual genes in the *E. coli* K-12 genome, there is a correlation between low position preference values and highly expressed genes.[49] This makes sense from a structural point of view, in that regions that are not preferentially localized in nucleosomes would tend to exist in more open conformations, and hence be more accessible to the RNA polymerase. This works even though in *E. coli* there are no histones or nucleosomes; however, the *E. coli* chromosome is compacted roughly 7,000-fold, so there must still be a need for some sort of chromatin structure, and the physical chemical properties are likely to be the same. The *E. coli* chromosome contains clusters of highly expressed genes, localized to certain regions of the chromosome.[50] Similarly, based on an extensive series of gene expression experiments in human cells, it has been shown that highly expressed human genes cluster together in distinct regions of the chromosomes.[51,52] It is possible to predict such regions throughout the whole genome using methods such as the nucleosomal position preference. As an example, Figure 5A shows a plot of the nucleosomal position preference along chromosome 1 of the yeast *Schizosaccharomyces pombe*.[53] A close-up (Fig. 5B) shows the marked region with a low position preference (dark green in lane C) corresponds to the *SPAC23A1.07* gene, which is a zinc finger protein that can be highly expressed in *S. pombe*. Other regions, such as highly

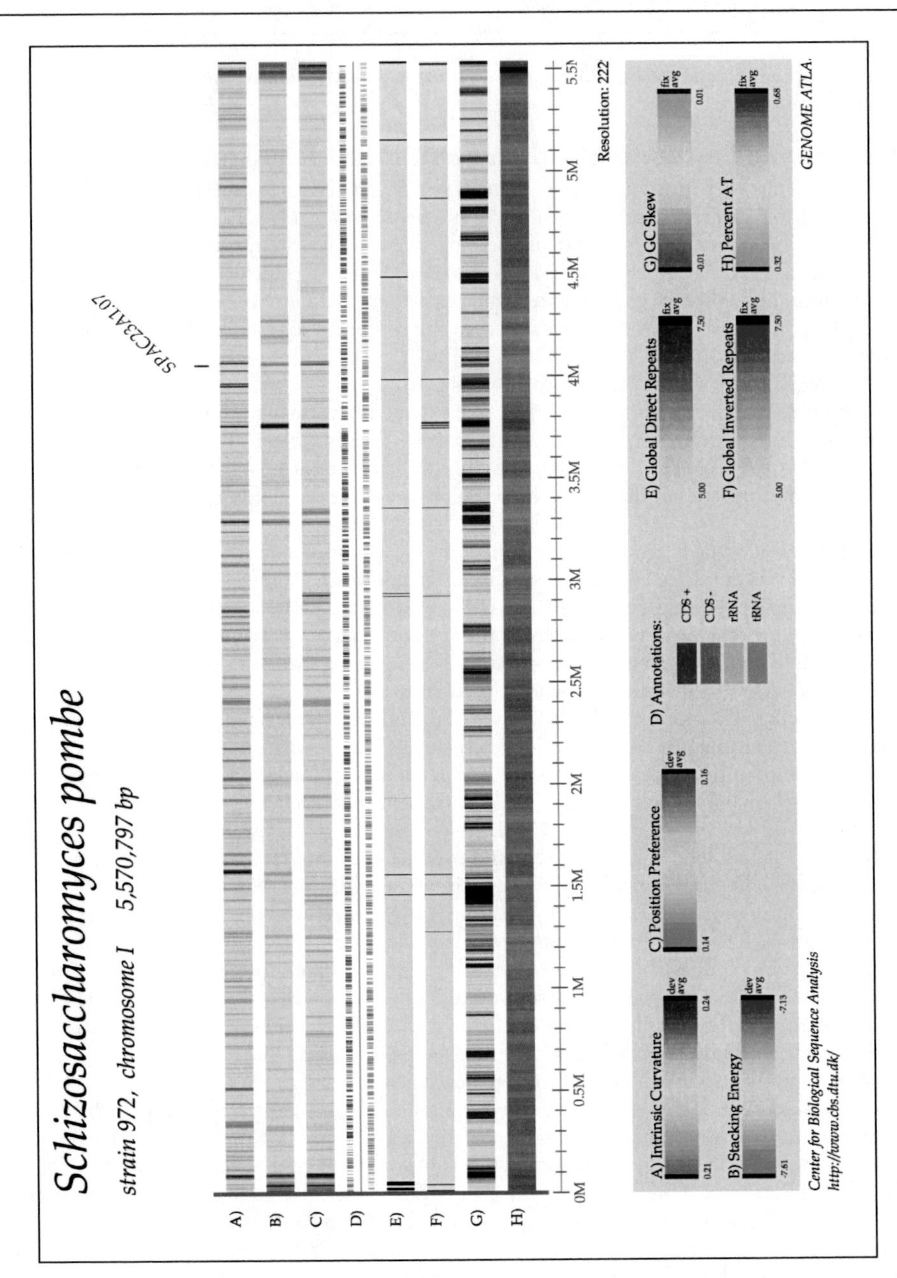
Schizosaccharomyces pombe
strain 972, chromosome I 5,570,797 bp
SPAC23A1.07
A)
B)
C)
D)
E)
F)
G)
H)
0M
0.5M
1M
1.5M
2M
2.5M
3M
3.5M
4M
4.5M
5M
Resolution: 222
A) Intrinsic Curvature
0.21
0.24
dev avg
B) Stacking Energy
-7.61
-7.13
dev avg
C) Position Preference
0.14
0.16
dev avg
D) Annotations:
CDS +
CDS -
rRNA
tRNA
E) Global Direct Repeats
5.00
7.50
fix avg
F) Global Inverted Repeats
5.00
7.50
fix avg
G) GC Skew
-0.01
0.01
fix avg
H) Percent AT
0.32
0.68
fix avg
Center for Biological Sequence Analysis
http://www.cbs.dtu.dk/

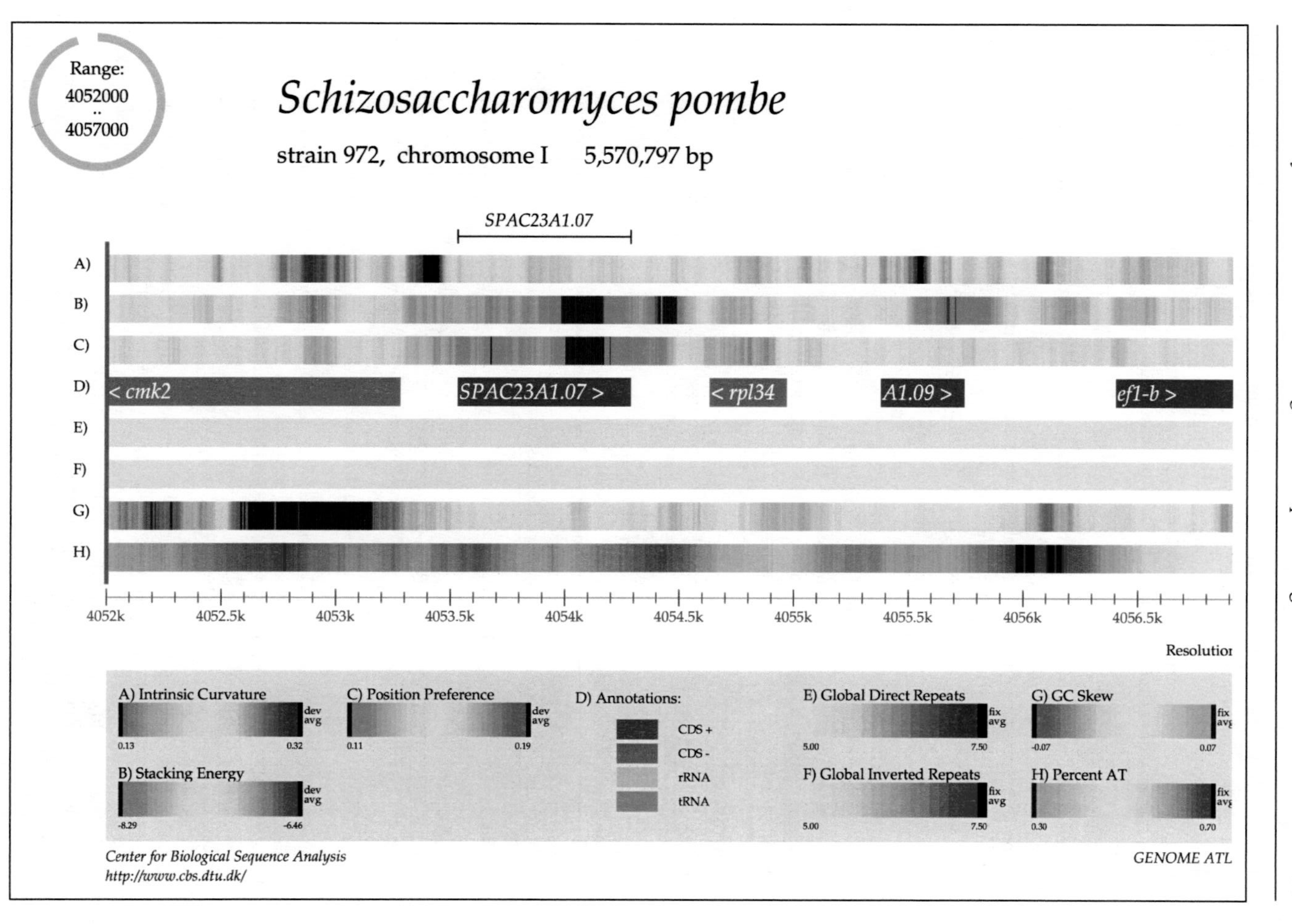

Figure 5B.
Continued.

expressed rRNA operons, consistently exhibit position preference values significantly lower than the chromosomal average in both eukaryotic and prokaryotic genomes (i.e., they are more likely to exclude chromosomes). In general, although different genes are expressed under different conditions, utilization of DNA structural properties can be helpful in finding regions along a chromosome that are potentially highly expressed.

Conclusions

Nucleosome positioning is governed by various sequence signals, including the differences between coding and non-coding DNA sequences imposed by evolutionary constraints. Expanding the alphabet of position signals beyond simple sequence, for example by using the extended set of DNA structure parameters in connection with machine learning methods, will further improve our ability to predict the role of nucleosome placement in transcriptional regulation.

Acknowledgements

This work was supported in part by NIH grant P20 RR-16455-01 from the BRIN Program of the National Center for Research Resources (M.D.) and by a grant from the Danish Research Foundation (D.W.U. and S.B.). M.D. is a Special Fellow of the Leukemia & Lymphoma Society.

References

1. van Holde KE. Chromatin. New York: Springer-Verlag, 1989.
2. Luger K, Mader AW, Richmond RK et al. Crystal structure of the nucleosome core particle at 2.8 Å resolution. Nature 1997; 389:251-260.
3. Trifonov EN, Sussman JL. The pitch of chromatin DNA is reflected in its nucleotide sequence. Proc Natl Acad Sci USA 1980; 77:3816-3820.
4. Zhurkin VB. Sequence-dependent bending of DNA and phasing of nucleosomes. J Biomol Struct Dyn 1985; 2:785-804.
5. Drew HR, Travers AA. DNA bending and its relation to nucleosome positioning. J Mol Biol 1985; 186:773-790.
6. Shrader TE, Crothers DM. Artificial nucleosome positioning sequences. Proc Natl Acad Sci USA 1989; 86:7418-7422.
7. Ioshikhes I, Bolshoy A, Trifonov EN. Preferred positions of AA and TT dinucleotides in aligned nucleosomal DNA sequences. J Biomol Struct Dyn 1992; 9:1111-1117.
8. Baldi P, Brunak S, Chauvin Y et al. Naturally occurring nucleosome positioning signals in human exons and introns. J Mol Biol 1996; 263:503-510.
9. Pedersen AG, Baldi P, Chauvin Y et al. DNA structure in human RNA polymerase II promoters. J Mol Biol 1998; 281:663-673.
10. Harrington RE, Winicov I. New concepts in protein-DNA recognition: sequence-directed DNA bending and flexibility. Prog Nucleic Acid Res Mol Biol 1994; 47:195-270.
11. Werner MH, Gronenborn AM, Clore GM. Intercalation, DNA kinking, and the control of transcription. Science 1996; 271:778-784.
12. Dickerson RE, Chiu TK. Helix bending as a factor in protein/DNA recognition. Biopolymers 1998; 44:361-403.
13. Olson WK, Gorin AA, Lu XJ et al. DNA sequence-dependent deformability deduced from protein-DNA crystal complexes. Proc Natl Acad Sci USA 1998; 95:11163-11168.
14. Zhurkin VB, Lysov YP, Ivanov VI. Anisotropic flexibility of DNA and the nucleosomal structure. Nucleic Acids Res 1979; 6:1081-1096.
15. Ulyanov NB, Zhurkin VB. Sequence-dependent anisotropic flexibility of B-DNA. A conformational study. J Biomol Struct Dyn 1984; 2:361-385.
16. Ohyama T. Intrinsic DNA bends: an organizer of local chromatin structure for transcription. Bioessays 2001; 23:708-715.
17. Satchwell SC, Drew HR, Travers AA. Sequence periodicities in chicken nucleosome core DNA. J Mol Biol 1986; 191:659-675.
18. Travers AA, Klug A. The bending of DNA in nucleosomes and its wider implications. Philos Trans R Soc Lond B Biol Sci 1987; 317:537-561.
19. Brukner I, Sanchez R, Suck D et al. Sequence-dependent bending propensity of DNA as revealed by DNase I: parameters for trinucleotides. EMBO J 1995; 14:1812-1818.

20. Widlund HR, Cao H, Simonsson S et al. Identification and characterization of genomic nucleosome-positioning sequences. J Mol Biol 1997; 267:807-817.
21. Cao H, Widlund HR, Simonsson T et al. TGGA repeats impair nucleosome formation. J Mol Biol 1998; 281:253-260.
22. Widlund HR, Kuduvalli PN, Bengtsson M et al. Nucleosome structural features and intrinsic properties of the TATAAACGCC repeat sequence. J Biol Chem 1999; 274:31847-31852.
23. Roychoudhury M, Sitlani A, Lapham J et al. Global structure and mechanical properties of a 10-bp nucleosome positioning motif. Proc Natl Acad Sci USA 2000; 97:13608-13613.
24. Nishikawa J, Amano M, Fukue Y et al. Left-handedly curved DNA regulates accessibility to *cis*-DNA elements in chromatin. Nucleic Acids Res 2003; 31:6651-6662.
25. Suto RK, Edayathumangalam RS, White CL et al. Crystal structures of nucleosome core particles in complex with minor groove DNA-binding ligands. J Mol Biol 2003; 326:371-380.
26. White CL, Suto RK, Luger K. Structure of the yeast nucleosome core particle reveals fundamental changes in internucleosome interactions. EMBO J 2001; 20:5207-5218.
27. Suto RK, Clarkson MJ, Tremethick DJ et al. Crystal structure of a nucleosome core particle containing the variant histone H2A.Z. Nat Struct Biol 2000; 7:1121-1124.
28. Davey CA, Sargent DF, Luger K et al. Solvent mediated interactions in the structure of the nucleosome core particle at 1.9 Å resolution. J Mol Biol 2002; 319:1097-1113.
29. Richmond TJ, Davey CA. The structure of DNA in the nucleosome core. Nature 2003; 423:145-150.
30. Dickerson RE. DNA bending: the prevalence of kinkiness and the virtues of normality. Nucleic Acids Res 1998; 26:1906-1926.
31. Yanagi K, Prive GG, Dickerson RE. Analysis of local helix geometry in three B-DNA decamers and eight dodecamers. J Mol Biol 1991; 217:201-214.
32. El Hassan MA, Calladine CR. Two distinct modes of protein-induced bending in DNA. J Mol Biol 1998; 282:331-343.
33. Dickerson RE. Helix structure and molecular recognition by B-DNA. In: Neidle S, ed. Oxford Handbook of Nucleic Acid Structure. Oxford: Oxford Sci, 1999:145-197.
34. Goodsell DS, Kaczor-Grzeskowiak M, Dickerson RE. The crystal structure of C-C-A-T-T-A-A-T-G-G. Implications for bending of B-DNA at T-A steps. J Mol Biol 1994; 239:79-96.
35. Gorin AA, Zhurkin VB, Olson WK. B-DNA twisting correlates with base-pair morphology. J Mol Biol 1995; 247:34-48.
36. Dickerson RE, Goodsell DS, Neidle S. "...the tyranny of the lattice..." Proc Natl Acad Sci USA 1994; 91:3579-3583.
37. Dlakic M, Harrington RE. The effects of sequence context on DNA curvature. Proc Natl Acad Sci USA 1996; 93:3847-3852.
38. Levitsky VG, Podkolodnaya OA, Kolchanov NA et al. Nucleosome formation potential of exons, introns, and Alu repeats. Bioinformatics 2001; 17:1062-1064.
39. Worning P, Jensen LJ, Nelson KE et al. Structural analysis of DNA sequence: evidence for lateral gene transfer in *Thermotoga maritima*. Nucleic Acids Res 2000; 28:706-709.
40. Widom J. Short-range order in two eukaryotic genomes: relation to chromosome structure. J Mol Biol 1996; 259:579-588.
41. Frontali C, Pizzi E. Similarity in oligonucleotide usage in introns and intergenic regions contributes to long-range correlation in the *Caenorhabditis elegans* genome. Gene 1999; 232:87-95.
42. Fukushima A, Ikemura T, Kinouchi M et al. Periodicity in prokaryotic and eukaryotic genomes identified by power spectrum analysis. Gene 2002; 300:203-211.
43. Wittig S, Wittig B. Function of a tRNA gene promoter depends on nucleosome position. Nature 1982; 297:31-38.
44. Turner BM. Cellular memory and the histone code. Cell 2002; 111:285-291.
45. de Ruijter AJ, van Gennip AH, Caron HN et al. Histone deacetylases (HDACs): characterization of the classical HDAC family. Biochem J 2003; 370:737-749.
46. Cunliffe VT. Memory by modification: the influence of chromatin structure on gene expression during vertebrate development. Gene 2003; 305:141-150.
47. Geiman TM, Robertson KD. Chromatin remodeling, histone modifications, and DNA methylation-how does it all fit together? J Cell Biochem 2002; 87:117-125.
48. Formosa T. Changing the DNA landscape: putting a SPN on chromatin. Curr Top Microbiol Immunol 2003; 274:171-201.
49. Pedersen AG, Jensendagger LJ, Brunak S et al. A DNA structural atlas for *Escherichia coli*. J Mol Biol 2000; 299:907-930.

50. Ussery D, Larsen TS, Wilkes KT et al. Genome organisation and chromatin structure in *Escherichia coli.* Biochimie 2001; 83:201-212.
51. Caron H, van Schaik B, van der Mee M et al. The human transcriptome map: clustering of highly expressed genes in chromosomal domains. Science 2001; 291:1289-1292.
52. Versteeg R, Van Schaik BD, Van Batenburg MF et al. The human transcriptome map reveals extremes in gene density, intron length, GC content, and repeat pattern for domains of highly and weakly expressed genes. Genome Res 2003; 13:1998-2004.
53. Wood V, Gwilliam R, Rajandream MA et al. The genome sequence of *Schizosaccharomyces pombe.* Nature 2002; 415:871-880.
54. Berman HM, Olson WK, Beveridge DL et al. The nucleic acid database. A comprehensive relational database of three-dimensional structures of nucleic acids. Biophys J 1992; 63:751-759.
55. Lavery R, Sklenar H. The definition of generalized helicoidal parameters and of axis curvature for irregular nucleic acids. J Biomol Struct Dyn 1988; 6:63-91.
56. El Hassan MA, Calladine CR. The assessment of the geometry of dinucleotide steps in double-helical DNA; a new local calculation scheme. J Mol Biol 1995; 251:648-664.
57. Bhat TN, Bourne P, Feng Z et al. The PDB data uniformity project. Nucleic Acids Res 2001; 29:214-218.

Index

D

E

F

G

H

I

J

K

L

M

N

O

P

Q

R

S

T

U

V

W

X

Y

Z